Gallium Arsenide IC Applications Handbook

*To our friend and colleague,
Dr. Edward L. Griffin,
for his outstanding leadership and support
in developing GaAs IC applications.*

Gallium Arsenide IC Applications Handbook

VOLUME 1

Edited by
Dennis Fisher
Inder Bahl
ITT GaAs Technology Center
Roanoke, Virginia

ACADEMIC PRESS
San Diego New York Boston London Sydney Tokyo Toronto

This book is printed on acid-free paper. ∞

Copyright © 1995 by ACADEMIC PRESS, INC.

All Rights Reserved.
No part of this publication may be reproduced or transmitted in any form or by any means, electronic or mechanical, including photocopy, recording, or any information storage and retrieval system, without permission in writing from the publisher.

Academic Press, Inc.
A Division of Harcourt Brace & Company
525 B Street, Suite 1900, San Diego, California 92101-4495

United Kingdom Edition published by
Academic Press Limited
24-28 Oval Road, London NW1 7DX

Library of Congress Cataloging-in-Publication Data

Gallium arsenide IC applications handbook / edited by Dennis Fisher.
  Inder Bahl.
        p.   cm.
   Includes index.
   ISBN 0-12-257735-3 (alk. paper)
   1. Microwave integrated circuits. 2. Gallium arsenide semiconductors. I. Fisher, Dennis. II. Bahl, I. J.
   TK7876.G35   1995
   621.381'32--dc20   95-30712
                                          CIP

PRINTED IN THE UNITED STATES OF AMERICA
95 96 97 98 99 00 MM 9 8 7 6 5 4 3 2 1

Contents

Contributors ix
Preface xi

1 MMIC Technology Overview

Inder Bahl and Dennis Fisher

1.1 Introduction 1
1.2 Brief History of MMICs 3
1.3 Benefits of Monolithic Integration 4
1.4 Advantages of GaAs for MMICs 7
1.5 MMIC Active Devices 8
1.6 MMIC Design 11
1.7 GaAs IC Fabrication 14
1.8 MMIC Packaging 20
1.9 Summary 23
    References 24

2 MMIC Application Overview

Inder Bahl and Dennis Fisher

2.1 Introduction   29
2.2 Role of GaAs MMICs in Systems   30
2.3 Historical Market Projections   30
2.4 Military System Applications   31
2.5 MIMIC Program   35
2.6 Commercial System Applications   40
2.7 Summary   53
   References   53

3 Digital GaAs Integrated Circuits

John Naber

3.1 Introduction   57
3.2 Logic Design   59
3.3 Trade-offs between Silicon and GaAs   63
3.4 Digital GaAs Product Insertions   70
3.5 Summary   76
   References   77

4 Phased-Array Radar

Inder Bahl and Dave Hammers

4.1 Introduction   79
4.2 Phased-Array Radar Architectures   96
4.3 Subsystem Functions and GaAs IC Applications   100
4.4 Transceiver Module Technology   105
4.5 Summary and Future Trends   132
   References   133

5 Electronic Warfare I: *Transmitters*

Ron Schineller

5.1 Introduction   137
5.2 EW Subsystems   142
5.3 Generic EW Chips   161
5.4 Summary   165
5.5 Future Trends   171
    References   172

6 Electronic Warfare II: *Receivers*

Sanjay B. Moghe, S. Consolazio, and H. Fudem

6.1 Introduction   174
6.2 The Crystal Video Receiver Structures   182
6.3 Superheterodyne Receiver Structures   184
6.4 Channelized Receiver Structures   193
6.5 IFM Receiver Structures   199
6.6 Integrated EW Transmit/Receive Modules   206
6.7 Other EW Receiver Types   220
6.8 Summary and Future Trends   220
    References   221

7 Instrumentation

Val Peterson

7.1 Introduction   225
7.2 Technology Choices and Merits of Using GaAs ICs   226
7.3 Typical Instrumentation Block Diagrams and GaAs IC Opportunities   232
7.4 Application Examples   250
7.5 Conclusions and Future Trends   253
    References   253

8 Personal Communications Service

J. Mondal, Sanjay B. Moghe, and S. Ahmed

8.1 Introduction 255
8.2 Requirements for a PCS Voice/Data System 259
8.3 MMIC Components for PCS 267
8.4 Future Trends in PCS 283
References 291

9 Satellite Communications

Ramesch K. Gupta

9.1 Introduction 293
9.2 Evolution in Communications Satellites 294
9.3 MMIC Technology in Satellite Transponders 302
9.4 Space-Qualified MMIC Components 303
9.5 Space-Qualification of MMICs 307
9.6 MMIC Subsystem Design Examples 308
9.7 Future Direction 319
References 320

10 Direct Broadcast Satellite Receivers

Charles Huang

10.1 Introduction 323
10.2 System Descriptions 324
10.3 Receiver Components 329
10.4 GaAs MMIC Development for DBS 332
10.5 Commercially Available GaAs MMICs 335
10.6 Designing Receiver Components with MMICs 344
10.7 Application Issues 350
10.8 Future Trends 352
References 353

Appendix A 355
Appendix B 357
Appendix C 359
Index 363

Contributors

Numbers in parentheses indicate the pages on which the authors' contributions begin.

S. Ahmed (255) Advanced Microwave Technology, Northrop Defense Systems Division, Rolling Meadows, Illinois 60008

Inder Bahl (1, 29, 79) ITT Gallium Arsenide Technology Center, Roanoke, Virginia 24019

S. Consolazio (173) Advanced Microwave Technology, Northrop Defense Systems Division, Rolling Meadows, Illinois 60008

Dennis Fisher (1, 29) ITT Gallium Arsenide Technology Center, Roanoke, Virginia 24019

H. Fudem (173) Advanced Microwave Technology, Northrop Defense Systems Division, Rolling Meadows, Illinois 60008

Ramesh K. Gupta (293) Comsat Laboratories, Clarksburg, Maryland 20871

Dave Hammers (79) ITT Gilfillan, Van Nuys, California 91409

Charles Huang (323) Anadigics, Inc., Warren, New Jersey 07059

Sanjay B. Moghe (173, 255) Advanced Microwave Technology, Northrop Defense Systems Division, Rolling Meadows, Illinois 60008

J. Mondal (255) Northrop Defense Systems Division, Rolling Meadows, Illinois 60008

John Naber (57) ITT Gallium Arsenide Technology Center, Roanoke, Virginia 24019

Val Peterson (225) Hewlett-Packard, Santa Rosa, California 95403

Ron Schineller (137) ITT Avionics, Nutley, New Jersey 07110

Preface

During the past 20 years tremendous progress has been made in GaAs integrated circuit technology. Conversely, the progress in applications of the technology has not been so great, until recently. The recent explosion in the commercial wireless communications market has given new life to the GaAs industry, which was starving for this kind of application, i.e., a very high volume requirement. Most microwave applications over the past 20 years have been served by Microwave Integrated Circuits (MICs), where active and passive discrete components such as transistors, inductors, capacitors, and resistors are attached externally to an etched circuit on an alumina substrate. As such, MICs are not well suited to high-volume, low-cost applications. In contrast, Monolithic Microwave Integrated Circuits (MMICs), wherein all circuit components, active and passive, are fabricated simultaneously on a common semi-insulating semiconductor substrate, thereby eliminating discrete components and wire bond interconnects, have the advantage of being well suited to high-volume production. MMICs have significant benefits over MICs in terms of smaller size, lighter weight, improved performance, higher reliability, and, most importantly, lower cost in high-volume applications.

In the United States significant credit goes to the Microwave/Millimeter Wave Monolithic Integrated Circuit (MIMIC) Program for MMIC technology maturity in terms of CAD tools, design, layout, materials, fabrication, packages, and testing in high volume. The MIMIC program is being managed by ARPA, a U.S. Department of Defense agency, and is now in the final phase of an 8-year effort.

At present, there is no single text that provides a comprehensive treatment of GaAs IC applications in commercial and military microwave systems. This book addresses various issues for MMIC insertions into both types of systems. The topics include (i) a brief introduction to various applications; (ii) technology factors such as GaAs vs Si, monolithic vs hybrid, and reliability; (iii) performance benefits with GaAs ICs; (iv) cost factors; (v) MMIC component selection and implementation; (vi) critical circuit parameters and their impact on system performance; (vii) systems architecture; (viii) hardware implementation; and (ix) long-term outlook for GaAs technology.

The book is divided into 10 chapters, each of which is self-contained. For material of less interest or not within the scope of this book, the reader is referred to appropriate, easily accessible references. References for future research are also listed and potential future applications are included.

Chapter 1 provides an introduction to the GaAs technology and the general subject area, including a brief history of MMICs; benefits of monolithic integration; advantages of GaAs for MMICs; MMIC active devices; and IC design, fabrication, and packaging. Chapter 2 gives an overview of GaAs IC applications both in the military and in commercial systems. Several military system applications are given along with a description of the MIMIC Program, its objectives, and its accomplishments. Commercial system applications include wireless communications, consumer electronics, automotive electronics, and traffic control.

Chapter 3 deals with digital GaAs IC applications. Several types of digital logic families are discussed. Prime factors such as performance and cost are considered in trade-offs between silicon and GaAs technologies for digital ICs. Finally, digital GaAs product insertions into several systems are described.

The remainder of this book deals with MMIC applications, beginning with Chapter 4 on phased-array radar applications of GaAs MMICs. This application has been the largest driving force for MMIC technology. Major topics described in this chapter include (i) an introduction to phased-array antennas, phased-array radar architectures, subsystem functions, and GaAs applications; (ii) transmit/receive (T/R) module technology; and (iii) performance, reliability, and cost considerations of T/R modules.

The next two chapters (5 and 6) describe electronic warfare (EW) applications of GaAs ICs. Chapter 5 deals with transmitters, covering technology requirements and challenges, EW subsystems, and generic EW chips. Chapter 6 discusses EW receivers such as crystal video, superheterodyne, channelized, IFM, integrated EW receiver modules, and other EW receiver types.

Chapter 7 treats applications of GaAs ICs in the instrumentation area. Technology choices and merits of using GaAs ICs are discussed and several examples of GaAs IC insertions into instruments are illustrated.

Potential applications of GaAs in personal communication services (PCS) are the topics of Chapter 8. Historical perspective, PCS system requirements, and MMIC components for PCS are treated in this chapter.

Chapter 9, which treats GaAs MMICs for satellite communications, describes the evolution in communications satellites. MMIC technology in satellite transponders, space qualification of both MIC and MMIC components, MMIC subsystem design examples, and future trends are discussed.

Chapter 10, the final chapter, describes direct broadcast satellite (DBS) receivers, the most recognized current high-volume application of GaAs IC. System descriptions, receiver components, GaAs MMIC development for DBS, receiver design with MMICs, and application issues are discussed.

In summary, this book deals with all aspects of MMIC technology insertions into commercial and military systems and contains enough material for a one-semester course at the senior or graduate level. The text can also be used by practicing engineers, technology professionals, and managers working in the microwave technology as well as for 3- to 5-day courses and workshops organized by professionals.

As with many edited books, this project has required the cooperation and coordination of the many authors who have contributed to this book. Their timely support and patience when they were very busy at work, and the understanding of their families, in preparing, correcting, and rewriting their chapters to result in an updated and comprehensive book are greatly appreciated. The editors have tried their best to minimize the overlap between chapters and also have attempted to use common symbols throughout the text.

Many other individuals have been involved in the background in bringing this book to publication. Though it is not possible to mention each of them individually, we thank our management for their support. We also thank most warmly Ms. Linda Blankenship, who aided in the typing of several parts of this book. Finally, we appreciate the patience and support of our wives, Glenna Fisher and Subhash Bahl, throughout the preparation of this book.

Dennis Fisher
Inder Bahl
Roanoke, Virginia
January 1995

1
MMIC Technology Overview

Inder Bahl and Dennis Fisher
ITT Gallium Arsenide Technology Center, Roanoke, Virginia

1.1 Introduction
1.2 Brief History of MMICs
1.3 Benefits of Monolithic Integration
1.4 Advantages of GaAs for MMICs
1.5 MMIC Active Devices
1.6 MMIC Design
1.7 GaAs IC Fabrication
1.8 MMIC Packaging
1.9 Summary
    References

1.1 INTRODUCTION

The current trend in microwave technology is toward circuit miniaturization, high-level integration, improved reliability, low power consumption, cost reduction, and high-volume applications. Component size and performance are prime factors in the design of electronic systems for satellite communications, phased-array radar systems, electronic warfare, and other military applications, while small size and low cost drive the consumer electronics market. Monolithic microwave integrated circuits (MMICs) based on gallium arsenide (GaAs) technology are the key to meeting the above requirements. They will play an increasing role in consumer electronics dealing with information transfer, communications, automotive applications, and entertainment. With MMIC technology a typical microwave subsystem can be produced on a single chip at costs of less than $100 while simpler single function chips cost less than $10. Some very simple function chips are now produced at costs as low as $1. While most MMICs currently in production operate in the 0.5- to 30-GHz microwave range, there are increasing applications in the millimeter-wave (mmW) spectrum (30–300 GHz) as higher frequency transistors mature. Monolithic technology is particularly beneficial to mmW applications through the elimination of the parasitic effects of bond wires which connect discrete components in conventional hybrid structures.

In MMICs, as shown in Fig. 1, all active and passive circuit elements or components and interconnections are formed in or on the surface of a semi-insulating substrate (usually GaAs) by some deposition scheme such as epitaxy, ion implantation, sputtering, evaporation, diffusion, or a combination of these processes. Typically MMICs use microstrip for the transmission medium at microwave frequencies and coplanar waveguide at millimeter wave frequencies. Via holes, metal-filled holes from the bottom of the substrate (ground plane) to the top surface of MMICs, provide low-loss and low-inductance ground connections. The most common active devices are metal semiconductor field effect transistors (MESFETs) and diodes, while high electron mobility transistors (HEMTs) find a niche in low-noise and high-frequency applications. Heterojunction bipolar transistors (HBTs) are gaining popularity as power devices. The pseudomorphic HEMT (or PHEMT) is a higher performance transistor which utilizes multiple epitaxial III–V compound layers other than GaAs to improve lattice matching. The MESFET is commonly referred to simply as a FET, and HEMTs are also known as modulation doped FETs (MODFETs). For reference, the conventional workhorse of the microwave industry has been the hybrid approach (also commonly referred to as the microwave integrated circuit (MIC) approach) in which discrete components such as FETs, resistors, capacitors, and inductors are mounted on alumina substrates containing a printed interconnect pattern; wire bonds connect the individual components to the interconnect pattern. In some cases resistors and capacitors are "printed" on the alumina substrate along with the interconnect pattern.

Over the past decade MMICs have matured from laboratory curiosities

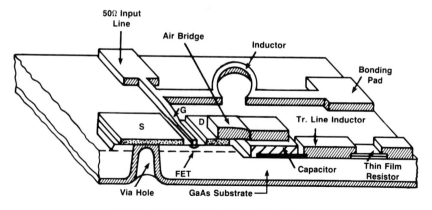

FIG. 1 A MMIC three-dimensional view.

to qualified production hardware for both commercial and military markets. The success of MMICs can be attributed to advances in materials, processing technology, devices, computer-aided design (CAD) tools, automated on-wafer testing, and packaging. Advantages of MMICs include low cost, small size, low weight, circuit design flexibility, broadband performance, elimination of circuit tweaking, high-volume manufacturing capability, package simplification, improved reproducibility, radiation hardness, improved reliability, and multifunction performance on a single chip. Indeed, the concept of implementing a "system on a chip" is now becoming a reality through monolithic microwave technology.

As noted above, an advantage of MMICs is the absence of the need to tweak the circuit to obtain the specified performance. Conversely, a disadvantage is that one cannot tweak the circuit. As a consequence, it is generally more difficult to achieve spec-compliance on the first design pass, thus increasing the nonrecurring engineering (NRE) cost of a MMIC. For high-volume applications the NRE cost is spread over many units and is by far offset by the much lower manufacturing costs. For low-volume applications, MICs may be the technology of choice unless the other advantages of MMICs are important.

1.2 BRIEF HISTORY OF MMICs

The first MMIC results for T/R modules using low-frequency silicon technology were reported in 1964. The results were not promising because of the low resistivity of silicon substrates [1] which provides insufficient isolation between the individual devices in the monolithic circuit. Furthermore, early microwave solid-state circuits require a large variety of solid-state devices, such as transistors, mixer diodes, pin diodes, varactors, impact avalanche transit-time diodes (IMPATTs), and Gunn diodes, and it is not easy to standardize the process specifications so that several kinds of devices can be produced simultaneously in an optimum fashion. In 1968 Mehal and Wacker [2] revived the approach and attempted to fabricate a 94-GHz receiver front-end by using Schottky-barrier diodes and Gunn diodes on a semi-insulating GaAs substrate. The results were poor due to the lack of adequate high-temperature processing techniques required for GaAs.

It was not until 1976, when Pengelly and Turner [3] applied the monolithic approach to an X-band amplifier based on the GaAs MESFET, that the technology began to show promise. By 1980 many MMIC results using MESFETs for various circuits had been reported. Since that time, tremendous progress has been made both in MMIC developments and in system applications [4–25]. The outstanding progress in the MMIC technology is attributed to the following:

1. Rapid development of GaAs material technology, including starting wafers, epitaxial growth, and ion implantation into semi-insulating substrates.
2. Advanced photo- or E-beam lithography technology developed for Si integrated circuits (ICs) which is directly applicable to GaAs ICs.
3. Excellent microwave properties of semi-insulating GaAs substrates which permit easy isolation of devices for high-level integration (high dielectric constant, $\epsilon_r = 12.9$ and low loss tan $\delta = 0.0005$).
4. Development of low-noise MESFETs and power MESFETs operating up to ~60 GHz have provided MMIC designers with versatile active circuit components.
5. Virtually any microwave solid state-circuit can be realized using combinations of MESFETs, dual-gate MESFETs, Schottky-barrier diodes, and switching MESFETs, each of which can be fabricated simultaneously using the same or similar process.
6. Development of good models for characterizing active and passive devices.
7. Availability of commercial CAD tools for accurate simulation and optimization of microwave circuits.
8. Availability of on-wafer high-frequency test probes which permits either low-cost screening or a large amount of statistically significant data to be measured without incurring the cost and variability of packaging.
9. Government funding for technology development and maturation.
10. Expanding military and commercial applications.

1.3 BENEFITS OF MONOLITHIC INTEGRATION

Hybrid MICs using discrete transistors, lumped elements, and thin or thick film matching circuits have been used extensively in microwave systems for nearly two decades. Conversely, MMICs have been used in systems for only the past few years. The relative merits of MMICs vis-à-vis hybrid MICs and conventional microwave components are in the areas of cost, size, weight, high-volume production capability, design flexibility, broadband performance, high level of integration, reproducibility, and reliability. These merits are compared in Table 1 and discussed below.

In the modern technology manufacturing environment, cost of electronic components is of prime importance in order to displace existing technologies and make feasible new advanced systems. The manufacture of hybrid circuits tends to be very labor intensive due to large part counts and wire bonds, whereas monolithic circuits are fabricated on wafers in batches, and hundreds and thousands can be manufactured at the same time. A monolithic IC can be designed to be quite tolerant of assembly variations. Thus,

TABLE 1 Comparison between Monolithic and Hybrid MICs

Feature	Monolithic	Hybrid
Substrate	Semi-insulator	Insulator
Interconnections	Deposited	Wire-bonded/deposited
Distributed elements	Microstrip or coplanar waveguide	Microstrip and/or coplanar lines
Lumped elements	Deposited	Discrete/deposited
Solid state devices	Integrated	Discrete
Controlled parasitics	Yes	No
Labor intensive	No	Yes
Repairability	No	Yes
Equipment costs	High	Low
Mass production	Yes	No
Debugging	Difficult	Easy
Integration with digital and electropotic ICs	Possible	N/A
NRE cost	Very high	Low
Production cost in high volume	Low	High
Size and weight	Small	Large
Design flexibility	Very good	Good
Circuit tweaking	Impractical	Practical
Broadband performance	Relatively good	Limited
Reproducibility	Excellent	Fair to good
Reliability	Excellent	Fair to good

monolithic circuits have a great advantage in terms of the manufacturing cost per unit when produced in large quantities, with an acceptable yield. In very large quantities, the recurring manufacturing cost will vary from less than \$1 per mm^2 for relatively small (1–5 mm^2) and simple function ICs to several dollars per mm^2 for larger (5–50 mm^2) and more complex function ICs. At these cost levels, widespread application of MMICs has become feasible. Continued cost reduction is expected to follow the pattern set by the silicon IC industry, making widespread application a practical reality.

Many of today's applications using MICs are in airborne systems and satellites, where size and weight are at a premium. Here monolithic ICs clearly have an edge over bulky hybrid MICs. It is worth mentioning that the weight of an individual chip resistor or a chip capacitor used in hybrid MICs is typically more than that of an entire MMIC chip. In general, MMIC-based subsystems will have 5 to 20 times the advantage in term of size and weight over hybrid MICs.

Since MMICs are produced in batches, they are well suited for high-volume production with reproducible performance. If the throughput of a foundry is 400 4″-diameter wafers/week, with 50% end-to-end product yield, the capacity for 10 mm² ICs is about 120,000 chips per week. Furthermore, since small signal MMICs can be fully characterized on-wafer, with high throughput, test costs [26] in a production environment are greatly reduced.

An important aspect of MMIC development to date has been the way in which MESFETs have been used to perform virtually all microwave functions. Furthermore, with the monolithic integration approach, the number of FETs required to perform a given function is essentially irrelevant. Thus, while a function requiring a relatively large number of FETs would be prohibitively expensive in MIC technology, it can be relatively inexpensive in MMIC technology. Such freedom to use active devices enables novel circuit solutions not practically possible in conventional technology. For example, owing to very low power consumption, MMICs using GaAs FETs are much better suited for use as microwave control circuits, such as switches, phase shifters, or attenuators, than are the discrete p-i-n diodes used in MICs. Indeed, millions of MMIC control circuits are now being used extensively for low-cost and high-volume commercial applications.

MMICs have demonstrated good performance over more than a decade bandwidth. A 5- to 100-GHz distributed amplifier [27] with 5 dB gain demonstrates how well one can design and fabricate broadband monolithic amplifiers. The elimination of parasitics associated with the lumped elements, active devices, and connecting bond wires or ribbons permits excellent broadband performance of MMICs, a distinct advantage over hybrid MICs. This advantage becomes increasingly important as the frequency of operation increases, particularly into the mmW spectrum.

A high level of integration at the MMIC chip level reduces the number of chips and results in low test and assembly costs, which in turn reduces the subsystem cost. Furthermore, combining digital circuitry with microwave circuits reduces the number of interconnects and increases reliability. Several different types of multifunction ICs have been developed using MMIC technology over the past 8 years, including front-end receivers, transceivers, and radar subsystems.

One of the most important advantages of the MMIC is that of reproducibility of performance. Reproducibility of MMICs mainly results from the excellent definition and repeatability of passive and active components and elimination of component interconnect parasitic variability associated with wire bonds. Well controlled processes and good statistical designs that accommodate the inevitable small variations in process from batch to batch result in reproducible MMIC chips.

Both hybrid and monolithic MIC technologies are considered reliable. However, a well-qualified MMIC process can be more reliable because of the much lower part counts and far fewer wire bonds.

1.4 ADVANTAGES OF GaAs FOR MMICs

Any assessment of MMIC technology options available to the microwave designer will generally be in terms of chip size, weight, reliability, reproducibility, cost, maximum frequency of operation and availability of a wide range of active devices for design flexibility. Various substrate materials used for MMICs are bulk silicon, silicon-on-sapphire (SOS), GaAs, and InP. Their electrical and physical properties are compared in Table 2. The semi-insulating property of the base material is crucial to providing higher device isolation and lower dielectric loss for MMICs. For example, while bipolar silicon devices are capable of operating up to about 10 GHz, the relatively low isolation property of bulk silicon precludes monolithic integration for frequencies above S-band (2–4 GHz). Sapphire substrates with their extremely high resistivity virtually eliminate the substrate frequency limitation for silicon devices, but the epitaxially grown devices themselves (usually FET structures in this technology) are of lower quality and hence are limited to about 6 GHz operation.

The GaAs FET as a single discrete transistor has been widely used in hybrid amplifiers (low noise, broadband, medium power, high power, high efficiency), mixers, multipliers, switching circuits, and gain control circuits. This wide utilization of GaAs FETs can be attributed to their high frequency of operation and versatility. All these benefits are automatically realized in MMICs as well. GaAs semi-insulating substrates provide sufficient isolation up to about 100 GHz. This, combined with much higher electron mobility (five to six times that of silicon), enables GaAs MMICs to be produced which operate up to 60 GHz. Further, MMICs at 94 GHz have been demon-

TABLE 2 Comparison of Monolithic Integrated Circuit Substrates

Property	Silicon	Silicon-on-sapphire	GaAs	InP
Semi-insulating	No	Yes	Yes	Yes
Resistivity (Ω-cm)	10^3–10^5	$>10^{14}$	10^7–10^9	$\sim 10^7$
Dielectric constant	11.7	11.6	12.9	14
Electrical mobility[a] (cm^2/V-s)	700	700	4300	3000
Saturation electrical velocity (cm/s)	9×10^6	9×10^6	1.3×10^7	1.9×10^7
Radiation hardness	Poor	Poor	Very good	Good
Density (g/cm^3)	2.3	3.9	5.3	4.8
Thermal conductivity (W/cm-°C)	1.45	0.46	0.46	0.68
Operating temperature (°C)	250	250	350	300
Ease of handling	Very good	Excellent	Good	Poor

[a] 10^{17} cm^{-3} doping level.

strated [28] using highly specialized HEMT and PHEMT devices epitaxially grown on semi-insulating GaAs. Hence, GaAs has been the technology of choice for most MMIC applications.

InP has been used [28] for millimeter-wave monolithic integrated circuits using HEMTs, but very little work has been done on InP MMICs using MESFETs. The low Schottky-barrier height of metals on n-type InP is an impediment to the development of an InP MESFET technology of equivalent performance to that of GaAs.

At the lower end of the microwave spectrum for new emerging telecommunication applications, GaAs power FETs are more suitable, compared with bipolar transistors, due to their high gain, low noise figure, high power with good efficiency, and low battery voltage (3 to 6 V) operation.

1.5 MMIC ACTIVE DEVICES

Since the first reported GaAs MMIC, the MESFET and the Schottky diode have been the primary devices used for analog ICs. Another derivative of MESFET being used is the dual-gate MESFET, which consists of two closely spaced gates. MESFET technology commonly uses 0.25- to 1.0-μm gate length for microwave applications and 0.5 to 2 μm for digital applications. MESFET low-noise and power MMICs demonstrate excellent performance at microwave frequencies. However, increasing emphasis is being placed on new devices for even better performance and higher frequency operation. HEMT and HBT devices [29, 30] offer potential advantages in microwave, millimeter-wave, and high-speed digital IC applications, arising from the use of heterojunctions to improve charge transport properties (as in HEMTs) or p–n junction injection characteristics (as in HBTs). HEMTs appear to have a niche in ultra-low-noise and high-frequency (mmW) applications. The MMICs produced using exotic structures such as pseudomorphic, lattice-matched HEMTs (PHEMTs) have significantly improved the noise performance and high-frequency (up to 94 GHz) operation. PHEMTs have shown excellent millimeter-wave power performance from Ku- through W-band. HBTs are vertically oriented heterostructure devices and are gaining popularity as power devices for high-efficiency, larger bandwidth applications, and single power supply operation. The current-gain cutoff frequency (f_T) for all these transistors lies in the 100- to 150-GHz range for state-of-the-art devices, with logic gate switching delays of under 15 ps at room temperature. A cross-sectional view of the three basic device types (MESFET, HEMT, and HBT) is shown in Fig. 2.

MESFETs, HEMTs, and HBTs have been used in both microwave and digital ICs to develop state-of-the-art circuit functions. Of these, MESFET technology is the most mature and is widely used in production applications. Nearly all microwave circuit functions have now been realized as

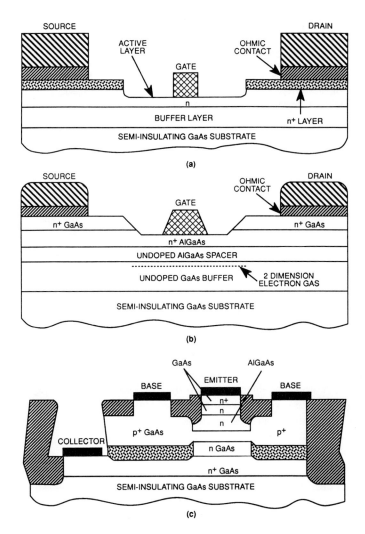

FIG. 2 Schematic cross section of (a) MESFET, (b) HEMT, and (c) HBT.

MMICs, including low-noise amplifiers, power amplifiers, high-efficiency amplifiers, broadband amplifiers, mixers, detectors, oscillators, multipliers, switches, phase shifters, attenuators, and modulators. Also, many passive circuits elements used in receiver or transmitter applications have been realized in MMICs including filters, power splitters and combiners, inductors, isolators, and circulators, as shown in Fig. 3. Many of these functions have been demonstrated (in different narrowband chips) over the entire 1- to 100-GHz frequency range. Furthermore, many of these functions have been combined on a single chip to form portions of a microwave system. Exam-

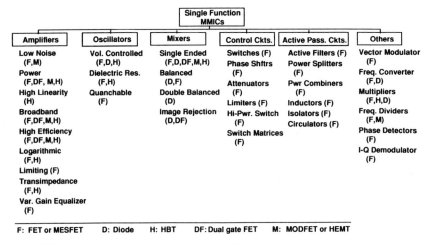

FIG. 3 Examples of single-function monolithic microwave integrated circuits.

ples of such multifunction ICs (as shown in Fig. 4) are receiver front-ends [21, 31–34], upconverters/downconverters [35, 36], transmitters [37], transceivers for active aperture radar applications [38–41], transceivers for wireless communication [42–45], and radar subsystems [46, 47]. All three technologies, MESFET, HEMT, and HBT, are suitable for both high-performance and high-level integration. The challenge remains that of performing the necessary integration with acceptable manufacturing yield and cost for affordable system applications.

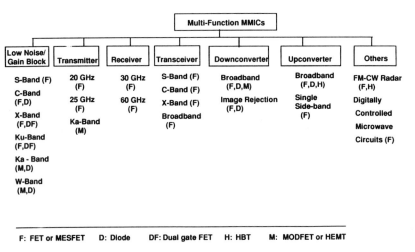

FIG. 4 Examples of multifunction monolithic microwave integrated circuits.

1.6 MMIC DESIGN

The design of MMICs which meet low cost targets requires state-of-the-art CAD tools. The need for increased design sophistication arises from the fact that the postfabrication tuning flexibility available in conventional hybrid microwave circuits is no longer present in the monolithically fabricated circuits. Consequently, a new philosophy for design methodology of MMICs is required. Key features of a successful design methodology include development of accurately characterized standard library cells including subcircuits, accurate models for linear and nonlinear active devices, accurate passive device models, use of circuit topology and circuit elements which are more tolerant to process variations, tolerance centering of designs, proximity effect models, comprehensive simulation of complete circuits, and automatic radio frequency (RF) testing of ICs on wafer. The latter is needed in order to obtain sufficient statistical characterization data without having to do expensive mounting or packaging.

An illustration of a microwave CAD approach showing various interactive features [48] is given in Fig. 5. A complete computer aided engineering (CAE) tool [49] as shown in Fig. 6 consists of device, circuit and system simulators, their accurate models (including physics based and electromagnetic), statistical design feature, and a link between the CAD, computer-aided test (CAT), and computer-aided manufacturing (CAM). The next generation MMIC CAD tool [50] which is based on a workstation is conceptually shown in Fig. 7. This interactive system will provide efficient coupling between circuit simulation, schematic captive/text editor, and the layout generator, greatly improving overall accuracy and reducing design cycle time. With such a system, first pass design success for simple microwave functions should be virtually guaranteed.

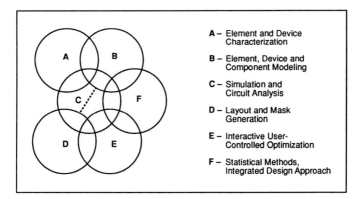

FIG. 5 Pictorial representation of the methodical aspects of microwave CAD and their linkage.

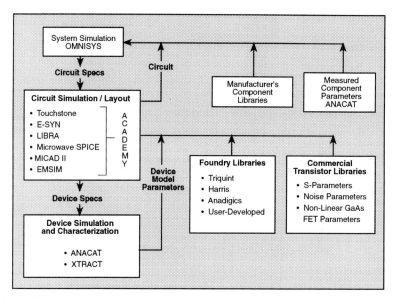

FIG. 6 EEsof's hierarchical microwave design system. (Reprinted with permission from O. Pitzalis [48], and Microwave Journal.)

In most cases, the ultimate purpose of MMICs is insertion into modules in a cost-effective manner which requires simple, noniterative assembly and test and elimination of labor-intensive tuning and testing. To a large extent, the MMIC chip performance-to-specification determines the manufacturing yield of the module. Figure 8 shows various steps of a module development methodology [51] from system definition to module completion. Given all the interactive effects between chips, components, and module, the chip design cannot be done effectively in isolation. Thus, the chip(s) and module must be designed according to concurrent engineering principles utilizing performance and cost models.

The evolution of a typical small signal MMIC design generally follows the flow diagram depicted in Fig. 9. The design starts with the customer specifications, which depends on the system requirements. System requirements also dictate the circuit topology along with the types of active elements to be used, for example, single- or dual-gate FETs and low-noise or -power FETs. In systems, the extent to which integration can be accomplished is limited by yield and cost. Since it is impossible to tune GaAs MMICs (without incorporating special on-chip programmable circuitry), an accurate and comprehensive modeling of each device and circuit component is required in order to save expensive and time-consuming iteration of mask fabrication, IC fabrication, and evaluation. The individual circuit elements must be electrically characterized by measuring RF and dc parameters. The microwave properties (e.g., S-parameters) are generally measured over the frequency band of interest at the required bias conditions. The final design is

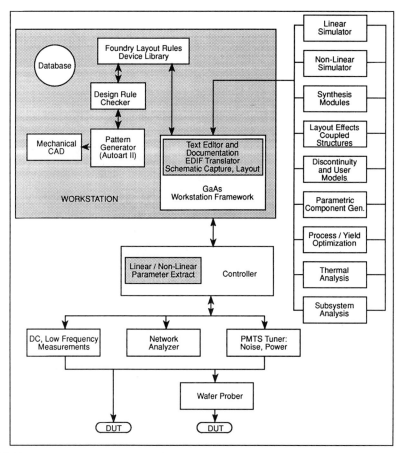

FIG. 7 Next-generation MMIC workstation concept. (Reprinted with permission from U.L. Rohde [50], and Microwave Journal.)

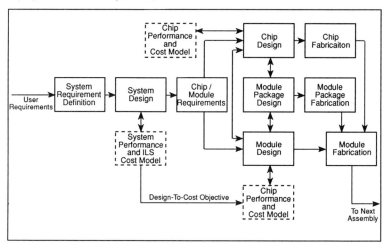

FIG. 8 Illustration of various steps of a module development methodology. (Reprinted with permission from A.K. Sharma [51], and IEEE, © 1990.)

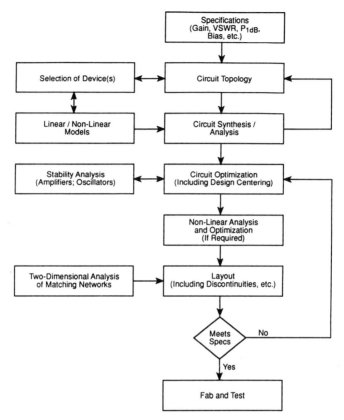

FIG. 9 Typical flowchart for a MMIC design.

completed by taking into account layout discontinuities, interaction between the components, stability analysis in case of amplifiers, and circuit yield analysis by considering process variations. In the case of nonlinear circuit design, (e.g., power amplifier, oscillator, mixer) an accurate nonlinear model for each device utilized is essential in order to design the circuit accurately.

Low-cost commercial MMIC applications (in the UHF-, L-, and S-bands) use low-loss inductors in the matching networks in order to reduce the chip size and improve RF performance and/or self-biasing techniques to operate with a single bias supply.

1.7 GaAs IC FABRICATION

There are many ways to fabricate MMICs. MESFET MMICs are most commonly fabricated with a recessed gate process as illustrate in Fig. 2a [4–

11, 52, 53], but the self-aligned gate process [54, 55] is gaining popularity because of its ability to efficiently process devices optimized for different functions, such as microwave small signal, microwave power, and digital, on the same wafer at the same time. The self-aligned gate process has demonstrated superior performance uniformity in a manufacturing environment. One particular embodiment of such a process that has shown state-of-the-art power performance is the multifunction self-aligned gate (MSAG) process which is shown in Fig. 10, along with salient features.

It is important for designers to have an appreciation for the influence of MMIC processing. At microwave frequencies, subtle changes in process variables can have significant effects on RF performance owing to electromagnetic interactions. (This is in contrast to lower frequency devices, usually implemented in silicon technology, where knowledge of the process is less important for the designer.) Thus, it is virtually impossible to tailor specific process changes to satisfy a particular microwave parameter change that a designer might desire without a high risk of having other parameters change in an unpredictable manner. Accordingly, if possible, designers should work with a very stable and predictable process, using innovative design techniques to overcome particular process deficiencies to achieve desired performance. In particular, the design must accommodate the normal statistical distribution of device parameters that will occur over the life of the MMIC manufacturing run. Otherwise manufacturing yield-to-RF specs may be unacceptably low.

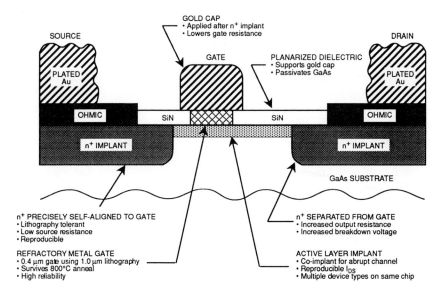

FIG. 10 Multifunction self-aligned gate (MSAG) FET cross section showing features to improve manufacturability.

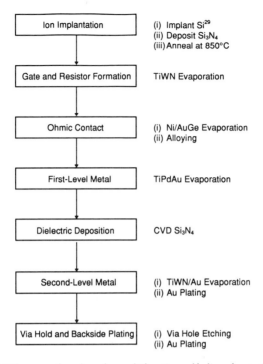

FIG. 11 MMIC process flowchart for multifunction self aligned gate (MSAG) process.

GaAs MMIC manufacturing processes are relatively complex, including over 250 individual process steps. These steps can be combined to show a simplified flow, as given in Fig. 11, for the MSAG process. The process for recessed gate MMICs has many similarities. The process includes the fabrication of active devices, resistors, capacitors, inductors, distributed matching networks, air bridges, and via holes for ground connections through the substrate.

It should be noted that, difficult as it is, MMIC processing is in fact less complex than silicon processing for devices operating at the low end of the microwave spectrum. Since silicon has inherently lower frequency capability and poorer isolation properties for integration purposes, more exotic processing is required to compete in the frequency region of overlap with GaAs applicability (~1 to 2 GHz). For example, a silicon bipolar complimentary metal oxide semiconductor (BiCMOS) process for such IC applications may require two to three times as many mask layers, adding significantly to the cost. With the more extensive process, there is often a penalty in added cycle time.

Active Layers

The MMIC process starts with the formation of an active layer on/into qualified semi-insulating GaAs substrate. There are basically two methods of forming an n-type active layer: ion implantation and epitaxy. In the ion-implantation technique, the dopant ions bombard the GaAs substrate in an area specified by a photolithographic photoresist pattern or mask. A suitable combination of energy and dose is used for the particular FET characteristics desired. During ion implantation, the crystal lattice of GaAs is damaged and the implanted atoms come to rest at random locations with the material. A high-temperature (850°C) annealing step is performed to heal the lattice damage and allow the implanted atoms to move onto lattice sites. With this technique, different active device types (with different active layer properties) can be readily fabricated on the same wafer through respective selective implants defined by photoresist masks. This technique is well suited to high-volume production since the methods and equipment are nearly identical to those used in the silicon industry.

As discussed earlier, epitaxial-based devices are sometimes required for particular high-performance applications. In the epitaxial technique, additional GaAs (or other III–V compound) material layer(s) are grown on the surface of the GaAs substrate in a manner that preserves the crystal structure. There are four basic types of epitaxy that have been used for GaAs: liquid-phase epitaxy (LPE), vapor-phase epitaxy (VPE), molecular-beam epitaxy (MBE), and metal–organic chemical vapor deposition (MOCVD). LPE is the oldest technique used to grow epitaxial layers on GaAs crystals, but it is not suitable for MMIC fabrication because of inherent surface roughness. VPE is typically used for GaAs power FETs. MBE is the most recent and powerful technique. MBE's advantages are that it can produce almost any epitaxial layer (III–V compound) composition, layer thickness, and doping with the highest possible accuracy and uniformity across a wafer. MOCVD has a similar flexibility with the added advantage of being better suited to low-cost manufacturing. However, the material electrical quality is not yet as good as that for MBE. The specified active areas are isolated either by mesa etching or by bombarding with ions, the latter of which increases resistivity by creating damage to the crystal lattice. A disadvantage of these techniques, relative to ion implantation, is that different device types generally require different epitaxial layers, requiring not only multiple expensive growth runs, but relatively costly processing to isolate the different device types.

Schottky or Gate Formation

The quality and placement of the gate metal is critical to FET performance in both low-noise and power FETs. The choice of the gate metal is generally based on good adhesion to GaAs, electrical conductivity, and thermal stability. Recessed gate FETs utilize evaporated materials such as TiPdAu or TiPtAu. MSAG FETs utilize a TiWN material, which forms a thermally stable, refractory Schottky gate in order to withstand the high-temperature annealing step which is performed while the gate is in place. It is deposited by reactive sputtering.

Gate formation can be defined by optical lithography techniques for critical dimensions down to about 0.5 μm. Below 0.5 μm the direct write electron beam method is often used. While quite expensive due to the high cost of the equipment, and having relatively slow throughput, the election beam method provides a high degree of precision to these small geometrics. The MSAG technique uses lower cost 1.0-μm optical lithography, along with a plasma underetch, to achieve a 0.4-μm or smaller gate dimension.

Ohmic Contact

Device ohmic contacts are made next. The purpose of an ohmic contact on a semiconductor material is to provide a good contact between the interconnect metal and the active channel at the semiconductor surface. The most common approach in the industry to fabricate ohmic contacts on GaAs is by alloying gold and germanium (88% Au and 12% Ge by weight, with a melting point of 360°C). A thin layer of AuGe alloy is evaporated on a thin layer of Ni. The total layer thickness is about 2000 Å. The ohmic contact pads are defined by a photoresist mask and chemical lift-off of metal on the photoresist regions. This step is followed by alloying at 400°C in a hydrogen ambient.

First-Level Metal

Next, a thick Ti/Au metal is overlayed on the gate by evaporation and lift-off. This metal reduces the gate resistance and also serves as a first-level metallization for MMIC fabrication, e.g., as a capacitor bottom plate or interconnect a first-level component.

Dielectric Deposition

Dielectric films are used in GaAs MMICs for passivation of active areas of FETs, diodes, and resistors, for metal–insulator–metal (MIM) capacitors, and for cross-over isolation. Silicon nitride (Si_3N_4) is commonly used as the dielectric material, which is easily deposited by either plasma-assisted chemical-vapor deposition or sputtering. The thickness of the dielectric film determines the capacitance per unit area of the MMIC capacitor. The thickness is usually between 1000 and 3000 Å and is optimized to have minimum pin holes, large breakdown voltage, and maximum possible capacitance. Typical values for capacitance and breakdown voltage are 300 pF/mm² and 60 V, respectively.

Second-Level Metal

Interconnection of components, airbridges, and the top plate of MIM capacitors is formed with the second-level layer TiWN/Au metal system. In order to achieve low-resistance connections, gold plating (4.5 μm thick) is added to provide maximum current capability of about 10 mA per μm of width and a sheet resistance of less than 10 mΩ/square. Typical line widths for microstrip line interconnects vary from 60 to 200 μm.

Backside Processing

Backside processing consisting of thinning by grinding or lapping, via hole source ground contact finalization and plating, is an important and cost-sensitive part of the processing. In a production environment, a significant investment has been made in the wafer by the time the frontside processing is completed and the backside processing started. Also, several of the backside operations critically affect the circuit function and the yield as a whole. After the frontside process, the wafer is thinned by a lapping technique from ~600 μm to the required thickness, typically 125 μm for small signal MMICs and 75 μm for power MMICs (to maximize heat dissipation). High-performance MMICs require low-inductance ground connections to the FET source and good thermal dissipation paths from the FET to its ground. In via-hole technology, holes are etched through the GaAs substrate under each FET source connection. Then the backside and the via-hole sidewalls are metallized. This provides a good connection from the frontside FET source metallization to the backside ground plane. This also eliminates the need for separate wire bonds to ground for each FET. Sputtering is the best technique with which to ensure that the metal enters the holes and covers the slopes as shown earlier in the MMIC diagram in Fig. 1. MMICs working below 6 GHz generally do not require via-hole technology and hence, their cost is considerably lower than that of higher frequency MMICs.

Dicing

After all processing operations are completed successfully, it remains to dice and select the chips. The first check for a good circuit is automatic testing on-wafer with microwave probes. After identifying the RF good ICs, the wafer is diced into chips. This can be done in several ways: scribing (diamond or laser), sawing, or by etching along defined lines. The RF-good and visually-good chips are then selected for mounting and/or packaging.

Transition to Commercial Applications

The MSAG process evolution represents a good example of the application of military-based technology to commercial production. Although the high-performance GaAs MSAG technology had provided an effective foundation for high-volume MMIC production, the costs associated with wafer production and cycle time were high for most commercial applications. Recognizing the growing importance of wireless communications below 6 GHz, the MSAG process was simplified by eliminating the high-cost drivers of the process, such as airbridges and backside vias, to address several low-

TABLE 3 Features of the Dual Use MSAG Processes

Feature	MSAG	MSAG-lite	MSAG-switch
Max. f(GHz)	<30	<6	<6
Drain voltage (V)	2 to 10	1.5 to 7	3 to 5
Metal 1 (μm)	0.6	0.6	None
Metal 2 (μm)	4.5	4.5	1.0
Glassivation	Polyimide	Polyimide	SiON
Airbridges	Yes	No	No
Backside vias	Yes	No	No
Photooperations	12	8	6
Major operations	50	37	30
Cycle time (weeks)	7	5	3.5

frequency microwave applications. Consequently, the GaAs die area and manufacturing cycle time were reduced. The two resulting lower frequency processes are denoted MSAG-Switch for switching applications and MSAG-Lite for active device applications such as amplifiers and mixers. Table 3 summarizes the major differences between the three MSAG fabrication processes [56, 57].

1.8 MMIC PACKAGING

Microwave packages [58–67] and assembly techniques play a very important role in the performance, cost, and reliability of MMICs. Since MMICs represent state-of-the-art technology in terms of size, weight, performance, reliability, and cost, their performance should not be compromised by the packaging. The affordability requirement on packages mandates that their complexity be minimized. Minimizing both the number of dielectric layers and the overall size improves electrical performance and production yields and lowers costs. However, a trade-off exists between simplicity and the number of functional features in terms of cost. Some high-volume applications demand package costs as low as 25 cents, while high-performance, low-volume applications can tolerate package costs in the $10–50 range.

Many of the packaging considerations for MMICs are similar to those for hybrid MICs. Most ceramic/metal packages should meet the environmental requirements of MIL-S-19500 and test requirements of MIL-STD-750/883. The package must pass rigorous tests of hermeticity, thermal and mechanical shock, moisture resistance and resistance to salt atmosphere, vibration and acceleration, and solderability. In order to minimize the effect of the package on MMIC performance, electrical, mechanical, and thermal modeling of packages must be performed.

The most important electrical characteristics of microwave packages

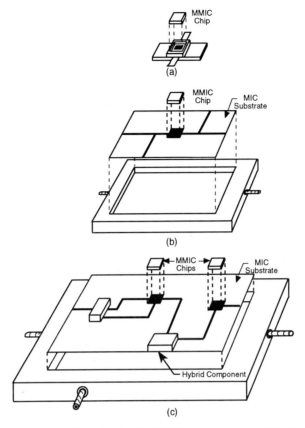

FIG. 12 Three packaging level details. (a) MMIC in package, (b) MMIC with support circuitry, and (c) MMICs with hybrid and support circuitry.

are low insertion loss, high return loss and isolation, and no cavity or feedthrough resonance over the operating frequency range. When a chip or chip set is placed in the cavity of a microwave package, there should be minimum degradation in the chip's performance. Generally, this cannot be accomplished without accurate electrical and electromagnetic modeling of the critical package elements. Microwave design must be applied to three parts of the package: RF feedthrough, cavity, and dc bias lines. Of the three, the design of the RF feedthrough is the most critical in determining the performance of packaged MMIC chips.

MMIC packaging can be performed at three levels as shown in Fig. 12. ICs can be mounted in individual packages; ICs can be packaged with support circuitry in a housing; or the ICs can be packaged at the subsystem level. The packaging requirements depend upon the application at hand. Table 4 summarizes [60] various packages in terms of performance and cost.

TABLE 4 Summarized Results of Packaging Survey

Package	Frequency (GHz) VSWR 1.2:1	VSWR 2.0:1	Loss per IO port	Isolation	Cost NRE	100	100K	Thermal resistance	Reliability	Adaptable
Cofired ceramic with Cu-W base	20 GHz	26 GHz	<0.25 dB	>40 dB	$ 15–40K	<$ 50	<$ 7	Low	Excellent	Boil or solder, coplanar or microstrip
Metal housing cofired feedthroughs	20 GHz	30 GHz	<0.25 dB	>40 dB	$ 10–15K	$ 50–100	$ 20–30	Low	Excellent	Feedthroughs adapt to your housing design
Microwave glass flatpack	N/A	20 GHz	<0.25 dB	>25 dB	$ 5K	<$ 6	<$ 4	Low when configured with a Cu-W Base	Good, usual handling precaution	Coplanar, microstrip version ok at lower freq.
Surface mount	12 GHz	20 GHz	<0.5 dB	>30 dB	$ 1–2K	$ 10–15	<$ 3	High, 35° C/W	Excellent	Poor, specific medium required
Metal housing glass feedthroughs	20 GHz	VSWR highly depends on transition	<0.2	>40	$ 1–5K	<$ 50	<$ 15	Low when configured with Cu-W base	Susceptible to mishandling	Requires Motherboard
Microwave quality T.O. cans	16 GHz	18 GHz	0.1 dB	>50	$ 1–5K	Not known	Not known	Dependent on attachment method	Susceptible to mishandling	Very inflexible
Nonhermetic chip carrier	Circuit and cavity dependent		Low	Topology dependent	<$500	<$ 2	<$ 2	Excellent, when made of Cu-W or Moly	Susceptible to mishandling	Best

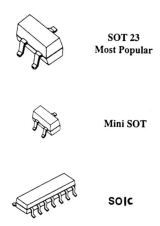

FIG. 13 Most popular plastic packages for RF and microwave applications—SOT package has 3 leads and SOIC package has 14 leads.

For wireless communications applications GaAs MMICs are being mounted into plastic packages in order to achieve low cost goals. Small outline transistor (SOT) and small outline integrated circuit (SOIC) packages are commonly used. These packages are shown in Fig. 13. SOIC packages have 8 to 16 pins and they work reasonably well up to 2 GHz. The measured dissipative loss in a SOIC 8 lead package is on the order of 0.2 dB at 2 GHz [68]. In order to improve the RF performance and power dissipation for power ICs, custom fused lead frames with low signal lead parasitics and reduced ground bond inductance are being used in custom plastic packages. Plastic molded IC packages are described in two packaging handbooks [63, 69]. The dielectric constant and loss tangent values of the organic moldling compound are about 3.7 and 0.01, respectively. The lead frame, which is the central supporting structure for ICs, is the backbone of a molded plastic package. Several different types of lead frame materials are being used such as nickel–iron and copper-based alloys, their selection for a particular application depends on factors such as cost, performance, and ease of fabrication. The ICs are packaged using surface mounting techniques. Over 10 billion SOIC packages are manufactured per year and the material and packaging labor cost all together is less than 25 cents per package in large quantities.

1.9 SUMMARY

The performance advantages of GaAs devices for microwave circuits have been realized for over two decades in the form of hybrid circuits utilizing discrete GaAs transistors. The further advantages of smaller size, lower weight, higher reliability, and lower cost in high-volume production are now realized routinely through MMIC technology. The MESFET has become the workhorse of the MMIC industry, while more complex and costly devices, such as HBT and HEMTs, are beginning to play important roles in high-performance niches. Two of the key factors in the maturation of MMIC technology are improved models and improved CAD tools to permit accurate simulation of performance for enhanced first pass design success. Although highly complex, MMIC processing has stabilized sufficiently that MMICs can be produced at reasonably high yields and at affordable cost. Several GaAs foundries are now processing 4" wafers which will further reduce MMIC costs. Finally, a wide variety of packages is available to cover the range of MMIC packaging requirements, but further work must be done to reduce costs.

In conclusion, MMIC technology has reached a state of maturity that makes high-volume systems applications practical. As will be seen in the succeeding chapters, with the advantages of small size, light weight, low cost, and multifunction performances, MMICs are now providing the enabling technology for a multitude of new applications, both military and commercial.

REFERENCES

[1] T. M. Hyltin, Microstrip transmission on semiconductor substrates. *IEEE Trans. Microwave Theory Tech.* **MTT-13**, 777–781, Nov. 1965.
[2] E. Mehal and R. W. Wacker, GaAs integrated microwave circuits. *IEEE Trans. Microwave Theory Tech.* **MTT-16**, 451–454, July 1968.
[3] R. S. Pengelly and J. S. Turner, Monolithic broadband GaAs FET amplifiers. *Electron. Lett.* **12**, 251–252, May 13, 1976.
[4] J. V. Dilorenzo and D. D. Khandelwal (Eds.), *GaAs FET Principles and Technology.* Artech House, Norwood, MA, 1982.
[5] R. S. Pengally, *Microwave Field-Effect Transistors—Theory, Design and Applications.* Wiley, New York, 1982.
[6] R. Soares, J. Graffeuil, and J. Obregon (Eds.), *Applications of GaAs MESFETs.* Artech House, Norwood, MA, 1983.
[7] R. E. Williams, *Gallium Arsenide Processing Techniques.* Artech House, Norwood, MA, 1984.
[8] R. A. Pucel (Ed.), *Monolithic Microwave Integrated Circuits.* IEEE Press, Piscataway, NJ, 1985. [Reprint volume]
[9] D. K. Ferry (Ed.), *Gallium Arsenide Technology.* Howards, Sams, Indianapolis, IN, 1985.
[10] N. G. Einspruch and W. R. Wisseman, *GaAs Microelectronics.* Academic Press, New York, 1985.
[11] I. J. Bahl and P. Bhartia, *Microwave Solid State Circuit Design,* Chap. 15. Wiley, New York, 1988.
[12] R. Soares (Ed.), *GaAs MESFET Circuit Design.* Artech House, Norwood, MA, 1989.
[13] J. Mun (Ed.), *GaAs Integrated Circuits: Design and Technology.* McMillan, New York, 1988.
[14] P. H. Ladbrooke, *MMIC Design: GaAs FETs and HEMTs.* Artech House, Norwood, MA, 1989.

[15] R. Goyal (Ed.), *Monolithic Microwave Integrated Circuits: Technology and Design*. Artech House, Norwood, MA, 1989.
[16] K. Chang (Ed.), *Handbook of Microwave and Optical Components*, Vol. 2. Wiley, New York, 1990.
[17] R. A. Pucel, Design considerations for monolithic microwave circuits. *IEEE Trans. Microwave Theory Tech.* **MTT-29**, 513–534, June 1981.
[18] D. G. Fisher, GaAs IC applications in electronic warfare, radar and communications systems. *Microwave J.* **31**, 275–291, May 1988.
[19] D. N. McQuiddy, Jr., High volume applications for GaAs microwave and millimeter-wave ICs in military systems. *IEEE GaAs IC Symp. Dig.* 3–6, 1989.
[20] D. N. McQuiddy, Jr., et al., Transmit/receive module technology for X-band active array radar. *Proc. IEEE* **79**, 308–341, Mar. 1991.
[21] M. I. Herman, et al., Multifunction W-band MMIC receiver technology. *Proc. IEEE* **79**, 342–354, Mar. 1991.
[22] J. J. Komiak and A. K. Agrawal, Design and performance of octave S/C-band MMIC T/R modules for multi-function phased arrays. *IEEE Trans. Microwave Theory Tech.* **39**, 1955–1963, Dec. 1991.
[23] E. D. Cohen, Military applications of MMICs. *IEEE Microwave Millimeter-Wave Monolithic Circuits Symp. Dig.* 31–34, 1991.
[24] R. Rosenzweig, Commercial GaAs MMIC applications. *IEEE Microwave Millimeter-Wave Monolithic Circuits Symp. Dig.* 59–60, 1991.
[25] D. Willems and I. J. Bahl, Advances in monolithic microwave and millimeter wave integrated circuits. *IEEE Circuit and System Dig.* 783–786, 1992.
[26] I. Bahl, G. Lewis, and J. Jorgenson, Automatic testing of MMIC wafers. *Int. J. Microwave Millimeter-Wave Computer-Aided Eng.* **1**, 77–89, Jan. 1991.
[27] R. Majidi-Ahy, et al., 5–10 GHz InP CPW MMIC 7-section distributed amplifier. *IEEE Microwave Millimeter-Wave Monolithic Circuits Symp. Dig.* 31–34, 1990.
[28] H. Wang et al., High performance W-band monolithic InGaAs pseudomorphic HEMT LNAs and design/analysis methodology. *IEEE Trans. Microwave Theory Tech.* **40**, 417–428, March 1992.
[29] F. Ali, I. Bahl, and A. Gupta (Eds.), *Microwave and Millimeter-Wave Heterostructure Transistors and Their Applications*. Artech House, Norwood, MA, 1989.
[30] F. Ali and A. Gupta (Eds.), *HEMTs and HBTs: Devices, Fabrication and Circuits*. Artech House, Norwood, MA, 1991.
[31] L. C. T. Liu et al., A 30 GHz monolithic receiver. *IEEE Microwave Millimeter-Wave Monolithic Circuit Symp. Dig.* 41–44, 1986.
[32] W. W. Nelson and A. F. Podell, High volume, low cost, MMIC receiver front end. *IEEE Microwave Millimeter-Wave Monolithic Circuit Symp. Dig.* 57–60, 1986.
[33] P. Wallace et al., A low cost high performance MMIC low noise downconverter for direct broadcast satellite reception. *IEEE Microwave Millimeter-Wave Monolithic Circuit Symp. Dig.* 7–10, 1990.
[34] R. M. Herman, et al., Highly integrated mixed-signal L-band configurable receiver array. *IEEE Microwave Millimeter-Wave Monolithic Circuit Symp. Dig.* 23–26, 1994.
[35] A. Y. Umeda et al., A monolithic GaAs HBT upconverter. *IEEE Microwave Millimeter-Wave Monolithic Circuit Symp. Dig.* 77–80, 1990.
[36] K. W. Chang et al., A W-band monolithic downconverter. *IEEE Trans. Microwave Theory Tech.* 1972–1977, Dec. 1991.
[37] A Gupta et al., A 20-GHz 5-bit phase shift transmit module with 16 dB gain. *IEEE GaAs IC Symp. Dig.* 197–200, 1984.
[38] L. C. Witkowski et al., A GaAs single-chip transmit/receive radar module. *GOMAC Dig.* 339–342, 1986.
[39] W. R. Wisseman et al., X-band GaAs single-chip T/R radar module. *Microwave J.* **30**, 167–173, Sept. 1987.

[40] M. J. Schindler et al., A single-chip 2–20 GHz T/R module. *IEEE Microwave Millimeter-Wave Monolithic Circuit Symp. Dig.* 99–102, 1990.
[41] C. Andricos et al., 4 Watt monolithic GaAs C-band transceiver chip. *GOMAC Dig.* 195–198, 1991.
[42] L. M. Delvin et al., A 2.4 GHz single chip transceiver. *IEEE Microwave Millimeter-Wave Monolithic Circuit Symp. Dig.* 23–26, 1993.
[43] G. Dawe et al., A high performance 2.4 GHz transceiver chip-set for high volume commercial applications. *IEEE Microwave Millimeter-Wave Monolithic Circuit Symp. Dig.* 11–14, 1994.
[44] T. Apel et al., A GaAs MMIC transceiver for 2.45 GHz wireless commercial products. *IEEE Microwave Millimeter-Wave Monolithic Circuit Symp. Dig.* 15–18, 1994.
[45] M. S. Wang et al., A single-chip MMIC transceiver for 2.4 GHz spread spectrum communication. *IEEE Microwave Millimeter-Wave Monolithic Circuit Symp. Dig.* 19–22, 1994.
[46] L. Reynolds and Y. Ayasli, Single chip FM-CW radar for target velocity and range sensing applications. *IEEE GaAs IC Symp. Dig.* 243–246, 1989.
[47] L. R. Whicker et al., A new approach to active phased arrays through RF wafer scale integration. *IEEE MTT-S Int. Microwave Symp. Dig.* 1223–1226, 1990.
[48] O. Pitzalis, Microwave to mm-wave CAE: Concept to production. *Microwave J.* 15–47, 1989.
[49] R. H. Jansen, Computer-aided design of hybrid and monolithic microwave integrated circuits—State of the art, problems and trends. *13th Eur. Microwave Conf. Procs.* 67–78, 1983.
[50] U. L. Rohde et al., MMIC workstations for the 1990s. *Microwave J.* 51–77, 1989.
[51] A. K. Sharma, Considerations in producibility engineering of MMICs. *IEEE Trans. Microwave Theory Tech.* 38, 1242–1248, Sept. 1990.
[52] R. L. Van Tuyl et al., A manufacturing process for analog and digital gallium arsenide integrated circuits. *IEEE Trans. Microwave Theory Tech.* **MTT-30**, 935–942, July 1982.
[53] T. Andrade, Manufacturing technology for GaAs monolithic microwave integrated circuits. *Solid State Technol.* 199–205, Feb. 1985.
[54] A. E. Geissberger et al., Refractory self-aligned gate technology for GaAs microwave FETs and MMICs. *Electron. Lett.* 23, 1073–1075, Sept. 1987.
[55] A. E. Geissberger et al., A refractory self-aligned gate technology for GaAs microwave power FETs and MMICs. *IEEE Trans. Electron. Devices* 35, 615–622, Dec. 1988.
[56] J. Bell and G. Studtmann, Successful commercialization of a high performance GaAs MMIC process. *GOMAC Dig.* 465–468, 1994.
[57] C. E. Lindberg and D. G. Fisher, MSAG-Lite: GaAs IC process technology addressing defense confersion. *IEEE MTT-S Microwave Symp. Dig.*, 1995.
[58] R. S. Pengelly and P. Schumacher, High-performance 20 GHz package for GaAs MMICs. *MSN Commun. Technol.* 18, 10–19, Jan. 1988.
[59] F. Ishitsuka and N. Sato, Low-cost, high performance package for a multi-chip MMICs module. *IEEE GaAs IC Symp. Dig.* 221–24, 1988.
[60] B. Berson, Strategies for microwave and millimeter wave packaging today. *19th Eur. Microwave Conf. Proc.* 89–95, 1989.
[61] H. Tomimuro et al., A new packaging technology for Gaas MMICs modules. *IEEE GaAs IC Symp. Dig.* 307–310, 1989.
[62] B. Berson, F. Rosenbaum, and R. A. Sparks, MMIC packaging. In *Monolithic Microwave Integrated Circuits* (R. Goyal, Ed.). Artech House, Norwood, MA 1989.
[63] R. R. Tummala and E. J. Rayaszewski (Eds.), *Microelectronic Packaging Handbook*. Van Nostrand Reinhold, New York, 1989.
[64] S. Chai et al., Low-cost package technology for advanced MMIC applications. *IEEE MTT-S Int. Microwave Symp. Dig.* 625–628, 1990.
[65] S. R. Smith and M. T. Murphy, Electrical characterization of packages for use with GaAs MMIC amplifiers. *IEEE MTT-S Int. Microwave Symp. Dig.* 131–134, 1993.

[66] D. Wein *et al.*, Microwave and millimeter-wave packaging and interconnection methods for single and multiple chip modules. *IEEE GaAs IC Symp. Dig.* 333–336, 1993.
[67] G. Strauss and W. Menzel, A novel concept for mm-wave interconnects and packaging. *IEEE MTT-S Int. Microwave Symp. Dig.* 1141–1144, 1994.
[68] F. Ndagijimana *et al.*, The inductive connection effects of a mounted SPDT in a plastic S08 package. *IEEE MTT-S Int. Microwave Symp. Dig.* 91–94, 1993.
[69] L. T. Manzione, *Plastic Packaging of Microelectronic Devices*. Van Nostrand Reinhold, New York, 1990.

2
MMIC Application Overview

Inder Bahl and Dennis Fisher
ITT Gallium Arsenide Technology Center, Roanoke, Virginia

2.1 Introduction
2.2 Role of GaAs MMICs in Systems
2.3 Historical Market Projections
2.4 Military System Applications
2.5 ARPA MIMIC Program
    2.5.1 Background
    2.5.2 Program Tasks
    2.5.3 Accomplishments
2.6 Commercial System Applications
    2.6.1 Wireless Communications
    2.6.2 Computers and Networks
    2.6.3 Equipment and Instrumentation
    2.6.4 Consumer Electronics
    2.6.5 Intelligent-Vehicle Highway System (IVHS)
2.7 Summary
    References

2.1 INTRODUCTION

Tremendous progress has been made in the development and insertion of monolithic microwave integrated circuit (MMIC) technology. MMIC technology was driven by military applications due to high performance, small size, light weight, and reliability requirements that could be met only through monolithic integration. More recently, high-volume, high-performance military applications of MMICs [1] such as active aperture arrays, missiles, electronic warfare, expendables, and communication systems are maturing. These applications benefit from the cost savings of MMIC manufacturing, where hundreds of thousands of virtually identical parts can be expected to be made at a cost of $100 or less at the subsystem level. The realization of even lower cost high-volume components with less stringent requirements has opened up commercial markets [2]. Indeed, commercial applications are rapidly becoming the major market for MMICs; examples are mobile communications, direct broadcast satellite (DBS), global positioning satellite (GPS), personal communications, and automotive object detection. As the MMIC industry is maturing, both military and commercial customers are demanding continuously higher performance and lower cost for MMICs, with primary focus on performance for military and low cost for commercial.

2.2 ROLE OF GaAs MMICs IN SYSTEMS

One of the major advantages of gallium arsenide (GaAs) ICs is that many types of microwave circuits (e.g., low-noise amplifiers, high-power amplifiers, mixers, phase shifters, etc.) can be combined on a single chip along with high-speed digital circuits. In other words, a complete subsystem, such as a transmit/receive (T/R) module which was previously implemented in hybrid technology, can now be realized on a single chip. For example, a complete multifunction monolithic transceiver chip containing 16 microwave circuits has been developed on a GaAs substrate measuring only 65 mm^2 [3]. A high-level integration approach minimizes part counts and saves labor-intensive packaging, resulting in lower cost subsystems with enhanced reliability.

2.3 HISTORICAL MARKET PROJECTIONS

Market projections, conferences, seminars, workshops, and foundry courses over the years have continually predicted that the GaAs industry has great potential and that there are numerous applications which require thousands to millions of GaAs ICs. What happened to those predictions during the past 10 years? With each passing year the projections have slipped, sometimes by more than a year. What have been the missing ingredients? Some of the main factors for slow growth in the GaAs IC industry are:

(i) Technology immaturity. The relatively slow emergence of the GaAs IC market can be attributable in part to long development times due to multi-iteration designs, and the difficulty in manufacturing multifunction chips with uniform, repeatable performance, high reliability, and good yield. As a consequence, development costs and manufacturing costs have remained high.

(ii) Lack of standard high-volume products. Available products are application specific. Specific products in moderate volume are expensive due to high nonrecurring engineering (NRE) costs, unavailability of reliability data, etc.

(iii) High-volume military applications such as active aperture antennas, electronic warfare, expendables, and smart weapons have not fully developed, owing in part to high MMIC costs.

(iv) Lack of commercial confidence resulting from factors such as unpredictable MMIC supply and weak links between users and MMIC vendors.

Chapter 1 highlighted some of the important technology progress and accomplishments which have eliminated the technology immaturity barrier.

- Stable GaAs substrate material quality
- Automated and high-yield MMIC processing
- Accurate models and integrated "first pass design" CAD tools
- Integrated automatic testing and prescreening of power ICs on wafer
- Single- and multichip packages for high-performance and low-cost applications
- Automatic assembly of chips into packages and modules

The remainder of this chapter will show that the other barriers to widespread MMIC system insertion are also being eliminated. The Advanced Research Projects Agency (ARPA) MIMIC program, discussed later in this chapter, has played a key role in promoting system insertions in addition to helping to mature the technology. Since the major portion of this book covers military electronic equipment, their designated nomenclature is given in Appendix A for the reader's convenience.

2.4 MILITARY SYSTEM APPLICATIONS

GaAs MMIC technology offers tremendous potential for both improving existing systems through high performance, small size, and light weight integrated microwave components and making next-generation high-volume military microwave systems affordable [1, 4–16]. Table 1 shows some of the military system applications. High-volume military system applications of MMICs are in the areas of radar, electronic warfare, smart weapons, and communications. Many of these systems incorporate active array antennas which in turn depend heavily on MMICs. MMICs are targeted for use in over 50 United States Department of Defense (DoD) systems covering the above-mentioned applications from 1 to 100 GHz. For example, the unmanned aerial vehicle (UAV) is a potential candidate for MMIC insertion due to small size and light weight requirements. The system which performs navigation, reconnaissance, surveillance, and electronic countermeasures will have electronics parts that make up more than 60% of the total cost. Many of these parts will be MMICs.

TABLE 1 Military Needs for High Volume MMICs for 4 Major System Insertions [5]

Systems category	MMIC quantities	
	1994–1996	1997–2000
Radar	3.6M	7.3M
Electronic warfare	4.0M	8.0M
Smart weapons	700K	1.0M
Communications	500K	600K

Active Phased-Array Radars

The active array antenna which uses a large number of radiating elements (on the order of 1000) has the capability to form and steer multiple beams and beam nulls simultaneously. This is achieved by adjusting the phase and amplitude of the signal of individual antenna elements. This approach uses efficient free-space power combining of many low- to medium-power radiators. The system can function effectively even with the failure of 5 to 10% of the elements and, as such, is more robust than a conventional radar.

The phased-array radar applications benefit significantly from MMIC technology. The transceiver (T/R) antenna element is one of the primary cost drivers in these applications. Figure 1 shows a simplified block diagram of a T/R module whose size and cost can be reduced drastically by monolithically integrating virtually all the microwave functions except the circulator and the antenna.

MMICs are being used in the following radar systems: ground based, fuze, advanced tactical surveillance (ATS), advanced air traffic control (AATC), airborne multimode, firefinder, helicopter all-weather fire control (HAWFC), advanced tactical fighter, airborne shared aperture (ASA), airborne early warning and control system (AWACS), spacebased ballistic missile detection, and satellitebased. Chapter 4 deals with the active array topic in detail.

Communication

Several communication systems are being developed using MMICs. Major areas of interest are in high-performance phased arrays, low-cost radios and receivers, and portable communication terminals. Satellite communication systems at millimeter wave frequencies utilize MMICs: for example, 44-GHz uplinks and 20-GHz downlinks between the satellites and air, sea, and earth terminals, which incorporate frequency hopping over a 2-GHz range for interception security and antijam protection. Specific system applications include defense satellite communication system (DSCS),

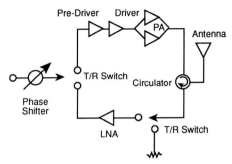

FIG. 1 A block diagram that illustrates the microwave functions of a T/R module.

MILSTAR Network, Mark XV, GPS, antijam data-links, joint tactical information distribution system (JTIDS), mobile intercept resistance radio (MIRR), integrated communication navigation identification avionics (ICNIA), and tactical radio.

Electronic Warfare

Electronic warfare (EW) consists of techniques, equipment, and tactics for minimizing the effectiveness of enemy's electronic systems that use electromagnetic sensing and transmitting devices. In order to achieve this, several functions of electronic support measures (ESM), electronic countermeasures (ECM), and electronic counter-countermeasures (ECCM) are required. The design of sophisticated electronic warfare systems (EW, ECM, ECCM, etc.) is influenced greatly by the increasing availability of low-cost, broadband, and reliable MMICs and digital ICs. Thus, MMIC technology is a critical factor in future systems that must satisfy electronic battlefield requirements while remaining within acceptable cost limits.

The EW application area provides a broad spectrum of GaAs IC system insertion opportunities ranging from form, fit, and function replacement of existing components to unconstrained designs that can utilize the full benefits of the GaAs technology, both analog and digital. With the exception of expendable decoys, the EW area GaAs IC applications are of modest volume by traditional integrated circuit standards. Quantities of a given chip type used in a production run are typically in the 1000 to 10,000 range. However, despite these relatively modest volumes, GaAs IC utilization can provide substantially reduced systems acquisition and life cycle costs.

A number of electronic warfare systems are expected to benefit from MMIC technology. Examples are integrated electronic warfare systems (INEWS), airborne jammers such as ALQ-126, ALQ-131, ALQ-135, ALQ-136, ALQ-161, generic expendable decoys, decoy (GEN-X), and next-generation systems including airborne shared aperture. MMICs are already in use in the GEN-X system. An example of a generic up/downconverter for EW

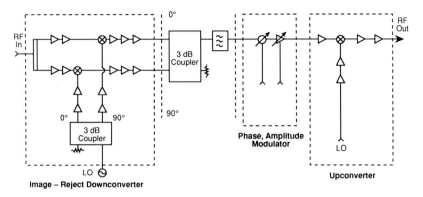

FIG. 2 The block diagram is shown for an up/downconverter for generic EW applications. Dotted lines denote typical individual MMICs.

applications [6] is shown in Fig. 2. The circuit functions shown in broken lines can easily be implemented as MMICs. The millimeter-wave spectrum is another area where MMIC technology will benefit EW systems in terms of low cost and light weight. Chapters 5 and 6 treat EW systems in detail.

Smart Weapons

Smart weapons for use against stationary and moving armored targets in all weather and terrain conditions must be affordable in large-volume production. MMICs are suitable for these applications. One of the earliest applications of MMICs in this area is the high-speed anti-radiation missile (HARM). Several other microwave and millimeter-wave smart weapons are currently under development. The use of MMIC chips is making these applications practical in terms of cost and size. Precision guided weapons (PGW) such as Longbow and sense and destroy armor (SADARM) system, both operate at Ka-band [12]. The Longbow millimeter-wave system includes a fire control radar (FCR) signaling a Hellfire anti-armor missile (HAAM) which has a millimeter-wave seeker. The millimeter-wave seeker, integrated in the front of HAAM provides fire-and-forget capability. The SADARM system consists of both millimeter-wave and infrared sensors to detect, discriminate, determine optimum aimpoint, and fire its warhead to destroy armored vehicles. Other smart weapons employing MMICs include advanced air-to-air missile (AAAM), precision-guided munitions such as the multiple-launched rocket systems and terminally guided warhead (MLRS/TGW), low-cost missile seeker, advanced medium-range air-to-air missile (AMRAAM), airborne adverse weather weapon system (AAWWS), multiple option fuse for artillery (MOFA), and X-Rod. Figure 3 shows the block diagram of a 94-GHz MMIC-based downconverter developed using high electron mobility transistor (HEMT) tech-

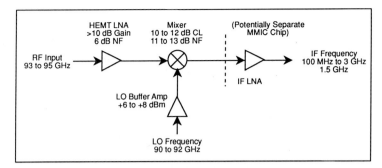

FIG. 3 A 94-GHz MMIC downconverter for missile seekers.

nology for missile seekers. The cost budget of a small kinetic energy weapon (X-Rod) operating at W-band (94 GHz) that will penetrate advanced armor at extended range can be realized only by using HEMT-based GaAs MMIC technology.

The MMIC-based MOFA is expected to provide the most important advance in fuze technology in terms of operating frequency and accuracy of setting the height at which the projectile bursts. A complete frequency modulation/continuous wave (FM/CW) radar costing less than $10 in large quantities was only possible with MMIC technology.

Specific MMICs for the above applications are described in the next section on the ARPA MIMIC program.

2.5 ARPA MIMIC PROGRAM

2.5.1 Background

In 1987, the United States Department of Defense (DoD) launched a major program called MIMIC, an acronym for microwave and millimeter-wave monolithic integrated circuits, to provide resources and structure in order to transition GaAs IC technology from a research and development mode to a mature manufacturing status. This program was designed to provide affordable, reliable, and reproducible microwave and millimeter-wave integrated ICs in the 1- to 100-GHz frequency range for a wide variety of defense applications, including upgrading of current systems and developing new affordable systems [1, 11].

The MIMIC program is sponsored by the ARPA and has full participation by the Air Force, Army, and Navy. The program is divided into four portions, designated Phases 0, 1, 2, and 3. The Phase 0, 1, and 2 programs started in January 1987, May 1988, and September 1991, respectively. The Phase 3 programs consist of several critical technology programs to supplement the efforts of Phases 1 and 2, running in parallel with both Phases 1 and 2. Altogether, the duration of the MIMIC Program, now nearing completion, is about 8 years and the budget is about $500M. The overall program objective is to [11]:

> Provide the needed microwave and millimeter-wave products at a price that will allow their use in fielded Department of Defense systems, that meet all required electrical, mechanical, and environmental parameters, and that continue to operate reliably for the time necessary to fulfill their intended application.

The MIMIC program's main focus is military system insertions; however, the impact of technology development on the nonmilitary applications is significant as well. The MIMIC program has established a solid infrastructure for MMIC technology.

Phase 0   The MIMIC Phase 0 Program was a study phase of 1 year undertaken by 16 cooperative teams consisting of 48 companies including foundries, CAD houses, material manufacturers, system houses, etc. During this study phase several critical technology development areas needed for successful development of MMICs were identified along with most promising system insertions for MMICs.

Phase 1   The MIMIC Phase 1 Program was a technology and hardware development phase undertaken by four teams consisting of 26 companies. The program was of 3 years duration and the total budget was about $225M. The primary objective of this phase was to develop the required technology to meet the goals of affordable ICs. The critical areas included GaAs material growth and wafer preparation, device and circuit modeling, CAD, computer-aided manufacturing (CAM), computer-aided test (CAT) equipment and tools, low-cost and high-performance packaging, and MMIC subsystem development. Another important aspect of this phase was the requirement that all circuits developed were targeted for specific system insertions to guarantee system relevancy for the development efforts.

Phase 2   The MIMIC Phase 2 Program is a 3- to 4-year effort which started in September 1991 and its primary objective is the hardware demonstrations for specific systems. The three teams (consisting of 31 companies) were awarded $230M to continue to develop MMICs and MMIC-based subsystems to demonstrate the readiness to manufacture affordable and reliable MMICs operating over the 1- to 100-GHz frequency range. In addition, Phase 2 MIMIC objectives include continued technology and product developments started in Phase 1, pursuit of new technology approaches, further demonstration of the affordability of MMIC devices and circuits, and foundry support.

Phase 3   Phase 3 consists of a large number of small but technologically critical programs which are being conducted concurrently with Phase 1 and Phase 2. These programs allow special and focused efforts which further advance the mainstream MIMIC activities. Examples of Phase 3 program activities are GaAs material development, multilayered multichip ceramic package development, on-wafer pulse power test capability development, process, device and circuit modeling, standard microwave hardware description language development, and merchant foundry qualification.

2.5.2 Program Tasks

The major tasks which have been and are being performed during this program are as follows:

(i) Establish a reliable, high-yield, and manufacturable process which can be used to fabricate MMICs as required to upgrade current and the next generation military systems.

(ii) Develop an integrated CAD system that provides an open architecture framework to use a wide variety of available software to design devices, MMICs, subsystems, and packages. Improved CAD capability is a key ingredient in achieving first pass success in MMIC designs. This, in turn, will reduce the time between the system concept and final implementation.

(iii) Develop a fully automated and integrated test capability which can be used for on-wafer testing and characterization of process control monitors (PCMs) and MMICs at early stages of fabrication. Both contact and noncontact probing techniques are included. On-wafer testing permits screening of substandard devices both at early stages of wafer fabrication and prior to mounting of the devices in packages, thus saving otherwise significant costs.

(iv) Develop low-cost and high-performance MMIC packages which will help in producing affordable subsystems.

(v) Establish a solid foundation for designing, fabricating, packaging, and testing for sustainable, low-cost, and high-yield production operations, making the technology transportable from one source to another.

(vi) Demonstrate chips in systems. The MIMIC Program is geared toward the successful development of military systems based on MMICs. In order to evaluate the success of this program, several system insertions were selected in all the four major disciplines, radar, electronic warfare, communications, and smart weapons. An extensive summary of hardware demonstrations for several systems is given in Figs. 4a and 4b.

2.5.3 Accomplishments

Substantial progress has been made in the MIMIC Program to advance and improve the performance of GaAs monolithic technology, including

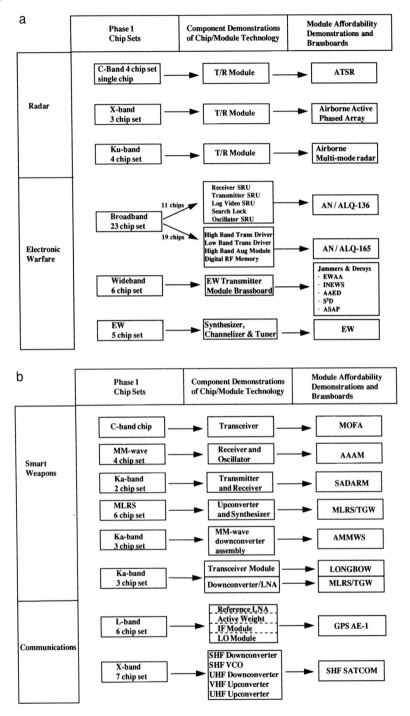

FIG. 4 (a and b) Examples of MMIC insertions into military systems—MIMIC Phase I.

device modeling, CAD tools, GaAs materials, fabrication, CAM, test development on CAT and packages. These advancements have resulted in improved performance, reliability, yields and unit costs. Computer controlled and high-throughput manufacturing facilities have been established. New techniques have been developed that can ultimately reduce the cost of chips, testing, and packaging. Progress is also evident from the fact that many chips are being used in production units and are increasingly scheduled for use in future military systems.

During the 1988 to 1994 period, more than 170 MMIC chips for use in numerous demo brassboards were designed, fabricated, tested, and evaluated in a brassboard environment. They span all four major systems areas as described earlier, namely radar, EW, communications, and smart weapons: A summary of the specific system demonstrations in the MMIC program is given below.

Radar A set of chips for COBRA (counter battery radar), another set for airborne phased-array radars, and several chips for Ku-band T/R modules were developed. Potential platforms for airborne phased-array radars (PARs) include the Air Force's advanced tactical fighter (ATF), the F-15 and F-16 aircrafts. A complete transceiver chip consisting of 16 microwave functions for C-band phased array radar/communication use was also developed. The first chip operated in the 5.25- to 5.85-GHz frequency range and demonstrated over 3.5 W power out with 35% power added efficiency (PAE) and 40 dB gain in the transmitting path. The receiver's noise figure was less than 4.0 dB with 22 dB minimum gain. A significantly reduced size T/R chip of 65 mm^2 was successfully developed for active phased-array antennae operating at C-band. MMIC chips are being used in systems such as cooperative engagement capability (CEC), Patriot radar, airborne active array radar, and groundbased radar (GBR). Over 10,000 MMIC-based T/R modules have been produced for the GBR program.

EW Chips were developed and demonstrated for several EW systems. Examples are AN/ALQ-136 (helicopter-mounted jammer), AN/ALQ-165 (tactical fighter jammer), GEN-X (jamming during the terminal guidance phase of approaching missiles), and other EW channelized receivers. Even for systems where the quantities are relatively low, significant cost savings are projected, using MIMIC program-developed chips. A dual 6.8- to 10.7-GHz T/R module using 72 MMICs was developed [17] for EW systems.

Communications In the communication area MMIC chips for several systems were developed. These include L-band GPS receivers, X-band DSCS, and a Q-band (millimeter-wave) communication system. Other applications of MMICs include single-channel advanced man portable (SCAMP) terminal and satellite extremely high frequency (EHF) data link.

Smart Weapons   One of the major applications of MMICs in military systems is smart weapons (SW). Examples demonstrated in the MIMIC program are HARM, MOFA, SADARM, AMRAAM, AAAM, X-Rod, MLRS/TGW, smart target activated fire and forget (STAFF), and Longbow. At present about 15,000 MMIC chips are being produced per year for the advanced air-to-air missiles.

2.6 COMMERCIAL SYSTEM APPLICATIONS

GaAs monolithic IC technology is finding wide applications in upgrade versions of existing and next-generation high-volume commercial systems up to 100 GHz as shown in Fig. 5. The widespread usage is taking place due to high-yield manufacturing of GaAs monolithic ICs, high performance, high reliability, and small size. A number of potential applications of monolithic ICs in the mmW frequency range (30 to 100 GHz) have also been identified [18–25]. MMICs have an important advantage over hybrid MIC technology at these high frequencies in having minimal performance degradation due to many fewer wire bonds. Also MMICs are far less labor-intensive and hence, are much lower in cost.

Commercial applications of GaAs ICs [2, 26–46] can be classified into five categories: wireless communication, computer and networks, equipment and instrumentation, consumer electronics, and intelligent-vehicle highway systems (IVHS). These applications are briefly described in the following subsections.

2.6.1 Wireless Communications

The communications applications area of MMICs is quite diversified. As shown in Fig. 6 it can be further subdivided into (a) satellite based

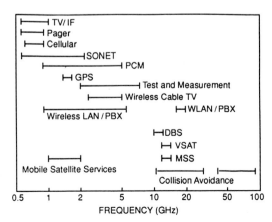

FIG. 5   Commercial application frequencies of GaAs MMICs.

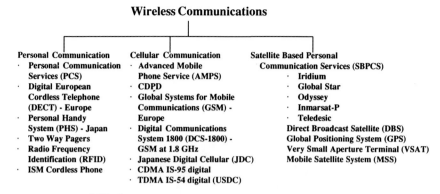

FIG. 6 MMIC applications in wireless communications.

communication and navigation and (b) cellular type wireless personal communication. Some applications span both of these areas. Satellite-based communications and navigation applications include direct broadcast satellite (DBS), television receive only (TVRO), global positioning satellite (GPS), very small aperture terminal (VSAT), and cellular telephone. Cellular personal communications applications include cellular telephones, cordless telephones, local cellular radio networks at 60 GHz, hand-held pagers, digital cellular telephones, and RF identification (RFID) card readers, and meter readers. The cellular area is characterized by the majority of the applications having operating bands in the RF or low microwave region (generally 0.8 to 2.4 GHz).

Satellite-based Communication

Satellites have become an integral part of our lives as they carry television programs, telephone connection, facsimile transmissions, and video conferences. GPS (navigation, guidance, and mapping) and data transmission using VSAT systems are also carried through satellites. Cellular telephones which connect anybody anywhere in the world through low Earth-orbit (LEO) satellites will bring another dimension to our dependency on satellites. VSAT offers privately owned communication networks to handle data transfer, facsimile, video, and voice. This satellite-based network will be used for data gathering and transfer, data applications, file retrieval, electronic mail and banking, real-time inventory control, message store and forward, point of sale, credit-card verification, etc. [26]. Earth terminals operate in the 16- to 16.5-GHz and 11.7- to 12.2-GHz bands for the uplink and downlink, respectively. Other potential areas of MMIC usage are in mobile satellite systems (MSS). Figure 7 shows the mobile satellite system which will be launched in 1994 for commercial applications. For example, the Omni TRACs system [34] receives its signal at X-band (10.95–12.75 GHz) and transmits at Ku-band (16–16.5 GHz), providing two-way satellite communication and position location for trucks.

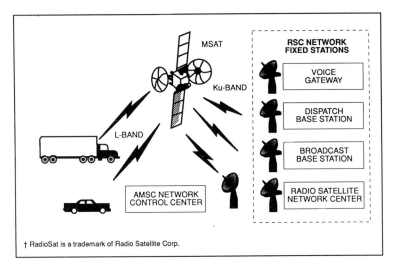

FIG. 7 The RadioSat system. (Reprinted with permission from G.K. Noreen [35], and Microwave Journal.)

MMICs are also finding use in millimeter-wave communication satellites at 30/20 GHz. A complete MMIC-based multichip receiver module at 30 GHz for use in the transponders on the Japanese Engineering Test Satellite XI has been developed. The advantages of millimeter-wave satellite communications are large bandwidth, small antenna size, compact earth terminal, and less interference between communications systems caused by atmospheric absorption.

Personal satellite communications and intersatellite communications links are envisioned for the future [18]. Personal satellite communications will provide portable video phone services, portable news gathering and distribution systems, medical and health consultation systems, and information exchange. Intersatellite communication systems will provide links between geostationary orbit satellites, spacecraft to geostationary orbit satellites and low earth orbit to geostationary orbit satellites.

We now describe several specific commercial satellite applications for MMICs.

Global Positioning Satellite (GPS) The Navstar GPS is based on a constellation of 24 satellites [29, 30, 38] orbiting Earth at very high altitude and is designed to provide highly accurate navigation 24 h a day. The GPS system is developed and managed by the U.S. Department of Defense, with investments of over $12 billion. The current system consists of 24 satellites (21 operating and 3 spares) bringing the GPS network to full operational capability. The GPS satellites which orbit Earth twice a day transmit precise time and position (latitude, longitude, and altitude) information via time-delay measurements from a minimum of three sets. With a GPS receiver, users can determine their location anywhere on Earth.

Satellites are transmitting two signals: (1) a clear/acquisition (CA) code for civilian use, the GPS receiver can determine position within 15 to 25 mm accuracy, and (2) a protected code used by U.S. military provides additional accuracy for military operations. For civilian use, the signal is transmitted at 1575.42 MHz. The GPS satellite differs from the geosynchronous communications satellites. The GPS constellation is grouped into six orbits, each containing four evenly spaced satellites. It takes 12 h for each satellite to complete one revolution around Earth at an altitude of 10,900 nautical miles. Regardless of the time or the location, there are at least four satellites passing overhead which are required to determine three-dimensional position. The GPS is designed to provide precise position, velocity, and time information to both civilian and military users with GPS receivers. The GPS will also be a key technology in recreational, personal security, automotive navigation, and marine navigation applications. With the help of GPS receivers, campers and hikers can find their route and their positions; many cars will have GPS-based in-car navigation systems featuring en route listings of restaurants, shops, and hotels and an optional television receiver. Other uses of GPS are in aircraft navigation.

The GPS industry is looking for MMICs in large volume at a cost between $2 and $10. Figure 8 shows a block diagram of a commercial GPS

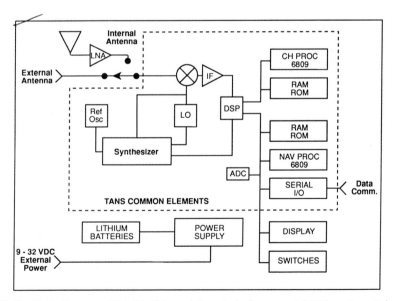

FIG. 8 Block diagram of Trimble Trimpack is typical of commercial GPS receivers. L-band spread-spectrum signal is downconverted and sampled before VLSI digital signal processing extracts positioning information [30].

receiver. A five-chip GPS receiver, based on one MMIC chip and four silicon very large scale integrated (VLSI) chips, has been developed [30]. The MMIC chip consists of several functions from low-noise amplifier (LNA) through analog/digital (A/D) converters, including voltage-controlled oscillators (VCOs), and has a noise figure (NF) less than 3.5 dB.

Satellite Cellular Telephone In order to provide worldwide cellular telephone and portable phone service, Motorola is working on the 66-Satellite Iridium System. Both the satellite system and the hand-held portable radio telephones will use large numbers of MMICs [36, 47]. The data link between the satellites will be at 23 GHz, the satellite communication link to gateway ground stations (which tie into existing telephone communication systems and which also provide command, control, and telemetry) is at 20 GHz, and the uplink is at 30 GHz. The communication link between satellites and portable cellular radio phones is at 1.6 GHz. The Iridium system will use active phased-array antennas using MMIC-based T/R modules at both L- and K-bands. Satellite-based personal communication services (SBPCS) will provide global communication with a common handset which will carry voice, paging, data, e-mail, fax, and video. An overview of the Iridium communication system is shown in Fig. 9. The system will be in operation in 1998 and will open up many new opportunities for MMIC applications.

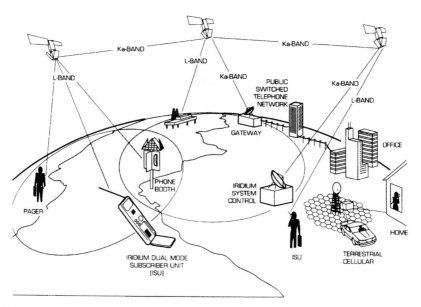

FIG. 9 Overview of Iridium communications system. (Reprinted with permission from R.J. Leopold and A. Miller [47], and IEEE, © 1993.)

TABLE 2 Comparison of Various Satellite-based Personal Communication Systems

	TRW Odyssey	Loral Globalstar (system B)	Motorola Iridium
Space segment			
Satellite altitude	5600 NM or 10370 km	750 NM or 1389 km	420 NM or 780 km
No. of orbital planes	3	8	6
No. satellites/initial	9 (3 Satellites per orbit)	24 (3 Satellites per orbit)	66 (11 satellites per orbit)
No. satellites/final	12 (4 Satellites per orbit)	48 (6 Satellites per orbit)	
No. beams per satellite	37 (need 15 beams to cover CONUS)	6 (Cover CONUS)	48 (need 59 beams to cover CONUS)
No. users per satellite	3000 Users per satellite	2800 Users per satellite	2990 Users per satellite
Frequency plan			
User-to-satellite	1.61 to 1.6256 GHz/BW = 16.5 MHz, LHCP	1.61 to 1.6256 GHz/BW = 16.5 MHz CP Divided 13 bands/1.25 MHz each	1.616 to 1.6265 GHz/BW = 10.5 MHz, RHCP Max 250 bands/31.5 kHz occupied BW
User-to-satellite	2.4835 to 2.5 GHz/LHCP	2.4835 to 2.5 GHz/CP Divided 13 bands/1.25 MHz each	1.616 to 1.6265 GHz/BW = 10.5 MHz, RHCP Max 250 bands/31.5 Khz occupied BW
Satellite-to-gateway	19.70 to 19.885 GHz/BW = 185 MHz LHCP & RHCP	5.185 to 5.216 GHz LH & RHCP	18.8 to 20.2 GHz/Center = 20 GHz RHCP
Intersatellite link	N/A	N/A	22.53 to 23.55 GHz/vertical
TT&C	C-band: transfer orbit Ka-band: on orbit	C-band	Ka-band
System performance			
Access method	CDMA	CDMA	FDMA/TDMA (50 kbps)

There are several other SBPCs in the development stage. These are Globalstar (14 satellites), Odyssey (12 satellites), Inmarsat-P (12 satellites), Teledesic (840 satellites), Orbcomm (36 satellites), etc. Table 2 compares Odyssey, Globalstar, and Iridium systems [48]. All three systems are expected to be fully operational by 2000. It is estimated that over the next 10 years more than one LEO satellite on average will be launched per month. Each satellite will cost about $30M.

Terrestrial Mobile Communications

Cellular Telephone The mobile communications market is a fast-growing business. GaAs MMICs are increasingly used in cellular products to make them smaller, lighter, cheaper, and more DC (usually battery) power efficient. For example, GaAs MMIC switches used in these products consume less power as compared with silicon p-i-n diode counterparts. There are several approaches in which GaAs ICs can be implemented into wireless communication systems. One of the approaches where GaAs ICs are used for all of the RF functions [49] is shown in Fig. 10. Low-noise amplifiers, power amplifiers, and switches are implemented using GaAs technology because of lower noise, higher PAE, and negligible power con-

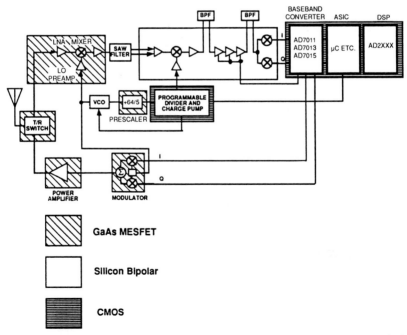

FIG. 10 GSM transceiver showing partitioning into ICs using GaAs, a moderate-speed silicon process, and CMOS. (Reprinted with permission from B. Clarke [49], and Microwaves and RF.)

sumption, respectively. GaAs mixer ICs work as well or better than any other technology. Future systems will use active array antennas employing MMIC-based T/R modules to increase functional capability, to improve system reliability, and to reduce size, weight, and cost.

Personal Communication Services (PCS) PCS will be the cellular phone of tomorrow. In PCS, both cordless and cellular phone services will merge and feature built-in electronic data and video transmission services. PCS handsets will carry voice, data, e-mail, fax, and video and will provide these services anywhere on the globe. PCS phones will be smaller, lighter, and lower powered than are current cellular phones [50]. These systems operate over different bands and will be based on digital techniques. Figure 11 shows the PCS/PHP (personal handy phone) handset block diagram using Motorola GaAs RFICs. Chapter 8 deals with PCS topics in detail.

RF Identification (RFID) RFID is another candidate for MMIC insertions into microwave commercial products. RFID is one of the automatic identification and data collection technologies which has applications in transportation (e.g., toll collection, traffic management, and automatic vehicle identification), factory automation, inventory control, material handling, and personnel management [43]. RFID tags are being used to collect tolls from vehicles and will become a leading edge technology for future IVHS. This technique has several advantages over the bar coding method in terms of range, accuracy, no line-of-sight requirements, and independence from environmental factors such as heat snow, rain, and dirt. Further, it can be used to track animals, people, packages, vehicles, and manufactured items. Basically a RFID system consists of an identification transponder

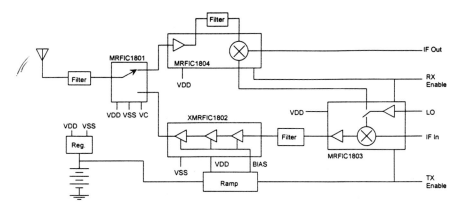

FIG. 11 Block diagram of the Motorola PCS/PHP RF IC 1800 series.

called "tag," and a fixed transceiver or "reader." Tags are equipped with memory capability and miniaturized antennas. A received signal triggers the tag to transmit its information to the reader, where it is decoded and passed to the host computer. The several frequency bands assigned to this application are 908–920, 1812–1830, and 2450 MHz.

2.6.2 Computers and Networks

Superfast computers, high-speed signal processors, medical image processors, and modeling and simulation systems primarily use high-speed digital GaAs ICs, while wireless local area network (WLAN) and fiber optic synchronous optical network (SONET) for the computer office use both MMICs and digital GaAs ICs. A WLAN provides high-speed data communications for computer networks and other purposes within an office or within a building [40]. Applications of WLAN include data transfer, printer sharing, e-mail, fax, and peer-to-peer communications.

A MMIC-based local area network (LAN) system which works at 18 GHz is a good compromise for wireless communication within a building (replacing existing hard-wired systems), since it attenuates rapidly outside the building permitting reuse of the carrier frequency. The safe distance between such buildings is about 40 m or more. Effective radiated power (ERP) for such systems is about 1 W. A block diagram of an 18-GHz WLAN receiver using MMICs is shown in Fig. 12.

2.6.3 Equipment and Instrumentation

A survey [51] projected that the world's microwave test equipment market, which included eight major products from 1 GHz through millime-

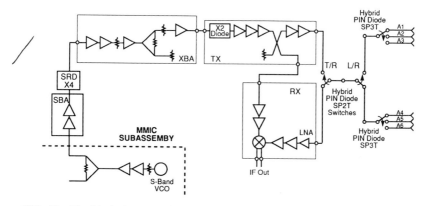

FIG. 12 The block diagram of the MMIC, 18-GHz transceiver subassembly showing the selected partitioning of the WLAN system. (Reprinted with permission from D.J. Mathews and C.L. Fullerton [40], and Applied Microwave.)

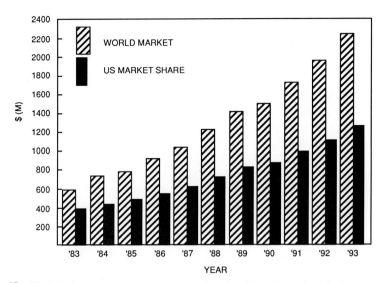

FIG. 13 World microwave test equipment market showing U.S. market share. (Reprinted with permission from L.D. Resnick and M.R. Stiglitz [51], and Microwave Journal.)

ter wave frequencies, will steadily increase. For analysis purposes the market forecast has been grouped into seven end-user categories: military, communications, aviation, electronic components, radar, test and measurement, and consumer products. As can be seen from Fig. 13, the United States is by far the largest market for microwave test equipment, driven by the large amount of military and communication equipment. MMIC insertions in this equipment will lower the cost and improve the systems' performance. Reduced part counts, fewer interconnections, and monolithic technology results in less sensitivity to assembly variables and the ability to meet stringent performance and reliability requirements for broadband components [52]. MMIC applications in several types of equipment, such as microwave equipment, microwave automatic test equipment, and data acquisition systems, are described in depth in Chapter 7.

2.6.4 Consumer Electronics

Consumer electronics has the highest potential volume of GaAs ICs due to larger customer markets. Each year there are an ever increasing number of new models of TVs, phones, cars, etc., which will require advanced electronics. GaAs MMICs which have the great potential for low cost in high-volume production are finding increasing application in cellular phones and mobile radios, TVRO downconverters, DBS receivers, high-definition TVs (HDTV), TV tuners, and VCRs. (Consumer MMIC applications in cellular phones and mobile radios were discussed in Section 2.6.1,

"Wireless Communications.") TVs are now requiring the capability for larger numbers of channels to accommodate expanded cable service. The low-noise property of GaAs MMICs is an important advantage in increasing the channel density for a given frequency bandwidth allocation. This requirement, plus the advent of HDTV, is creating new design-in opportunities for MMICs which should result in very high volume production. Some of the above applications are discussed in Chapter 10.

TV is the second largest application of consumer electronics. More than 60% of the households in the world have a TV set [53]. MMICs are being used in TVROs, DBS, VHF/UHF tuners, "wireless cable TV" or multichannel multipoint distribution service (MMDS), local multipoint distribution service (LMDS), etc. DBS has the largest application of GaAs MMICs in 12-GHz receivers. A 2.5-GHz rooftop receiver for wireless cable TV is another good example of MMIC application. Another potential MMIC application in TVs is in VHF/UHF or cable TV tuners to convert 950- to 1450-MHz signals to a fixed intermediate frequency (IF). To date, TV has consumed a greater number of MMICs than any other single application.

2.6.5 Intelligent-Vehicle Highway System (IVHS)

IVHS is composed of communication systems, radars, electronics, and computers to improve the highway safety and efficiency [54–57]. Numerous activities are taking place in the United States, Europe, and Japan in IVHS implementation. The U.S. Department of Transportation is expected to spend more than $200 billion to develop and implement IVHS infrastructure over the next 20 years. In major metropolitan cities in the world, during rush hours and heavy traffic, traffic congestion is increasing with time while the road expansions are space-limited. This results in traffic delays, wasted fuel, more air pollution, and more frequent accidents, costing billions of dollars every day. The solution is to use the existing roads more efficiently and safely through IVHS. The program is supported by the federal government, state governments, insurance industry, auto industry, trucking industry, electronics industry, etc.

The emergence of IVHS will lead to a major market opportunity for MMICs. At present several different types of MMICs are being developed for this application over the 1- to 100-GHz frequency range. IVHS will be addressing the following major areas.

(i) Travel information—car navigation and route guidance through GPS, yellow pages information, etc.

(ii) Vehicle control and safety—adaptive smart cruise control, collision warning/avoidance radar system, automatic breaking, etc.

(iii) Traffic management and safety—efficient traffic light timing, automatic vehicle identification for toll collection on busy highways, electronic warning signs, etc.

(iv) Public transport—delivering public transit information to homes, offices, bus stops, airports, and railway stations, tripping traffic lights to allow buses to stay on time, parking information, vehicle weighing, etc.

The above applications will save billions of dollars in lost time and revenues. Reduced emission from nonstopped vehicles will create less pollution and will help in maintaining a clean environment. Here also PCS and mobile satellite communication will play a major role in implementing several applications in IVHS by transmitting and receiving IVHS-related data, e.g., local traffic conditions and yellow pages information. The development of several GaAs MMICs for PCS and IVHS sensors is underway.

A number of microwave and millimeter-wave monolithic IC-based automotive and traffic systems are currently in development. Traffic control systems using vehicle speeds, vehicle lengths, and distance between vehicles derived from Doppler frequency shifts can determine an optimum traffic pattern in congested areas. Examples of such traffic control systems are Doppler radar systems for true motion control and traffic monitoring, car collision avoidance systems, beacon systems for road toll tax, and road infrastructures. Table 3 lists microwave and millimeter-wave frequency allocations for automotive and traffic systems. The development of Doppler radar for automotive collision avoidance or intelligent cruise control has been in progress for the past 25 years. Several laboratories, institutions, and companies have studied microwave and millimeter pulse and/or FM-CW radar systems for this application. These systems use a small antenna fitted into the front of a car and the system works up to about 200 m under adverse weather conditions. Traffic control systems using vehicle speeds, vehicle lengths, and the distance between the vehicles (derived from Doppler frequency shift data), can determine the optimum traffic pattern in congested areas. In Europe, two millimeter-wave frequency bands have been allocated for car-to-station and car-to-car communications (63–64 GHz)

TABLE 3 A Summary of Automotive Microwave and Millimeter-Wave Applications

Application	Frequency (GHz)	Approximate transmission power (dBm)
Electronic toll collection	5.8	20
Radar detector	24	—
Microwave doppler sensor	60	10
Mobile broadband system	60–66	17–27
Intervehicle and vehicle/beacon communication	60–66	13–23
Collision avoidance radar	60–77	7–13

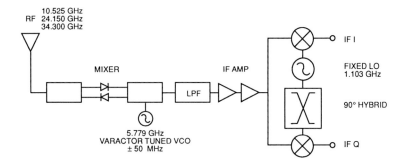

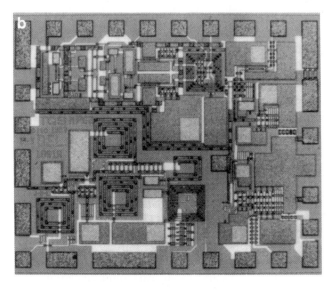

FIG. 14 (a) Simplified RF block diagram of an MMIC version of a radar detector front-end. (b) MMIC for a radar detector application. (Reprinted with permission from G.D. Martinson and M.M. Burin [31], and Applied Microwave.)

and collision avoidance radars (76–77 GHz). In the United States, vehicle-based radars operate at 77 and 94 GHz. One of the major reasons the millimeter-wave systems have limited commercial use is its high cost. With MMICs, the high cost barrier for large-scale use will no longer exist.

Devices for the detection of police speed-measuring radars have been in use for several years. The next generation of these radar devices seeks multi-functionality at affordable cost. Figure 14 shows a simplified RF block diagram of a radar detector and its MMIC version which is in production [31].

2.7 SUMMARY

Military and commercial applications are using increasing numbers of MMICs at increasing levels of integration. This will lead to lower costs, higher yields, and better performance, reproducibility, and reliability. An optimal combination of monolithic and hybrid circuits is a common approach. Millimeter-wave technology above 30 GHz is an important area whose growth is dependent on the monolithic approach, the use of MMICs will reduce circuit interconnection parasitics problems compared with conventional MIC technology. A major trend in MMIC technology is large-scale integration, that is, a complete subsystem on a single chip. Cost, size, and performance factors are driving this trend. MMIC technology using HEMTs will dominate low noise and millimeter-wave applications.

MMICs have made good progress toward meeting the requirements for active array antennas, especially low cost. Active array antennas are being developed using T/R modules based on single-function chips, multifunction chips and integrated T/R module chips. Several systems are in pilot production or limited rate production.

In the satellite communication area, GaAs MMICs offer important advantages such as small size, light weight, and higher reliability. The low cost of MMICs will make possible the development of active array antenna-based communication systems. Such systems with multiple-beam transmissions and time division multiple access (TDMA) beam routing will open up a new era. A growing application for MMICs in portable phones is high efficiency, low cost, small size power amplifiers.

The future for MMICs is very bright. Their insertion into military systems will grow steadily, while growth in the emerging PCS and IVHS applications is expected to be very rapid.

REFERENCES

[1] E. D. Cohen, Military applications of MMICs. *IEEE Microwave Millimeter-Wave Monolithic Circuits Symp. Dig.* 31–34, 1991.
[2] R. Rosenzweig, Commercial GaAs MMIC applications. *IEEE Microwave Millimeter-Wave Monolithic Circuits Symp. Dig.* 59–60, 1991.
[3] *Active Array Technology for Radars of the Future.* ITT Defense and Electronics, 1993, Van Nuys, CA.
[4] D. G. Fisher, GaAs IC applications in electronic warfare, radar and communications systems. *Microwave J.* **31**, 275–291, May 1988.
[5] D. N. McQuiddy, Jr., High volume applications for GaAs microwave and millimeter-wave ICs in military systems. *IEEE GaAs IC Symp. Dig.* 3–6, 1989.
[6] A. Podell, S. Moghe, and D. Lockie, GaAs MMICs designed for EW applications. *MSN&CT* **17**, 8–19, 1987.
[7] E. R. Schineller, A. Pospishil, and J. Grzyb, Insertion of GaAs MMICs into EW systems. *Microwave J.* **32**, 93–102, Sept. 1989.

[8] R. A. Gilson, MIMIC system insertion opportunities. *GOMAC Dig.* 45–48, 1989.
[9] E. D. Cohen, The U.S. MIMIC program—Status and expectations. *Appl. Microwave* **1**, 14–26, Nov/Dec. 1989.
[10] D. N. McQuiddy, Jr., High volume GaAs MMICs for military systems. *Appl. Microwave* **1**, 35–43, Nov/Dec. 1989.
[11] E. D. Cohen, MIMIC from the Department of Defense perspective. *IEEE Trans. Microwave Theory Tech.* **38**, 1171–1174, Sept. 1990.
[12] C. Rudolph and C. Fingerman, Ka band MIMIC modules for the longbow and SADARM smart weapons systems. *GOMAC Dig.* 13–16, 1990.
[13] R. W. Perry, MMIC insertion into SHF Satcom system. *GOMAC Dig.* 17–20, 1990.
[14] G. Johnson, F. Flores, and K. Redding, EW channelizer for airborne platforms—MMIC is an enabling technology. *GOMAC Dig.* 21–24, 1990.
[15] E. R. Schineller *et al.*, MIMIC EW brassboards. *GOMAC Dig.* 25–29, 1990.
[16] M. R. Stiglits and C. Blanchard, Phase I of the DoD MIMIC program: A status report. *Microwave J.* **35**, 90–100, Jan. 1992.
[17] G. Garbe *et al.*, A 6.8–10.7 GHz EW module using 72 MMICs. *IEEE MTT-S Int. Microwave Symp. Dig.* 1329–1332, 1993.
[18] H. Kato, etc., A 30 GHz MMIC receiver for satellite transponders. *IEEE Trans. Microwave Theory Tech.* **38**, 896–902, July 1990.
[19] H. H. Meinel, Millimeter-wave technology advances since 1985 and future trends. *IEEE Trans. Microwave Theory Tech.* **39**, 759–767, May 1991.
[20] S. Kitazume and H. Kondo, Advances in millimeter-wave subsystems in Japan. *IEEE Trans. Microwave Theory Tech.* 775–781, 1991.
[21] H. Meinel, Millimeter-wave commercial applications and technologies. *Microwave Eng. Eur.* **1**, 9–20, Nov. 1991.
[22] L. Raffaelli and E. Stewart, Millimeter-wave monolithic components for automotive applications. *Microwave J.* **35**, 22–32, Feb. 1992.
[23] R. Leyshon, Millimeter technology gets a new lease on life. *Microwave J.* **35**, 26–35, March 1992.
[24] D. A. Williams, Millimeter wave radars for automotive applications. *IEEE MTT-S, Int. Microwave Symp. Dig.* 721–724, 1992.
[25] Y. Takimoto and M. Kotaki, Automotive anticollision radar. *Appl. Microwave* **4**, 70–82, Fall 1992.
[26] N. K. Osbrink, Earth-terminal design benefits from MMIC technology. *MSN&CT* **16**, 68–77, Aug. 1986.
[27] K. B. Bhasin and D. J. Connolly, Advances in gallium arsenide monolithic microwave integrated-circuit technology for space communications systems. *IEEE Trans. Microwave Theory Tech.* **34**, 994–1001, Oct. 1986.
[28] J. Gladstone, Commercial applications of GaAs ICs, *IEEE Microwave Millimeter-Wave Monolithic Circuits Symp. Dig.* 103–107, 1988.
[29] J. Hurn, *GPS—A Guide to the Next Utility.* Trimble Navigation, Sunnyvale, CA, 1989.
[30] C. Sherod, GPS positioned to change our lives and boost our industry. *MSN* **19**, 24–40, July 1989.
[31] G. D. Martinson and M. M. Burin, Radar detector technology. *Appl. Microwave* **2**, 68–85, Summer 1990.
[32] T. Noguchi, Commercial applications of GaAs ICs in Japan. *IEEE GaAs IC Symp. Dig.* 263–360, 1990.
[33] H. I. Ellowitz, The roaring DBS receiver market. *Microwave J.* **34**, 24–35, July 1991.
[34] J. Browne, GaAs MMICs combine for Ku-band mobile SATCOM transceiver. *Microwave & RF*, **30**, 159–167, Aug. 1991.
[35] G. K. Noreen, Mobile satellite communications for consumers. *Microwave J.* **34**, 24–34, Nov. 1991.

[36] H. R. Malone et al., High volume GaAs MMIC applications. *IEEE GaAs IC Symp. Dig.* 135–138, 1991.
[37] A. L. Fawe, European progress in satellite communications. *Microwave Eng. Eur.* **1**, 23–29, Nov. 1991.
[38] R. Schneiderman, GPS becomes a high flying market. *Microwaves and RF* **30**, 30–40, December 1991.
[39] R. Rosenzweig, Crossroads for the U.S. GaAs IC industry. *Appl. Microwave* **3**, 15–20, Winter 91/92.
[40] D. J. Mathews and C. L. Fullerton, Microwave local area network for the computer office. *Appl. Microwave* **3**, 40–50, Winter 91/92.
[41] C. Huang, The applications of GaAs MMICs to DBS receivers. *Microwave Eng. Eur.* **2**, 23–31, Jan. 1992.
[42] R. Schneiderman, Fighting to survive in turbulent 1992. *Microwaves RF* **31**, 33–41, Jan. 1992.
[43] M. Kachmar, RF technology expands into new ID markets. *Microwaves RF* **31**, 49–52, March 1992.
[44] S. Watanabe, Technology transfer of high frequency devices for consumer electronics concerns and expectations. *IEEE Microwave Millimeter-Wave Monolithic Circuits Symp.* 5–6, 1992.
[45] A. Colquhoun and L. P. Schmidit, MMICs for automotive and traffic applications. *IEEE GaAs IC Symp. Dig.* 3–6, 1992.
[46] S. Lande, Customer acceptance of GaAs ICs. *IEEE GaAs IC Symp. Dig.* 7–10, 1992.
[47] R. J. Leopold and A. Miller, The iridium communications system. *IEEE MTT-S Int. Microwave Symp. Dig.* 575–578, 1993.
[48] D. Renkowitz, Satellite systems for cellular telephones. *IEEE MTT-S Int. Microwave Symp. Dig.* 1619–1622, 1994.
[49] B. Clarke, Advanced RFICs for VHF and UHF communications. *Procs. Second Annu. WIRELESS Symp.* 55–65, 1994.
[50] B. J. Leff, Making sense of wireless standards and system designs. *Microwave RF* **33**, 113–118, Feb. 1994.
[51] L. D. Resnick and M. R. Stiglitz, The World Microwave Test Equipment Market. *Microwave J.* **31**, 40–47, Dec. 1988.
[52] V. Peterson, Instrument applications of GaAs ICs. *Microwave J.* **32**, 46–52, Aug. 1989.
[53] R. B. Gold, Applications of GaAs ICs in video distribution. *IEE GaAs IC Symp. Dig.* 151–155, 1994.
[54] T. Rose, Intelligent vehicle highway systems: Going places fast. *Microwave J.* **36**, 172–178, May 1993.
[55] R. Schneiderman, Intelligent-vehicle market moves into the fast lane. *Microwaves RF* 42–47, June 1993.
[56] W. C. Collier and R. J. Weiland, Smart cars, smart highways. *IEEE Spectrum* 27–33, April 1994.
[57] R. Schneiderman, Technology helps drive the intelligent-vehicle highway systems market. *Microwaves RF* **33**, 33–40, May 1994.

3
Digital GaAs Integrated Circuits

John Naber
ITT Gallium Arsenide Technology Center, Roanoke, Virginia

3.1 Introduction
3.2 Logic Design
    3.2.1 Device Types
    3.2.2 Logic Families
3.3 Trade-offs between Silicon and GaAs
    3.3.1 Performance Factors
    3.3.2 Cost Factors
3.4 Digital GaAs Product Insertions
    3.4.1 Convex/Vitesse Supercomputer
    3.4.2 Cray Computer's Supercomputer
    3.4.3 TriQuint's and Vitesse's Clock Driver ICs for Personal Computers and Workstations
    3.4.4 Fiber Optic Communication Chip Sets
    3.4.5 ARPA/University of Michigan/Vitesse RISC Processor
    3.4.6 ARPA's Digital GaAs Insertion Program
3.5 Summary
    References

3.1 INTRODUCTION

The first digital gallium arsenide (GaAs) circuit was fabricated and tested at Hewlett Packard in the early 1970s [1]. However, it was not until 1980 that Gigabit Logic was founded and the first commercial digital GaAs foundry came into existence. From the early to middle 1980s great expectations were placed upon the future capabilities of GaAs. There were even rumors of making major inroads into the silicon-dominated industry with talk about a digital GaAs central processing unit (CPU) for workstations or personal computers, which has been the "holy grail" for digital GaAs because of the possible large volume of ICs. In 1985 the digital GaAs integrated circuit (IC) industry was expected to reach over $1 billion by the early 1990s [2]. All this hype and overly optimistic predictions made it easy to obtain money and justification for investment. For example, nine companies offered noncaptive digital GaAs foundry services: Anadigics, Ford, Gain, Gigabit, Harris, MSC, Tachonics, TriQuint, and Vitesse. Of this original group of nine, three went bankrupt, one merged, two got out of the business, and only two remain that offer digital GaAs products and services (TriQuint and Vitesse). Some of the obstacles and pitfalls these companies did not overcome included difficulties in achieving high integration levels, foundry overcapacity, system users entrenched with silicon who were reluctant to work with GaAs, and improvements in silicon technology. Moreover, the future success of the remaining GaAs foundries will depend on recognizing and maintaining their current niche market success by taking advantage of the inherent benefits of GaAs for certain applications. Even though there has been an initial shakeup in the digital GaAs community, the future looks encouraging as evident by foundries reporting profits for the first time and by Motorola's recent announcement of a $100M commitment to a GaAs fabrication facility [3]. More realistic forecasts for digital GaAs predict sales of $400M by 1995 based on the $100M of existing sales in 1991 [4].

Now that the industry is well into its second decade of commercial sales, some trends are now clear. For example, the metal semiconductor field effect transistor (MESFET) using direct-coupled FET logic (DCFL) is the choice for high-integration, low-power circuits and gate arrays with 350,000 equivalent gates are now available [5]. This 350K gate array contains 1.2 million transistors, which is comparable to an Intel 486 microprocessor. A microphotograph of this circuit is shown in Fig. 1. However, for ultra-high-speed circuits, heterostructure bipolar transistors (HBTs) are the most logical choice, with commercially available flip-flops having toggle rates of 9 GHz at medium scale integrations (MSI) to large scale integrations (LSI) levels.

The purpose of this chapter on applications of digital gallium arsenide is to give enough background information to acquaint the first-time chip or system designer with the trade-offs among GaAs and all the various silicon digital technologies. The comparisons and trade-offs are sufficiently complicated that trying to develop a single metric can be very misleading. Therefore, to make a knowledgeable decision on which technology is best matched for a particular system requires evaluating a multitude of parameters, such as cost, speed, and power. Section 3.2 will briefly describe various types of GaAs transistors and logic approaches. Section 3.3 will take a close look at comparing the speed, power, and cost parameters with the different technologies and processes. Section 3.4 will describe some of the most popular system insertions of digital GaAs, which is the primary purpose of this chapter.

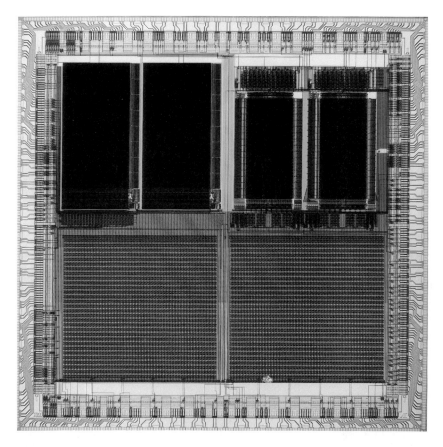

FIG. 1 A die photograph of Vitesse's 350,000-gate gate array. The die measures 15 mils on a side and contains 1.2 million transistors.

3.2 LOGIC DESIGN

In order to draw comparisons between technologies it is appropriate to elucidate and understand the physical properties of the natural occurring silicon and the man-made gallium arsenide. Table 1 shows a tradeoff between semiconductor properties of silicon and GaAs [6, 7].

GaAs has an inherent advantage over silicon in four of the seven listed physical properties in Table 1. The primary reason that GaAs has received as much attention and research as it has is due to its high-speed advantage over silicon. This advantage is realized from an electron mobility (the ability of the electrons to move through the crystal lattice of the semiconductor) that

TABLE 1 Comparison of Semiconductor Properties between GaAs and Silicon

Property	Silicon	GaAs	Advantage
Bandgap (eV)	1.12	1.43	GaAs
Electron mobility (cm^2/Vs)	800	5000	GaAs
Hole mobility (cm^2/Vs)	300	250	None
Native oxide	SiO$_2$	None	Silicon
Radiation hardness (rad)	1.0E + 6[a]	1.0E + 7[a]	GaAs
Thermal conductivity (W/cm^2-°C)	1.4	0.5	Silicon
Resistivity (ohm-cm)	4.0E + 5	4.0E + 8	GaAs

[a]Total dose only; GaAs shows similar performance to silicon for transient upset.

is over six times greater for GaAs than for silicon. Another advantageous property of GaAs is the capability of generating light for use in lasers and light-emitting diodes (LEDs) because of its direct bandgap. Moreover, the bandgap in GaAs is larger, which lends itself to higher temperature operation. The last advantageous physical property of GaAs from the table is the three orders of magnitude increase in the intrinsic substrate resistivity, which makes it semi-insulating. This means that the substrate capacitance of GaAs will be less than that of silicon, leading to higher frequencies of operation.

One of the disadvantageous properties of GaAs with respect to silicon is thermal conductivity. Silicon can dissipate three times the power per area than can GaAs. The last disadvantageous physical property of GaAs is that is has no native oxide, which limits the types of transistor devices possible in the technology. In contrast, silicon metal oxide semiconductor FET (MOSFET) devices use a layer of SiO$_2$ to isolate the gate from the channel. These devices can be formed with both p and n channels, thus permitting complementary (CMOS) operation for very low power consumption. The workhorse device for GaAs is the MESFET, which has a gate forward bias conduction limit of only 0.7 V, unlike MOSFETs which can have gate biases up to the power supply rail. This limited gate voltage swing of the MESFET can degrade noise margin, which adversely affects yield.

3.2.1 Device Types

There are four different types of transistors available for the design of GaAs digital logic gates, three being FET structures that modulate majority charge carriers and one being a bipolar structure that modulates minority charge carriers. The FET devices are: MESFET, high electron mobility transistor (HEMT), and junction field effect transistor (JFET). The bipolar device is the HBT.

The MESFET device is the most widely used for digital GaAs because of its simplicity. The MESFET uses a Schottky barrier created by the direct metal to semiconductor interface to control the FET channel [12]. The HEMT device is a further enhancement of the MESFET. The device uses an epitaxially grown III–V compound layer under the gate to create a region void of any dopant atoms, which significantly improves the speed and reduces the inherent noise of the device because of a longer mean free path of the charge carriers before a collision [13]. A drawback of this technology over the MESFET is its higher cost due to the requirement of epitaxial grown wafers. The last FET type of device is the JFET, which uses a p–n junction to modulate the channel. Considerable research has been done in this area in pursuit of a complementary logic device [14]. However, the speed of JFETs is significantly lower than that of MESFETs or HEMTs because of the low mobility of holes in GaAs, which is on the same order as that in silicon. Hence, p-channel devices will drastically reduce the overall speed performance of the device, but the potential for significant power savings drives continued research in this area [15].

The last type of GaAs transistor and most different from the previous three is the HBT that uses the injection of minority charge carriers to control the flow of current in the device. HBTs are current gain devices with gains (Beta) of approximately 100. On the other hand, MESFETs are voltage gain devices with voltage gains of approximately 10. HBT technology has made great strides over the past 5 years and is now commercially available. In addition to their large current gain, another benefit of HBTs is the lower offset voltage leading to higher resolution in combined analog/digital circuits, such as A/D and D/A converters. Another strong point of HBTs is their very high-speed and simplistic gate design using integrated injection logic (I^2L) in which a single transistor can implement an inverter gate [16]. Some drawbacks are high power consumption and high cost. The power is high due to the large currents required for the bipolar transistors. The cost is high due to the complexity of the process that requires more mask layers, which is further exacerbated by the required epitaxial grown wafers.

3.2.2 Logic Families

The first digital GaAs logic gate used buffered FET logic (BFL) because of its large noise margin and very high speed [17]. However, BFL is very power hungry, requiring diodes and two power supplies and uses only depletion or normally-on FETs. There has been a large assortment of logic families designed in digital GaAs over the past 20 years, but basically three are now used between Cray, TriQuint, and Vitesse. These are source-coupled FET logic (SCFL) (TriQuint), capacitor diode FET logic (CDFL) (Cray), and DCFL (Vitesse). These three logic families for a two-input NOR gate are shown in Fig. 2. The main attributes of the particular logic families

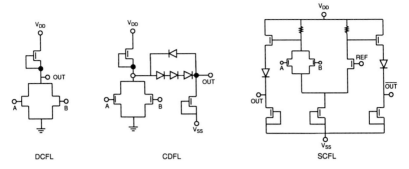

FIG. 2 A two-input NOR gate of the three most popular digital GaAs logic families used today.

will be described but for a much more thorough coverage reference [18] is recommended.

DCFL is the best choice for low power consumption and very large scale integration (VLSI) levels. This is the logic family Vitesse uses for its large gate arrays. Initial skeptism said DCFL could not be used for VLSI designs because the variation of the threshold voltages would be too large, affecting noise margin or circuit yield. However, Vitesse has proven the skeptics wrong with the introduction of their 350,000-gate gate array [5]. DCFL is similar in design to silicon *n*-channel metal-oxide semiconductor (NMOS) by using an active load depletion mode FET (DFET) and a normally off enhancement mode FET (EFET) for the switch. DCFL has the important advantage of requiring only a single voltage power supply. The power is low due to the few number of FETs needed to implement a logic function, similar to CMOS.

SCFL is used by TriQuint and Vitesse for achieving the highest possible speed from a logic gate. SCFL is a differential logic family similar to bipolar emiter-coupled logic (ECL). SCFL achieves its high speed by the gain increase afforded by a differential amplifier coupled with FETs which are maintained in their saturation state, thus reducing the FET capacitance to be charged. SCFL consumes the largest amount of power because of the complexity of the logic design, uses only DFETs, and requires two power supplies. However, this complexity compares more favorably with DCFL when macros such as flip-flops, adders, or multiplexers are considered.

CDFL is a compromise between SCFL and DCFL with regard to speed, power, and complexity. CDFL was used at GigaBit logic before their merger with TriQuint and is currently used by Cray Computer. CDFL is a two-stage logic design that has a first stage the same as that of DCFL, but uses a driving stage as well. The driving stage is AC coupled to the first stage by reverse biased Schottky diodes to reduce the static DC power dissipation.

TABLE 2 Summary of DCFL, SCFL, and CDFL Logic Families

Logic family	No. FETS[a]	No. diodes/resistors[a]	Typical power	Typical speed	Typical gates
CDFL	4	4/0	Med	Med	LSI
DCFL	3	None	Low	Low–med	VLSI
SCFL	8	2/2	High	High	MSI

[a]For a two-input NOR gate.

CDFL requires two power supplies and can accomodate large scale integration (LSI) levels of circuit integration.

Table 2 summarizes some of the characteristics of the aforementioned logic families. As can be seen from the table these logic families have distinct trade-offs for a particular type of design and run the gamut from low integration/high power and high speed to high integration/low power and lower speed.

3.3 TRADE-OFFS BETWEEN SILICON AND GaAs

There are a multitude of factors to consider when doing a digital design. However, the primary factors addressed here are performance and cost.

3.3.1 Performance Factors

The principal performance parameters addressed are speed and power dissipation as they relate to CMOS, bipolar CMOS (BiCMOS), ECL, and GaAs. The comparisons applied to CMOS will also be inferred to apply to BiCMOS as well unless where noted. This is based on the premise that most bipolar usage in large BiCMOS digital designs are circuits used to drive the outputs and large capacitive loads, such as clock lines. Hence, the majority of the logic will consist of CMOS not bipolar.

The transconductance with unity voltage gain (F_t) of an active device is a good figure of merit for the high-speed performance of a given semiconductor or process. CMOS has a much lower F_t (typically 1–5 GHz) than GaAs (typically 12–20 GHz) due to its larger substrate capacitance and smaller transconductance (gm). Transconductance is proportional to electron mobility, which as mentioned previously is much larger in GaAs [8]. The other performance factor, power dissipation, with regard to CMOS shows a large discrepancy between static and dynamic power. CMOS's static power dissipation is in the range of a few microwatts because the gate is totally isolated from the power supply rail for a logic low state.

However, the dynamic power dissipation is determined when the gate switches and it increases with increasing frequency. This is due to the logic gate having no precharge on the gate output as in the case of GaAs and ECL. (Conversely, GaAs and silicon ECL have constant current sources in their logic design and hence the static and dynamic power dissipations are equivalent.) The internal logic swing can have a significant impact on the power dissipation, since the time to charge a capacitor is directly proportional to the voltage. In other words, the larger the voltage swing the more time it takes to charge to that voltage. By combining the larger capacitance and larger voltage of CMOS one can realize that the power dissipation will scale linearly with frequency. This is the reason that CMOS power dissipation scales linearly with frequency and power dissipation is usually specified in microwatts per megahertz. Moreover, CMOS's large power dissipation at high speed is further exacerbated by the fact that its internal logic swings require 5 V (more recently 3 V in some cases) versus 1 V for ECL and 0.7 V for GaAs. Thus, the commonly held belief that CMOS is the lowest power consumption technology must include a caveat for higher speed applications. Indeed some of the new high-speed CMOS processes have power dissipations actually exceeding that of ECL or GaAs. For example, the new Alpha microprocessor from Digital Equipment Corp. runs at a 200-MHz clock rate, but dissipates considerable power (35 W) [9]. This chip is fabricated with a state-of-the-art CMOS process using 0.5-μm gate lengths.

ECL digital logic uses bipolar transistors instead of MOSFETs utilized in CMOS to clock flip-flops at frequencies up to 1.1 GHz. It significantly surpasses the high-speed performance of CMOS because the logic design uses a constant current source to steer current to complementary outputs and bipolar transistors supply more than an order of magnitude greater gain than do FET devices (current gain versus voltage gain) for driving capacitive loads at high speed. Moreover, the logic swing is only 1 V versus 3 or 5 V of CMOS. The static power dissipation is high due both to the constant current through the resistive loads and current sources and to the large number of transistors per gate. Hence, ECL derives its speed advantage from the constant current source logic design coupled with the larger gain of bipolar transistors and the smaller internal logic swings.

Table 3 lists key gate characteristics of CMOS, ECL, and GaAs [10]. Table 4 lists the specifications and capabilities of state-of-the-art CMOS, BiCMOS, ECL, and GaAs gate array processes. It can be observed from this table that chip specifications requiring speeds greater than 200 MHz force the designers to use ECL, BiCMOS, or GaAs. CMOS is the usual choice for designs with speeds less than 100 MHz, but designs between 100 and 400 MHz must be addressed on a design by design basis. Figure 3 shows the principal operating ranges and power dissipations for the three technologies

TABLE 3 Gate Characteristics of CMOS, ECL, and GaAs

Gate characteristics	CMOS	ECL	DCFL GaAs
No. of transistors inverter	2	7	2
No. of transistors D-FF	19	22	19
No. of resistors and diodes D-FF	None	14	None
Logic swing	3 to 5 V	1 V	0.7 V
Inverter static pwr	0.001 mW	1.2 mW	0.5 mW

based on the data from Table 4. As can be seen, GaAs is superior with respect to power consumption for frequencies above 300 MHz.

The performance trade-offs between CMOS, BiCMOS, ECL, and GaAs have been evaluated at a transistor or gate level. However, another critical variable in the performance equation is the design or architecture itself used in the circuit. The importance of architecture selection in the design when doing speed and power comparisons will be shown in the following paragraphs. In doing a technology trade-off between GaAs and silicon for a particular system upgrade or insertion, care must be taken not to force GaAs to be an exact gate for gate replacement of ECL, CMOS, or BiCMOS. Constraining a GaAs design to an existing configuration will lead to unfair comparisons and unoptimized designs. For example, in designing a GaAs microprocessor it would be extremely inefficient to do a direct design translation of an Intel 486 or Digital Corporation's Alpha CPU. These designs are pipeline intensive to maximize performance. However, in using GaAs, alternative architectures can be implemented to greatly simplify the overall design as discussed in Section 3.4 with respect to a GaAs reduced instruction set computer (RISC) CPU.

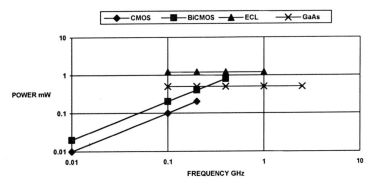

FIG. 3 Frequency and power dissipation per gate for CMOS, ECL, and GaAs.

TABLE 4 Specifications of Commercially Available Gate Arrays in CMOS, ECL, and GaAs

Specification	CMOS[a]	BiCMOS[b]	ECL[c]	GaAs[d]
Gate/emitter size	0.8 μm	0.8 μm	1.0 μm	0.6 μm
Total no. of gates	400,000	150,000	10,000	350,000
Total usable gates	250,000	112,000	7,848	175–245,000
Wafer diameter	150 mm	150 mm	100 mm	100 mm
Chip size (mm)	Not available	Not available	9.8 × 9.8	14.9 × 14.7
Metal/poly layers	4	4	3	4
Mask layers	≈17	≈24	≈17	13
Power/gate (mW)	Not available	Not available	1.2	0.14[e]
Max. clock (MHz)[f]	215	400	1100	2500

[a] Mitsubishi M6008X (Source: ASIC and EDA, pp. 22–27, January 1992).
[b] Texas Instruments TGB1000 (Source: ASIC and EDA, pp. 22–27, January 1992).
[c] Motorola MCA3 (Source: design manual).
[d] Vitesse VGFX350K (Source: data sheet).
[e] Unloaded gate.
[f] Maximum toggle rate of a D-type flip-flop.

The above concept is illustrated through a simple example of comparing GaAs with BiCMOS for a 16-bit adder that must clock at 100 and 150 MHz. BiCMOS was chosen over CMOS because these frequencies are outside the range for strictly CMOS processes, but are well within the limits of most BiCMOS processes. The 100-MHz adder in BiCMOS mandates using a pipelined or parallel architecture that requires 125 registers and four 4-bit adders and consumes 100 mW of power. However, it would be extremely inefficient to use a pipelined approach in GaAs. For example, if the same 16-bit pipelined adder were used in GaAs the circuit would clock at 485 MHz, but dissipate 365 mW of power. Thus, if this gate for gate translation were done, the obvious choice of technology would be BiCMOS instead of GaAs because the speed and power are both excessive for the given requirement. On the other hand, if a judicious choice of architectures are evaluated, the GaAs design can meet the 100-MHz design objective with lower power dissipation than BiCMOS by using a serial architecture. The serial architecture can be used because of the 2.5-GHz toggle rates of the GaAs registers. Hence, a 16-bit serial adder can be realized using 49 registers and no pipelined stages. This comparison is shown in Fig 4. A 61% reduction in registers for the GaAs design in comparison to the BiCMOS design translates to a 13% overall reduction in power dissipation at 100 MHz.

Another example showing the architectural trade-offs between GaAs

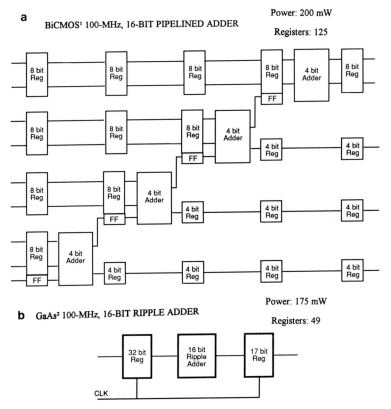

FIG. 4 Different architecture approaches in GaAs and BiCMOS for a 100-MHz 16-bit adder. a, BiCMOS gate array using 30% logic switching and no clock skew; b, Vitesse GaAs VIPER gate array using worst-case clock skew.

and BiCMOS is a 150-MHz 16-bit adder. In order to meet the 150-MHz performance in BiCMOS a dual-stage adder is required. However, a GaAs design can again be implemented with a much simpler and robust design by using a carry look ahead (CLA) adder. Both designs are shown in Fig. 5.

A summary of the technology comparison between GaAs and BiCMOS for a 100-, 150-, as well as a 200-MHz, 16-bit adder using the optimum architecture for each technology is shown in Table 5. Table 5 shows that GaAs can offer power reductions of up to 18% over BiCMOS at clock frequencies as low as 150 MHz. It also shows that each application requires the appropriate architecture and must be carefully evaluated when doing a technology comparison.

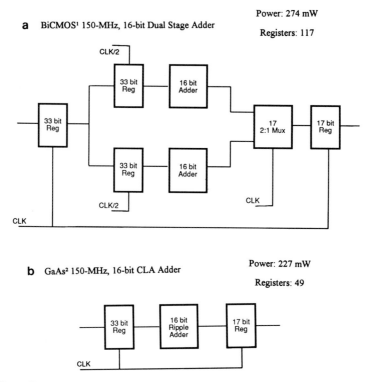

FIG. 5 Different architecture approaches in GaAs and BiCMOS for a 150-MHz, 16-bit adder. a and b, same as in Fig. 4.

TABLE 5 Technology Comparison between BiCMOS and GaAs for 100-, 150-, and 200-MHz 16-Bit Adders

Design goal	BiCMOS			GaAs			GaAs pwr reduction (%)
		MHz	mW		MHz	mW	
100-MHz adder	(1)	135	200	(2)	116	175	13
150-MHz adder	(3)	165	275	(4)	356	225	18
200-MHz adder	Not capable			Same as above			N/A

Note. (1) Pipelined, (2) ripple, (3) dual stage, (4) carry look ahead.

TABLE 6 Cost Factors of Various Technologies

Cost factor	CMOS[a]	BiCMOS	ECL	DCFL GaAs
		For 1 wafer		
Wafer cost	$30 (150 mm)	$30 (150 mm)	$15 (100 mm)	$300 (100 mm)
Mask cost[b]	17K	24K	17K	13K
Fab cost[c]	0.51K	0.69K	0.51K	0.39K
Cost/sq. mm	$0.99	$1.40	$2.23	$1.74
		For 1000 wafers		
Wafer cost	$30 (150 mm)	$30 (150 mm)	$15 (100 mm)	$300 (100 mm)
Mask cost[b]	17K	24K	17K	13K
Fab cost[c]	0.51K	0.69K	0.51K	0.39K
Cost/sq. mm[d]	$0.32	$0.42	$0.69	$0.90

[a]From Ref. [12] based on a 4M DRAM process using 0.8-μm technology.
[b]Assuming $1000 per mask.
[c]Assuming $30 per mask level processing cost [12].
[d]Wafer cost becomes much larger than mask cost.

3.3.2 Cost Factors

There are many applications where cost and not speed is a driving design factor. In these cases GaAs as well as silicon ECL and BiCMOS cannot compete with silicon CMOS. This becomes apparent when the cost factors of Table 6 are taken into consideration. The purpose of this section is to obtain typical values for the cost factors, which can be used as cost metrics to evaluate the different current production technologies. Actual values will of course vary from vendor to vendor. In order to develop an accurate cost model other factors need to be considered such as: process yield, test yield, packaging cost, and die size. Packaging cost and die cost is relatively technology transparent and should not have much impact on the results from Table 6. A significant cost factor not included in the table is test and process yield because this information is company proprietary. However, some general insight can be drawn from the data in Table 6.

CMOS is the most cost-effective technology. The data used to obtain the CMOS cost factors are from a typical production process for 4M dynamic random access memories (DRAMs) [11]. However, keep in mind that the next-generation 64M DRAM CMOS process will require 24 mask levels versus 17, 0.35-μm geometry versus 0.8-μm, and 200-mm wafers versus 150-mm of the 4M DRAM process. Therefore, if a CMOS process meets all performance goals for a given design it should be used without regard given to BiCMOS, bipolar, or GaAs alternatives because of its significantly lower price per square millimeter. BiCMOS is the next most cost-effective technology for designs outside the spectrum of CMOS perfor-

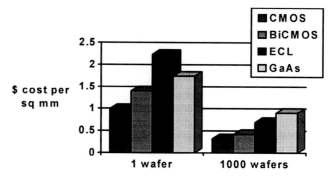

FIG. 6 Cost per square mm for different technologies at small and large quantities.

mance. However, as demonstrated in the previous subsection, alternative architectures can be used for GaAs that will significantly reduce size and power with respect to BiCMOS. For GaAs to be cost competitive with BiCMOS the GaAs chip size must be on the order of 20% less for small volumes and 50% smaller for large volumes of wafers. The increased size reduction of GaAs is required for larger volumes because the larger wafer cost of GaAs eliminates the low-volume advantages provided by the reduced mask and fabrication cost. A similar comparison holds for ECL with costs being somewhat closer in high volume. Figure 6 graphically depicts the results from Table 6.

In summary, CMOS is the lowest cost approach when high-speed performance is not the driving factor. If high-speed performance is required, GaAs can be cost competitive with silicon BiCMOS or ECL provided that an optimal architecture is chosen to minimize chip size.

3.4 DIGITAL GaAs PRODUCT INSERTIONS

This section will describe some of the more popular system and product insertions of digital GaAs in both military and commercial systems, with emphasis placed on the commercial side. The purpose is to give the reader brief descriptions of the assorted products and the reason GaAs was chosen over ECL or BiCMOS without getting into the specific details of the designs.

3.4.1 Convex/Vitesse Supercomputer

Convex Computer Corporation, a producer of supercomputers located in Richardson, Texas, and Vitesse Semiconductor Corporation, a GaAs IC foundry, located in Camarillo, California, joined forces over a 2-year period to develop the world's first commercially available GaAs computer called the C3800. The C3800 uses a parallel architecture with up to 8 CPUs to give 1 gigaflop performance at a 17-ns cycle time. Also, the C3800 can access 4 gigabytes of main memory at 500 megabytes per second. Each C3800 uses 550 GaAs ICs consisting of 30 different types that range in complexity from 15,000 to 45,000 gates. The total cost for the C3800 is $8M. For comparison, Convex's C3400 machine utilizing strictly BiCMOS has only 40% of the performance to the GaAs C3800, giving 400 Mflops using a 40-ns cycle time [19].

GaAs was chosen over BiCMOS and CMOS for the C3800 because of the demanded high speeds. For example, embedded static random access memory (SRAM) cycle time had to be less than 4 ns. The only competing silicon technology that could meet the speed requirement was ECL. However, ECL's high power consumption would have caused the system to exceed its thermal budget with air cooling. Hence, for Convex the only viable alternative was digital GaAs. Figure 7 shows four ICs that make up the core processor, which is referred to as the "Javelin" chip set. The complexity of these ICs range from 30,000 to 45,000 gates.

3.4.2 Cray Computer's Supercomputer

Cray Research, the source of all commercially available Cray computers, spun off Cray Computer to develop a supercomputer using GaAs called the Cray III. Cray Computer is located in Colorado Springs, Colorado. The project started in the mid 1980s but to date, no product has been sold. Cray Computer has delivered a system for evaluation for the National Center for Atmospheric Research in conjunction with Thinking Machines CM-5 parallel processor [20].

The Cray III as initially envisioned is a 16-processor vector machine containing 50,000 GaAs ICs with greater than 10 Gigaflops performance attained from a system clock rate of 500 MHz. The reason that GaAs was chosen for the next generation beyond the silicon-based Cray II is raw speed, which allows the system to clock at more than double the rate achievable with ECL gate arrays (244 MHz versus 500 MHz). Table 7 shows a comparison between the Cray II and Cray III and the improved performance resulting from utilizing GaAs. Note that the Cray III is only half the size of the Cray II, but offers eight times the performance.

3.4.3 TriQuint's and Vitesse's Clock Driver ICs for Personal Computers and Workstations

GaAs ICs are currently being used in higher-end 486 PCs and various workstations to distribute the high-speed system clock rate to other devices on the motherboard. The advent of faster CPUs with clock rates from 50 to 66 MHz exacerbate the timing requirements of the clock signals on the

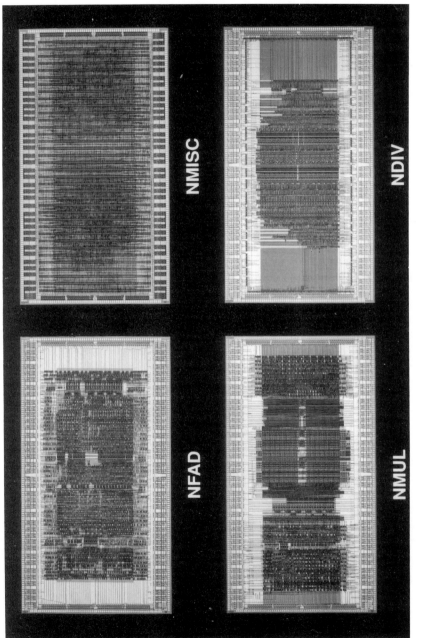

FIG. 7 Picture of the Javelin chip set for Convex's supercomputer core processor (used by permission).

TABLE 7 Comparison between the Cray II and the Cray III Systems[a]

Specification	Cray II	Cray III	Cray III improvement
Technology	Silicon	GaAs	N/A
Speed performance	1.2 Gigaflops	10 Gigaflops	8× Faster
Clock rate	244 MHz	500 MHz	2.1× Faster
Power	150 kW	150 kW	None
IC integration	16 Gates	200–2000 Gates	125× More gates
IC packaging	Chip carriers	Bare die	2×
Physical size–height	45 Inches	34 Inches	2× Smaller

[a] From [21].

printed circuit board going to the various support ICs. For example, Intel's clock driver specifications for its new superscalar (dual instruction) Pentium (P5) requires dual 60- to 125-MHz clocks with 1.5-ns rise and fall times, less than 90 ps of jitter, and less than 250 ps of skew between clocks [22]. Figure 8 shows the proposed architecture of the Pentium and the routing of the 33-MHz and 66-MHz clocks.

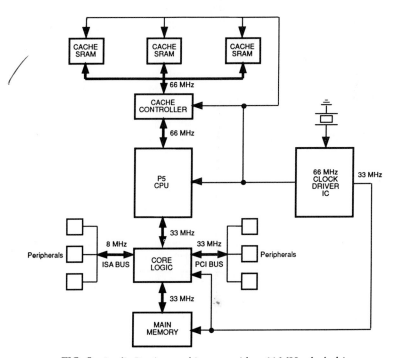

FIG. 8 Intel's Pentium architecture with a 66-MHz clock driver.

TABLE 8 Comparisons between Silicon and GaAs Clock Driver ICs

Company	Process	Skew/jitter (ps)	Rise time (ns)	Output/current (mA)	Power diss. (mW)
AMCC	BiCMOS	500	1.5	24	400
Cypress	BiCMOS	250	1.5	24	500
Pioneer	BiCMOS	250	1.5	24	—
TriQuint	GaAs	250/75	1.4	30	800
Vitesse	GaAs	500	1.5	4	500

This clock driver market has silicon offerings from AMCC, Cypress, Motorola and Pioneer Semiconductor. TriQuint and Vitesse both offer GaAs chips with these capabilities. Table 8 compares the performance of some of these clock driver ICs by listing a selection of pertinent specifications. However, additional factors such as flexibility and cost will be considered when choosing a clock driver from the various vendors.

3.4.4 Fiber Optic Communication Chip Sets

There are a host of new high-speed serial data networks and standards for data transfer rates in excess of the capabilities of Ethernet (10 Mbps) or the fiber distributed data interface (FDDI, 100 Mbps). These new and emerging high speed standards or formats consist of the fiber channel, high-performance parallel interface (Hippi), small computer system interface (SCSI) and the synchronous optical network (SONET). The data rates of these networks range from 266 Mbps to 1250 Mbps. Hence, GaAs, BiCMOS or ECL must be used because the required performance is beyond CMOS's capabilities. Most chip sets have a single transmitter and receiver pair that typically transmits the data using some form of mb/nb coding with the clock signal being recovered directly from the data. The two largest commercial GaAs houses (TriQuint and Vitesse) offer these types of chips as well as a silicon competitor (AMCC). Table 9 shows a comparison between

TABLE 9 Comparisons of Various Fiber Optic Communication Transmitter and Receiver Chip Sets for the Fiber Channel Format

Company	Process technology	Format	Speed (Mbps)	Cost (units)
AMCC	BiCMOS	Fiber channel	266, 531, 1060	$138 (1000)
TriQuint	GaAs	Fiber channel	194.4, 200, 266	$110 (100)
Vitesse	GaAs	Fiber channel	1060	$125 (1)

these three companies' product offerings for the fiber channel chip set only. This chip set was selected because of its growing popularity due to its smaller connectors and ability to use the previous listed formats.

As can be seen from this chart GaAs can indeed compete with silicon and is in fact 10 to 20% less expensive than its silicon counterpart. However, other parameters such as power dissipation, temperature performance, and noise immunity need to be considered before a chip set is selected for a particular system.

3.4.5 ARPA/University of Michigan/Vitesse RISC Processor

As alluded to earlier in the introduction a GaAs CPU would be the ultimate insertion for high-performance PCs and workstations. It initially appeared that with the relatively slow evolvement of GaAs to VLSI, this would never happen. Indeed, several attempts have been made, but have floundered or gone nowhere [23, 24]. However, with the CPU design expertise from the University of Michigan and using Vitesse's process on a ARPA-funded program, there is a real possibility of breaking the barrier to a GaAs CPU.

The program consists of three (Aurora) phases in developing a 32-bit RISC CPU based on the MIPS R3000. In phase I a 60,000-transistor (20,000 two-input NOR gates), 137-MHz CPU was demonstrated [25]. This integration level is approximately half the size of an Intel 80286. Phase II has demonstrated a 160,000-transistor (53,000-gate), 200-MHz CPU that consumes 24 W of power [26]. This power level is well within air cooling limits and is only 1.2 times larger than an Intel 80286. The goal of the last phase is to demonstrate a 400-MHz three-stage pipelined CPU. Some critical developments to this project in order to compete with next-generation silicon CPUs was first the high-yielding, 0.6-μm VLSI process of Vitesse followed by a suite of CAD design tools by Cascade Design Automation that contains an efficient compiler for circuit synthesis. Table 10 summarizes the three Aurora designs.

TABLE 10 Summary of the Three-Phase GaAs RISC Program

Phase	No. of transistors	Equal no. of 80286s[a]	Clock (MHz)	Power (W)	Yield
Aurora I	60,000	0.46	137	11	17%
Aurora II	160,000	1.23	200[b]	24[b]	N/A
Aurora III	500,000	3.85	400[b]	N/A	N/A

[a]Integration size with respect to a 130,000 transistor Intel 80286.
[b]Goal.

TABLE 11 Digital GaAs ARPA Insertion Contracts and GaAs Benefits

Company	System	GaAs benefit
E-Systems	Array processor for RC-135 reconnaissance	6× Speed improvement at 300 lbs less weight
E-Systems	Modem and synthesizer for AN/PRC-126 radio	Improved system performance
Grumman	Radar processor	Greater range and better resolution
Honeywell	Digital map computer for air navigation	Improved speed and system performance
ITT Avionics	Digital RF memory for AN/ALQ-136 jammer	Performance upgrade with same size/power
KOR electronics	Digital RF Memory for ULQ-21 drone jammer	Reduced cost
Martin Marietta	Signal processor for Hellfire Missle	Reduced cost, weight, and size
Martin Marietta	On board processor for a spacecraft	7× Improvement in speed
McDonnell Douglas	Image processor for OH-58D helicopter	Performance upgrade
Sanders Associates	Signal processor for AN/ALQ-126B jammer	Performance upgrade with same size/power
Texas Instruments	Signal processor for AN/APS-137 radar	Increased resolution

3.4.6 ARPA's Digital GaAs Insertion Program

The only application of GaAs digital circuits to be discussed with respect to the military markets will be those in the ARPA insertion program. This program was started in 1989 to demonstrate the cost and performance advantages digital GaAs brings to fielded military systems. Table 11 lists the original 11 contract awards to 9 different companies representing systems from all branches of the armed services [27].

This table shows four main benefits to a variety of systems by incorporating digital GaAs. These are improved speed or system performance, system enhancements or upgrades, reduced size and weight, and reduced cost. The cost reduction can by realized by increased reliability, meaning fewer total systems needed. Furthermore, the cost and performance benefits can be synergistic, such that the increased system performance results in fewer systems needed to accomplish a given task.

3.5 SUMMARY

This chapter has presented some of the trade-offs that must be addressed when doing a high-speed digital design. These trade-offs consist of evaluating cost and performance factors, which include circuit speed, power, and architecture. While an "apples to apples" comparison is difficult when choosing between high-performance digital technologies, it was shown that GaAs can compete at operating speeds as low as 100 MHz with silicon BiCMOS and high-speed CMOS. Furthermore, there are now several examples of systems in which GaAs was chosen over silicon. It is the author's belief that GaAs will never play a major role in this silicon-dominated semiconductor industry, but will continue to effectively compete in the high-performance niche markets.

REFERENCES

[1] P. Greiling, The historical development of GaAs FET digital IC technology. *IEEE Trans. Microwave Theory Tech.* MTT-32 1144–1156, Sept. 1984.
[2] G. Bylinsky, What's sexier and speedier than silicon. *Fortune Magazine* 74–76, June 24, 1985.
[3] M. Gold, Motorola makes $100M investment in GaAs. *Electronic Engineering Times* 1, April 22, 1991.
[4] S. Lande, Customer acceptance of GaAs IC's. *IEEE GaAs IC Symp.* 7–10, October 1992.
[5] R. Wilson and B. Fuller, Biggest GaAs gate array. *Electronic Engineering Times* 4, February 10, 1992.
[6] R. Eden, A. Livingston, and B. Welch, Integrated circuits: The case for gallium arsenide. *IEEE Spectrum* 20(12), 30–37, Dec. 1983.
[7] M. Shur, *GaAs Devices and Circuits*. Plenum, New York, NY, 1987.
[8] S. Sze, *Physics of Semiconductor Devices,* 2nd ed. Wiley, New York, 1985.
[9] D. Dobberpuhl *et al.*, A 200-MHz 64b dual-issue CMOS microprocessor. *IEEE J. Solid-State Circ.* 27(11), 1555–1567, November 1992.
[10] R. Cates, Gallium arsenide finds a new niche. *IEEE Spectrum* 25–28, April 1990.
[11] M. Penn, Economics of semiconductor production. *Microelectronics Journal,* 23(4), 255–265, 1992.
[12] M. Howes and D. Morgan, *Gallium Arsenide Materials, Devices and Circuits.* John Wiley and Sons, New York, 1985.
[13] N. Einspruch and W. Wisseman, VLSI electronics microstructure science, In *GaAs Microelectronics*, Vol. 11. Academic Press, New York, 1985.
[14] C. H. Vogelsang *et al.*, Complementary GaAs JFET 16K SRAM. *IEEE GaAs IC Symposium*, 75–78, November 1988.
[15] M. Wilson *et al.*, Process optimization of high performance ion implanted GaAs JFETs. *IEEE GaAs IC Symposium*, 169–172, October 1992.
[16] J. DiGiacomo, *VLSI Handbook.* McGraw–Hill, New York, 1989.
[17] R. VanTuyl and C. Liechti, High-speed integrated logic with GaAs MESFETs. *IEEE J. Solid-State Circ.* SC-9(5), 269–276, October 1974.
[18] S. Long and S. Butner, *Gallium Arsenide Digital Integrated Circuit Design*. McGraw-Hill, New York, NY, 1990.
[19] S. Ohr, Making the 400-MHz computer a reality. *Computer Design* 83–99, Aug. 1992.
[20] G. Zorpette, Large computers—Technology 1993. *IEEE Spectrum* 30(1), 37, Jan. 1993.
[21] D. Kiefer and J. Heightley, Cray-3: A GaAs implemented supercomputer system. *IEEE GaAs IC Symp.* 3–6, 1987.
[22] R. Wilson, Range of clock chips reveal P5 timing. *Electronic Engineering Times* 52, Nov. 30, 1992.
[23] E. Fox *et al.*, Reduced instruction SET architecture for a GaAs microprocessor system. *IEEE Comp.* 71–81, Oct. 1986.

[24] D. Harrington *et al.*, A GaAs 32-bit RISC microprocessor. *IEEE GaAs IC Symp.* 87–90, 1988.
[25] R. Brown *et al.*, GaAs RISC processors. *IEEE GaAs IC Symp.* 81–84, 1992.
[26] M. Upton *et al.*, *A 160,000-Transistor GaAs Microprocessor*. IEEE International Solid-State Circuits Conference. To be presented February 1993.
[27] A. Prabhakar, Digital gallium arsenide upgrades for military systems. *IEEE GaAs IC Symp.* 15–17, 1989.

4
Phased-Array Radar

Inder Bahl* and Dave Hammers†
*ITT Gallium Arsenide Technology Center, Roanoke, Virginia
†ITT Gilfillan, Van Nuys, California

4.1 Introduction
    4.1.1 Frequency Bands
    4.1.2 Radar Types and Applications
    4.1.3 System Description
    4.1.4 Electronically Scanned Radars
    4.1.5 Technology Trends: Active Phased-Array Radar (APAR)
    4.1.6 Cost Requirements
4.2 Phased-Array Radar Architectures
    4.2.1 Passive Phased-Array Radars
    4.2.2 Active Phased-Array Radars
4.3 Subsystem Functions and GaAs Applications
    4.3.1 Exciters
    4.3.2 Transmitters
    4.3.3. Antennas
    4.3.4 Receivers
    4.3.5 Signal Processing/Control
4.4 Transceiver Module Technology
    4.4.1 Transceiver Module Architectures
    4.4.2 Low-Noise Amplifiers
    4.4.3 Power Amplifiers
    4.4.4 Phase Shifters
    4.4.5 Switches
    4.4.6 Attenuators
    4.4.7 T/R Module Chips
    4.4.8 Circulators
    4.4.9 Housing
4.5 Summary and Future Trends
    References

4.1 INTRODUCTION

Since the second world war, tremendous progress has been made in the field of radar. Radar is an acronym derived from radio, detection, and ranging. The purpose of a radar is to detect a target, determine its position in terms of range, height, elevation, and azimuth and speed and, if possible, obtain information about its size and configuration. It has to work when mounted either on a moving platform, such as a tank, car, ship, submarine, aircraft, or space ship or permanently installed on the ground. It has to cover friendly as well as enemy areas in an environment of interfering reflections (called clutter) from sea, mountains, buildings, forests, adverse weather conditions such as rain, fog, and snow, jamming signals (intentional and unintentional) and, in many cases, must keep its location from being detected. In addition, it must be effective in sensing targets varying in size, speed, and surface reflectivity (called radar cross section). Major applications include vehicle speed sensors, airport traffic control, weather sensors, ground mapping, air/ground surveillance and target detection, altimation, and weapon guidance.

Radar's principles, system design, hardware implementation, signal processing, etc., have been thoroughly described in the literature [1–24]. An excellent review of radar technology is given in [15]. The evolution of radar technology includes microwave megnetron, high-power klystron, low-noise traveling wave tube (TWT), coherent signal processing, monopulse tracking, pulse compression, electronically steered arrays, high-speed digital processing and control, hybrid microwave integrated circuit (MIC) techniques, solid-state devices, and monolithic microwave and millimeter-wave integrated circuits (MMICs), and including over the horizon (OTH), phased-array (PA) and synthetic aperture (SA) radar systems.

Some of the many functions radars are designed to perform [23] along with a brief definition of each are given in Appendix B. Since the major portion of this chapter covers military radars, their designated nomenclature is given in Appendix A. This section provides an overview of radar frequency bands, their applications, and types of radars.

4.1.1 Frequency Bands

Depending upon the application, including function, platform, and range of detection, the frequency of operation for radars covers VHF to optical (30 MHz to 30×10^4 GHz). Table 1 shows the microwave radar frequency bands. In addition, infrared and optical frequency bands have been used for weather radars. In general, selection of frequency range for any radar depends upon a number of diverse considerations, including atmospheric attenuation, variation of radar cross section of target and clutter (including rain) with frequency, availability of power sources, target location accuracy, and antenna size.

Most microwave frequency radars are narrowband (< 10% bandwidth). The bandwidth of the transmitted signal determines the frequency agility, the range resolution, and the multifunctionality of the radar system. The trend in microwave frequency radars is toward increasing bandwidth (e.g., 25%).

TABLE 1 Radar Frequency Bands

Band designation	Nominal frequency range	New band designation (GHz)	Specific radiolocation (radar) bands based on ITU assignments for region 2 (North and South America)
HF	3–30 MHz	A	
VHF	30–300 MHz	A < 0.25	138–144 MHz
		B > 0.25	216–225 MHz
UHF	300–3,000 MHz	B < 0.5	420–450 MHz
		C > 0.5	890–942 MHz
L	1,000–2,000 MHz	D	1,215–1,400 MHz
S	2,000–4,000 MHz	E < 3	2,300–2,500 MHz
		F > 3	2,700–3,700 MHz
C	4,000–8,000 MHz	G < 6	5,250–5,925 MHz
X	8,000–12,000 MHz	H > 6	8,500–10,680 MHz
Ku	12.0–18 GHz	I < 10	13.4–14.4 GHz
		J > 10	15.7–17.7 GHz
K	18–26.5 GHz	J < 20; K > 20	24.05–24.25 GHz
Ka	26.6–40 GHz	K	33.4–36.0 GHz
EHF	30–300 GHz		

4.1.2 Radar Types and Applications

Several types of radars are being used for military and civilian applications. They can be identified in several ways: frequency of operation, mode of operation, type of waveforms used (pulse, pulsed repetition frequency, and CW), application, and by the method of scanning the beam. Radars can be classified into many application areas, e.g., surveillance (including long-range and battlefield), altimeters, doppler, air defense, weapons guidance, synthetic aperture, air traffic control and landing aid, weather monitor, terrain avoidance, speed monitoring/control, vehicle traffic sensors, and special purpose military. These radars are summarized in Table 2. This list excludes classified defense system radars.

Long range (more than 100 nautical miles) surveillance radars operate in the UHF-, L-, and S-frequency bands, while the battlefield surveillance radar's frequency is generally in the X- or Ku-band. Surface to air defense radars operate in the S-band, and the surface to air defense and missile guidance system's frequency range is in the C- and X-bands. X-band and Ku-band frequencies are also used for airborne multifunctional guided missile radars. Airport surveillance radars (ASRs) perform in the L-, S-, and C-bands. Altitude sensors used in all commercial and military aircraft and surfacebased weather radars operate in the C-band. Most automobile

TABLE 2 Types of Radars

Type	Main platform	Frequency band	Purpose	Application
Surveillance and search	Ground, shipborne, airborne	UHF, L, S, C	3-D air surveillance, long range search and track of missiles, aircraft, satellites	Military
Altimeters	Ground, airborne	S, C	Auto height finding	Military/commercial
Doppler weapons guidance	Airborne	X, Ku	To measure three-dimensional velocity of the aircraft mounted in the missiles and control its path to hit the target	Military/commercial
Air defense	Ground, shipborne	S	Locate target aircraft and direct weapons against it	Military
Missle defense	Vehicle	C, X	Locate incoming missile and fire counter missile to destroy it	Military
Synthetic aperture	Airborne and space based	X	Maps ground from the airplane/satellite	Military
Air traffic control	Ground	L, S, C, X	Helps in locating aircraft entering and leaving an airport. L-band is used for long range search	Commercial
Weather monitor	Ground, airborne, space-based	C, X	Monitor cloud formation in the flight path/predict weather conditions and forecast	Military/commercial
Terrain avoidance	Airborne	X	Locate mountains and other obstacles to avoid during the flight	Military/commercial
Speed monitoring and breaking	Vehicle	X, Ku, K, V, W	Cruise control, collision avoidance	Commercial
Vehicle traffic sensors	Ground	Q, V, W	Traffic monitoring	Commercial
Special purpose				Military
Docking		X, K	Guide a spacecraft to docking	
Marine radar		S, C, X	Locate other ships and seashore	
Spacebased radar		L, S		
Over the horizon		HF	Monitor the movement of ships and planes	

speed detection police radars are at 24 GHz. RF to X-band frequencies have been used for synthetic aperture radar applications. Missile seeker radars and collision avoidance radars are being developed at millimeter-wave frequencies.

While most radars operate in the RF and microwave portion of the spectrum, a growing number of radars operate in the millimeter-wave, infrared, and optical frequency bands. These radars are in different stages of evolution from mere concepts to fully operational systems. In many applications, millimeter-wave radars have distinct advantages over microwave, infrared, and optical counterparts. Millimeter-wave radars have the advantages of small size and weight, high resolution in both azimuth and range, and good scanning possibilities and can operate in adverse visibility and weather conditions with little degradation of performance in these applications. With the ever increasing capability of military firepower, mobility, multifunctionality, and the complexity of the battlefield scenario, it has become necessary to have effective surveillance, target acquisition, and identification of equipment. Millimeter-wave radars are well suited to perform the above functions effectively over a very short range (1–2 nautical miles).

4.1.3 System Description

A simplified block diagram of a conventional radar is shown in Fig. 1. The radar is composed of several main components: an antenna (for both transmission and reception of the signal), a duplexer or a circulator (which separates transmission and receiving paths with good isolation), a transmitter (consisting of a controlled microwave power source), a receiver (consisting of a low-noise amplifier (LNA), mixer, local oscillator (LO), intermediate frequency (IF) amplifier, detector, and video amplifier), a signal processor (consisting of processes which improve the signal to interference ratio), and a display. The LNA amplifies the weak received signal, mixer and LO combination downconverts the received signal to an intermediate frequency, and the detector section eliminates the IF frequency, achieves the

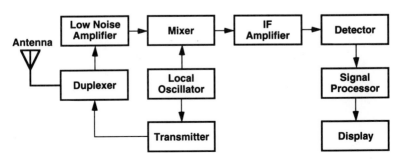

FIG. 1 Simplified radar block diagram.

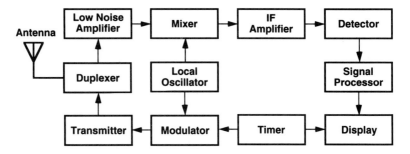

FIG. 2 Simplified block diagram of a conventional pulse radar.

baseband signal, and amplifies it. Note that the local oscillator is derived from the same frequency source as the transmitted signal. This is to provide coherency between the transmitted and the received signals, a characteristic very desirable for Doppler signal processing in modern radars. The signal processor and display units convert the radar-derived information for the operator. In an automatic system the signal processor and associated computers extract target data which automatically updates tracks and previous data for controlled actions.

Figure 2 shows a block diagram of the conventional pulse radar. The waveforms at send time are shown in Fig. 3. The transmitter signal is modulated with the required pulse width and repetition frequency.

Figure 4 shows a block diagram of a monopulse radar, in which the directional information of an incoming signal is obtained more accurately by a comparison of the relative amplitude of signals received from its sum and difference radiation patterns.

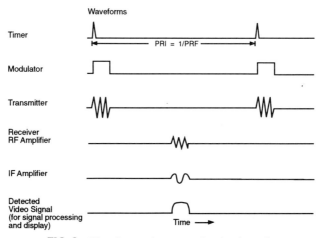

FIG. 3 Waveforms of a conventional pulse radar.

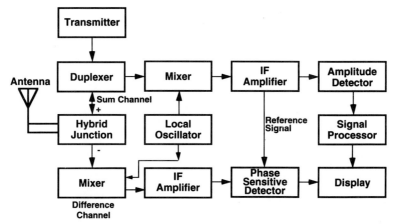

FIG. 4 Simplified block diagram of a monopulse radar.

4.1.4 Electronically Scanned Radars

Conventional radars have limited capabilities and response time due to mechanical inertia of the antenna rotating systems. Work on electronically scanned radars with enhanced capabilities started in the early 1960s. In this type of radar the beams are scanned in azimuth and elevation by modulating the signal fed into the antenna elements. Basically there are two types of electronically scanned radars: frequency and phase. The first practical electronically scanned radars were of the frequency scanned type.

Frequency Scanned Radars

Frequency scanned radars (FSR) belong to the simple class of electronically scanned radars in which an electromagnetic beam is positioned in space by changing the frequency of operation [3] without changing the antenna position. In this case the phase of the radio frequency (RF) signal fed to each radiating element in the antenna array is varied by changing the frequency of the source. The scan sensitivity, or the amount that the beam moves with respect to frequency change, depends on the extent of the frequency range and the propagation time of the RF signal across the elements. This scheme is relatively simple, reliable, and cost-effective and provides rapid detection and accurate position measurement of multiple targets. This technique has been successfully implemented in air surveillance and aircraft control radars by using mechanical control in azimuth and frequency scanning in elevation to provide three-dimensional search. Some of these radars in use are summarized in Table 3.

A block diagram of an FSR is shown in Fig. 5. The frequency scanned antenna consists of folded rectangular waveguides with radiating slots. The 3-D radars (AN/SPS-48 family) which operate at S-band and provide 45° beam scan in elevation are in extensive use.

TABLE 3 Electronically Scanned Radars

Radar	Platform	Frequency	Antenna size	Power source	Scan type	Range	Purpose
ANTARES	Ground, static	S-band	9 × 8 m	Magnetron	Azimuth–Mechanical Elevation–Phase	~370 km	Auto height finding or low elevation long range surveillance
Air defense radar (AR-3D)	Ground, static	S-band	4.9 × 7.1 m	Klystron	Azimuth–Mechanical Elevation–Stacked Beam	~550 km	Three-dimensional air defense
Tracking radar (AN/SPS-32)	Shipborne	S-band	7.6 × 6.0 m	—	Azimuth–Phase Elevation–Phase	—	—
Surveillance radar (AN/SPS-48)	Shipborne	S-band	Slotted array	CFA[a]	Azimuth–Mechanical Elevation–Frequency	—	3-D air surveillance and weapons support
Surveillance radar (AN/SPS-52)	Shipborne	S-band	Slotted array	—	Azimuth–Mechanical Elevation–Frequency	—	3-D air surveillance and weapons support
Surveillance radar (AN/TPQ-37)	Ground, trans-portable	X-band	Slotted array	TWT	Azimuth–Mechanical Elevation–Frequency	—	3-D battlefield surveillance of artillery
Surveillance radar (AN/TPS-59)	Ground, trans-portable	L-band	9.1 × 4.9 m	Solid state	Azimuth–Mechanical Elevation–Phase	—	Long range surveillance for U.S. Marine Corps and tactical air operations
Matador (TRS 2210)	Ground, mobile	S-band	—	Magnetron	Azimuth–Mechanical Elevation–Phase	—	3-D mobile air defense

[a] Cross-field amplifier.

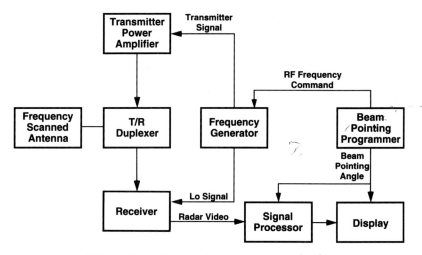

FIG. 5 Block diagram of a frequency scanned radar.

Phased-Array Radars

Phased-array radars (PARs), which is the topic of this chapter, form an electromagnetic beam in azimuth and elevation space by changing the phase of the RF power fed to each individual radiating element in the antenna array. The beam moves to any position in the zone of coverage within microseconds (μs) in proportion to the phase gradient achieved across all the radiating elements. Since no mechanical movement is involved, PARs can be made with larger aperture size, greater radiating power, faster scanning rates (as compared to conventional radars), and with the possibility of performing search, acquisition, and multitarget tracking in an interleaved manner. Since the beam can be steered within microseconds the system needs to be controlled by a digital computer to perform the various functions. Beam scheduling, related to the positioning of radar beam for functions like search, acquisition, and tracking, is one of the most important tasks performed by the system.

The block diagram of a typical PAR is shown in Fig. 6. The major difference between a PAR and a conventional radar lies in the beam formation system, e.g., front-end antenna—multielement antenna array vs single-element antenna. In the conventional PAR each antenna element is connected to phase shifter used in either transmit or receive operations. During the past 30 years the phase shifters have been manufactured using waveguide, ferrite, and silicon diode technologies. PARs are being used in groundbased, shipboard, airborne, spacebased, satellite and airport traffic control applications. Table 4 gives the history of phased-array radars developed during the 1960s and 1970s.

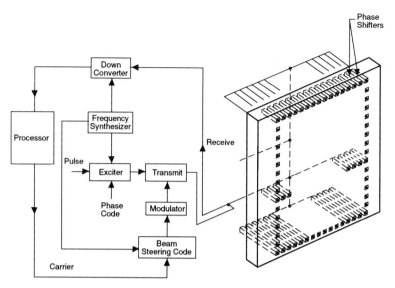

FIG. 6 The configuration is shown for a groundbased PAR containing hundreds or thousands of phase shifters.

Attempts were made early to place power sources at the radiating elements. The MERA (molecular electronics for radar application) system [25] was the first solid-state phased-array radar. It operated at X-band and had 604 elements, each having an output power of 0.5 W. This airborne system was demonstrated in 1968. The RASSAR (reliable advanced solid-state radar) program [25] followed MERA and had features similar to those of the AN/APQ-122 system. This air-ground radar operated in X-band, had 1648 elements (each transmitting 1 W), and was demonstrated in 1974.

Pave Paws radars [26], which are entirely solid state, were the first major PARs to join the U.S. Air Force protection system. These radars were used to watch around the clock for submarine-launched ballistic missiles and operate in three major modes: searching, verifying, and tracking. There are two systems installed, one on the east coast and the other on the west. These systems have two flat sides each of which has 1792 active and 885 dummy radiating elements. The system has 240° of horizontal coverage and 3° to 85° elevation coverage. Each system can track hundreds of missiles simultaneously. The system on the east coast is capable of tracking objects the size of a baseball over St. Louis or the size of a small car over Spain. Figure 7 shows the Pave Paws early warning radar.

Another powerful PAR (twice the power of Pave Paws) is the BMEWS (ballistic missile early warning system) phased-array radar [26]. This over the horizon radar can detect landbased missiles in far greater numbers than

TABLE 4 Development of Phased-Array Radar Systems

Radar	Year	Platform	Frequency	Antenna size (m)	Power source	Range	Purpose
ESAR	1960	—	—	—	Tetrode	—	—
Spacetrack (AN/FPS-85)	1968	Ground	VHF	~26.9 × 26.9	—	Satellite ranges	Search and track of missiles and satellites
Missile site radar (MSR)	1969	Ground	S-band	~8.5 Diameter	—	—	Short range search and track, and missile guidance
Precision approach radar (AN/TPN-19)	1971						
Perimeter acquisition radar (PAR)	1974	Ground	VHF	~30 m Octagonal	TWT	~400 km	Long range search and track of missiles and satellites
AEGIS (AN/SPY-1)	1974	Shipborne	S-band	~3.7 × 3.8 Hexagonal	—	—	Surface to air missile defense
Patriot (AN/MPQ-53)	1975	Ground/mobile	C-band	—	—	—	Search and track, and missile air defense
Pave Paws (AN/FPS-115)	1975	Ground	S-band	—	—	—	Tactical air defense system
COBRA DANE (AN/FPS-108)	1976	Ground	L-band	29 Diameter	TWT	~1850 km	Warning of submarine launched missile attacks
COBRA JUDY (AN/SPS-)	1980	Shipborne	S-band				Search and track of missiles and satellites

FIG. 7 AN/FPS-115 Pave Paws early warning radar array antenna (photograph courtesy of Raytheon Company).

can smaller seabased radars. This system operates at 51.28 MHz and can detect aircraft and cruise missiles between 800 to 3500 km from North American shores [26].

Another PAR example is the airborne joint surveillance target attack radar system (JSTARS). This radar detects, locates, identifies, classifies, tracks, and targets hostile ground movements in any weather, providing a real-time tactical view of the battlefield. The system consists of a 24′-long multimode phased-array antenna with two operating modes: synthetic aperture radar/fixed target indicator and wide area surveillance/moving target indicator. These modes can be performed simultaneously and the radar can track vehicles and other targets while directing weapons to ground targets.

Other examples of first-generation phased-array radar systems are the U.S. Navy's Aegis, Patriot, and Cobra Dane. The Aegis radar (AN/SPY-1D) as shown in Fig. 8 has four conventional array antennas (only two have been shown) each consisting of approximately 4000 passive elements arranged in an octagonal configuration 12′ across. This system is deployed on many cruisers. The Patriot (AN/MPQ-53) is an anti-aircraft and anti-missile multifunction PAR system. Its two large circular space-fed lens array antennas are shown in Fig. 9. Several other antennas are for identification friend or foe (IFF), electronic warfare (EW), and missile control functions. The radar operates at C-band and performs several functions: (a) target search and track, (b) missile search, track, and communications during mid-course guidance, and (c) target terminal guidance via sensors on the missile [27]. The L-band Cobra Dane, which has a single array of 29 m diameter and an average power of about 1 MW, is shown in Fig. 10. This radar has the

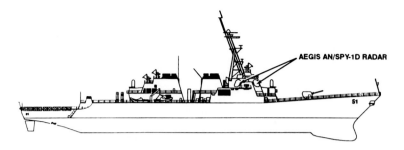

FIG. 8 USS *Arleigh Burke* with AN/SPY-1D (two arrays visible on slanted panels below and to the side of the bridge).

distinctive feature of high resolution (less than 1 m) and is a multimission system. It provides information regarding missile test firings within the former Soviet Union and firings of intercontinental ballistic missiles (ICBM).

Newer phased-array radars are AN/MPS-39 or MOTR (multiple object tracking radar), AN/TPQ-37 Firefinder radar, and AN/APQ-164 airborne multifunction radar. Figure 11 shows the AN/APQ-164 radar antenna.

FIG. 9 Photograph of the Patriot phased-array radar. This is a C-band multifunction radar that provides tactical air defense, including target search and tracking and missile fire control. The phased-array antenna uses 5000 ferrite phase shifters to electronically scan the antenna beam (photograph courtesy of Raytheon Company).

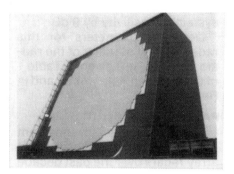

FIG. 10 AN/FPS-108, or Cobra Dana, a large phased-array radar on Shemya Island near the tip of the Aleutians.

4.1.5 Technology Trends: Active Phased-Array Radar (APAR)

The trend in advanced military radars is toward wider bandwidths, multifunction capability, rapid and accurate surveillance and tracking of multiple targets, protection in a harsh EW environment, detection of new threats such as low flyers and small cross-sectional targets, radiation hardness, high reliability, and, above all, low cost for affordable systems. There is a strong emphasis on decreasing the radar's susceptibility to intercept and jamming, thus providing electronic counter-countermeasures (ECCM) effectiveness and improving the radar's performance in an EW environment. The phased array-antenna, with capability of rapidly steerable multiple beams and nulls, provides a particularly important approach to meeting these requirements.

Through the 1960s, 1970s, and 1980s, the basic RF technology approach to phased-array radar centered on the application of high peak power transmitter tubes and centralized receivers. These assemblies would interface with thousands of phase shifters (one at each element) through corporate beamforming structures. However, in the late 1980s a new class of phased-array radar, the active phased-array radar (APAR), entered the scene. In an APAR, the large transmitter tubes of the conventional radar are replaced by thousands of elements, called transmit/receive (T/R) modules, which are distributed across the antenna and serve as independent miniradars. The modules are small, weighing only a few ounces, with all of the electronics imbedded in one or more microminiature chips. In combination, the output power of all of the miniature T/R modules results in the equivalent of the single high-power transmitter, but with significant advantages of wide bandwidth, reliability, and efficiency. Single and multichip MMIC-based T/R modules are being developed for APAR systems. Figure 12 shows a comparison of conventional phased-array and active phased-array radars.

One of the most difficult challenges for T/R modules is achieving high

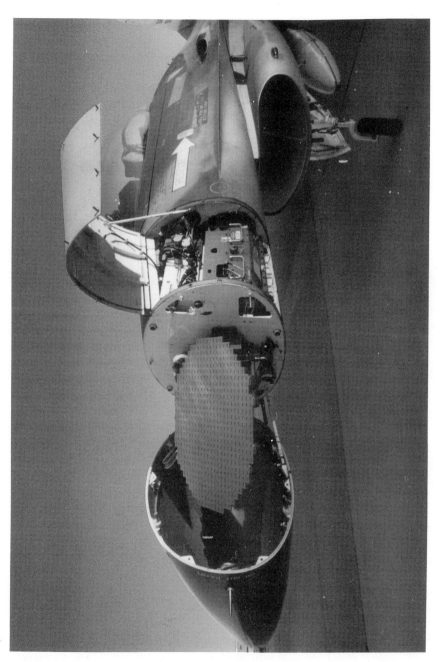

FIG. 11 AN/APQ-164 (B1-B) Antenna and elements (photograph courtesy of Westinghouse Electric Corporation).

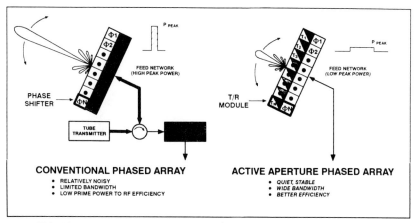

FIG. 12 Comparison of conventional and active aperture phased arrays.

transmitted power with high efficiency. Here the key figure of merit is power added efficiency (PAE) defined as the ratio of RF output power to DC power consumed. Significant progress has been made in recent years to improve the PAE of power ICs using Class-AB power devices and designs. For example, power ICs with 5 W output power and 60 PAE over 15% bandwidth at C-band have been developed. These highly efficient ICs drastically reduce the prime power requirements which, in turn, simplifies the array design. Other challenges include light material for T/R module housings and antenna elements and required cooling of the array.

4.1.6 Cost Requirements

Successful development of APAR for high-volume applications depends basically on two factors: cost and performance. The low-cost and large-scale usage of T/R modules is one of the driving forces for the success of GaAs MMIC technology, whereas performance, in terms of amplitude and phase tracking from module to module, is essential to achieve low sidelobe antenna performance and to take care of ECM threats. Also, high-efficiency T/R modules dissipate less heat. This reduces the cooling requirements which in turn reduces the system's cost and improves its reliability. High-efficiency T/R modules also require less prime power which enhances the transportability of the radar system from one place to another.

The cost of an APAR will depend upon its capability and complexity, and is composed of several subsystems such as T/R modules, mechanical structures to support and cool T/R modules, power supplies, RF feed and DC bias, signal processor and central computer, and displays. Some costs are fixed while other costs depend upon the life of the active array. The cost ratio of T/R modules to the total APAR for various active array sizes can

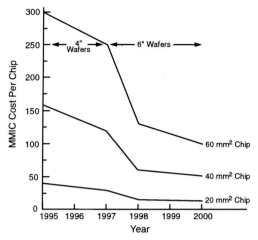

FIG. 13 MMIC costs vs time. Large chips are multifunction. Cost does not include NRE and is based on high production volume.

be assessed as 30% for 500–1000 modules, 40% for 1500–2000 modules, 50% for 2500–3500 modules, and 60% for above 4000 modules. Assuming a MMIC-based T/R module cost (when produced in large quantities) of $500 to $1000, and 1600 modules per APAR system, the average GaAs IC-based APAR cost will be $2–4M. Figure 13 shows the cost of MMICs as a function of time. The cost model is based on several assumptions and learning curves from the experience the GaAs industry achieved during microwave and millimeter-wave integrated circuit (MIMIC) programs. Also, the semiconductor industry has consistently demonstrated that a higher level of automation in fabrication and testing in high-volume production greatly improves the yield and reduces final cost.

Today's MMIC cost for T/R module applications is in the $1000 to $2000 range. The cost can approximately be split as follows: MMIC (eight chips), 30%; assembly, 15%; controller, 15%; other parts (connectors, feedthrus, resistors, capacitors, etc.), 25%, and testing, 15%. Assembly costs are moderately high due to packaging of large numbers of discrete devices and components which are on the order of 200; each requires about three to four wire bonds. High MMIC costs are due to large chip sizes, low end-to-end chip yield (wafer, RF, optical), and labor-intensive testing. With today's improved computer-aided design (CAD) tools, accurate device models, and electromagnetic (EM) simulators, the total GaAs chip size for C-band T/R modules can be reduced to about 100 mm^2 with transmitter power output of about 15 W and about 75 mm^2 with transmitter power output of about 8 W. By using only one or at the most only a few highly integrated MMIC chips, the part counts, die bonding, wire bond connections, and subassembly test can be reduced significantly (by at least a factor of 5). Using automated GaAs IC manufacturing technology with improved optical yield, RF testing, and low-cost packaging, the cost of MMIC-based T/R modules in production volume can be achieved with the $300 to $1000 range, depending upon the frequency of operation, bandwidth, power output, etc.

In summary, within the next 2 to 3 years, with high-volume production, 4″-diameter wafers, higher level of automation, and improved processes in terms of higher end-to-end wafer yields, two to four chip solution for T/R modules will cost about $2/mm^2. In other words, in large production volumes, APAR systems will contain about $200 worth of GaAs MMICs.

The following sections describe the phased-array radar architectures, subsystem functions and their realizations using GaAs technology, and the GaAs transceiver module technology.

4.2 PHASED-ARRAY RADAR ARCHITECTURES

Basically there are two phased-array radar configurations: passive and active. Currently used phased-array radars are mostly of the passive type, whereas the new low-cost multifunction and high-reliability phased-array radars being developed are of the active (or APAR) type just discussed. A brief description of the passive type architecture is first given to help in understanding the APAR type which is the primary subject for this chapter.

4.2.1 Passive Phased-Array Radars

Most PARs use high-power tubes such as magnetrons, klystrons, and travelling wave tubes as centralized transmitters to feed radiating elements each of which contain a controllable passive phase shifter as shown in Fig. 14. These systems use a corporate or row/column feed network to supply power to the radiating elements through the phase shifters.

In passive PARs, the total loss includes losses in the feed, phase shifters, circulator, and the radiating elements. These systems suffer from heavy losses in the feed network resulting in poor power efficiency. The total loss in the corporate feed is about 2 to 6 dB depending upon the frequency of operation and the number of elements used. Due to very high losses in the power distribution networks, system requirements mandate very high tube power output. Low power efficiency and high power tubes result in poor system reliability. The transmitters and the associated high-voltage power supply are the reliability limiting components in the passive PAR. Additionally, this technology has difficulty meeting the most demanding threats, particularly those at low altitudes such as anti-ship cruise missile (ASCM) and small targets, due to inadequate frequency stability.

The transmitters utilize hybrid solid-state amplifiers for moderate power

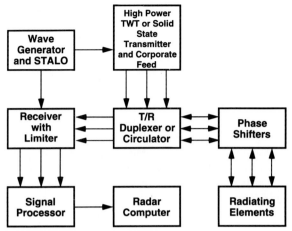

FIG. 14 Passive phased-array radar configuration showing main building blocks.

applications. They require low DC operating voltages and have inherent high reliability. However, for higher power applications, tubes are generally preferred because they convert prime power to microwave energy more efficiently than do solid-state amplifiers, which require lossy combiners. Traveling wave tubes are generally used as the RF power amplifiers in broadband applications, although they do not offer the required performance.

The antenna in a PAR is used for both transmitting and receiving the signal. Therefore, the phase shifters feeding the radiating elements have to be reciprocal. Since the array is switched digitally with the help of an electronic computer, digital phase shifters with very low switching times (<1 µs) are required. There are two different types of phase shifters: those using solid-state switching devices (i.e., p-i-n diodes or field effect transistors (FETs)) and those using ferrites. In diode/FET phase shifters, phase shift results from a change in phase values associated with the two states of switch reactance. Phase shifting with ferrites is usually accomplished by the change in magnetic permeability which occurs with application of a magnetic biasing field.

Diode/FET phase shifters have the advantages of fast switching, simplicity of the driver circuit, reciprocal operation, and lower temperature sensitivity. Ferrite phase shifters have the advantages of high power handling capability, lower insertion loss, and lower voltage standing wave ratio (VSWR).

The number of phase shifters required in a PAR generally equals the number of radiating elements. One PAR uses thousands of small, highly reliable low-cost phase shifters compatible with large-scale production. The current cost of the phase shifters is significant (~ 30%) compared to the total cost of the PAR.

4.2.2 Active Phase-Array Radars

Since 1965 there has been an increasing emphasis in APARs. At that time, the general consensus was that the electronically steerable antenna technique had great promise in terms of its capability, flexibility, maintainability, and reliability. However, its implementation into systems has been slow and plagued by high costs. The development of APARs has been hampered over the past 30 years by two factors: lack of low-cost high-speed computers that can provide the necessary control for the multimode operation and the multitarget tracking, and the availability of low-cost T/R modules. As these two major obstacles have now been essentially overcome, APARs are now feasible.

The APAR configuration, as shown in Fig. 15, uses solid-state transceiver modules at the element level. Since transmitters are located in the T/R modules and feed losses are minimum, the total amount of prime power required is reduced. In this configuration, the full transmitted power is generated very close to the radiators and the subsequent power combining in space is very efficient. The exciters for each module can be either low-power phase-locked oscillators providing a coherent wavefront with predictable phase, or a low-level oscillator driving subarray feeds connected to a group of modules. From a reliability point of view, the former is preferred; but costwise, the latter may be desired. Since the subarray feeds handle only low-power levels, power dissipated does not contribute significantly in the

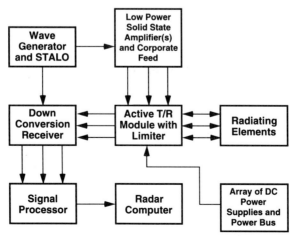

FIG. 15 Active phased-array radar (APAR) configuration showing main building blocks.

total power loss. Due to high-efficiency transmitters and lower distribution loss, the APAR system power efficiency is two to three times better than that of a passive PAR.

Another advantage of the APAR is the graceful degradation of system performance when a fraction of the modules fail, as contrasted with catastrophic shutdown with tube failure in a passive PAR. In addition to the increased reliability associated with solid-state transmitters and lower feed losses, provision for quick interchange of defective modules enhances the overall operational availability of APARs. Perhaps the most significant advantage of APARs is full beam and waveform agility. This provides approaches for flexible beam shaping and beam nulling, multiple beam generation, and adaptive energy management.

The APAR architecture trade-offs center on matching the active power–aperture product and the flexible waveform to the functional problem to be solved. In essence, this means determining the output power and the number of elements required relative to the sensitivity needed for target extraction in the various deployment environments. This evaluation is usually done at the highest frequency of the operating bandwidth since the minimum number of modules to be used for avoiding grating lobes in a $\pm 45°$ scan occurs when the elements are spaced $\lambda/2$. Other factors such as desired target resolution and accuracy in angle and range usually have less influence in the desired power–aperture product and hence, in the distribution of power and elements in the active array. This requirement sets the aperture and pulsewidth sizes, which again impacts the number of elements on a $\lambda/2$ spacing across the array. Furthermore, clutter and ECM control can heavily influence the choice of resolution and waveform, to such a degree that they become the overwhelming drivers in the power–aperture product and therefore the number of elements and T/R modules.

Usually phased-array radars are applied in situations where high-performance target tracking is needed since the data rate can be kept high or adaptive depending on the maneuvers of the target set. In these cases multiple antenna feed channels for monopulse or simultaneous tracking beams are needed. This complicates the active array since these feeds have to be packed in the array with all the T/R modules, cooling, control, and power distribution networks. However, these complications can be outweighed by the performance advantages of the APAR system.

4.3 SUBSYSTEM FUNCTIONS AND GaAs APPLICATIONS

4.3.1 Exciters

Exciters or oscillators represent the basic microwave energy source and solid-state devices have the advantages of light weight and small size. As shown in Fig. 16, a basic microwave oscillator consists of (a) a transistor

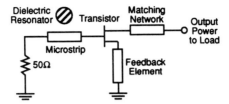

FIG. 16 Basic configuration of a dielectric resonator oscillator. The feedback element is used to make the active device unstable, the matching network allows transfer of maximum power to the load, and the dielectric resonator provides frequency stability.

as the active device (a diode can also be used), (b) a passive frequency-determining resonant element, such as a microstrip, surface acoustic wave (SAW), cavity resonator, or dielectric resonator for fixed tuned oscillators, and (c) a varactor or a yttrium–iron–garnet (YIG) sphere (not shown in Fig. 16) for tunable oscillators. These oscillators have the capability of temperature stabilization and phase locking. Dielectric resonator oscillators provide stable operation from 1 to 100 GHz as fixed frequency sources. In addition to their good frequency stability, they are simple in design, have high efficiency, and are compatible with MMIC technology. Gunn and impact avalanche transmit-time (IMPATT) diode oscillators provide higher power levels and cover microwave and millimeter-wave bands. Transistor oscillators using GaAs or other III–V compound metal semiconductor field effect transistors (MESFETs), high electron mobility transistors (HEMTs), and heterostructor bipolar transistors (HBTs) provide highly cost-effective, miniature, reliable, and low-noise sources for use up into the millimeter-wave frequency range, while silicon bipolar junction transistor (BJT) oscillators operate only to about 20 GHz. Compared to a GaAs MESFET oscillator, a BJT or a HBT oscillator typically has 6 to 10 dB lower phase noise very close to the carrier frequency. Figure 17 shows the performance of various solid-state oscillators. Higher power levels are obtained by connecting a high-power amplifier to the output of the medium-power oscillator. This amplifier also acts as a buffer between the load and the oscillator circuit which minimizes frequency pulling. The buffer amplifier stage also operates in saturation, which provides constant output power and minimizes output power variation with temperature. Since exciters are stand-alone components and only one is required for each system, either MMIC or MIC technologies meet the requirements which include good frequency and power stability over temperature, minimum frequency pulling, low single side band phase noise, and low bias voltages. Performance of a dielectric resonator oscillator (DRO)-stabilized MMIC oscillator is summarized in Table 5 [28].

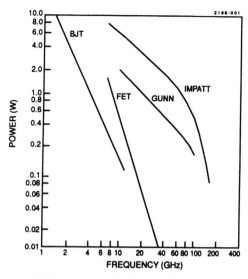

FIG. 17 Maximum CW power available from solid-state microwave oscillators.

4.3.2 Transmitters

The primary function of the transmitter is to provide a highly stable, phase- and amplitude-adjusted pulse of microwave energy to the radiating element. In conventional radars, centrally driven tube transmitters consume a large amount of prime power and dissipate much heat in generating high-

TABLE 5 Summary of DRO-Stabilized MMIC Performance [27]

Parameter	C-band	X-band	Ku-band	Units
Frequency	5.027	10.740	13.120	GHz
Output power	+12	+16	+10	dBm
SSB phase noise				
100 kHz	−115	−110	−100	dBc/Hz
10 kHz	−88	−80	−70	dBc/Hz
Frequency pulling	0.02	0.02	0.001	% (3 : 1 VSWR)
Bias voltage	8	4	11	V
Bias current	19	65	35	mA
Stability measured from −54 to +85°C				
Freq. stability	±2	±4	±5	ppm/°C
Power stability	±0.75	±1.0	±0.5	dB

peak low-duty cycle pulsed energy. Furthermore, tube transmitters usually limited to narrowband operation where the bandwidth is less than 10% of the operating frequency. These factors dictate the undesirable size, weight, cost, and reliability of conventional radars. In APARs, low-peak high-duty cycle energy allow wideband operation especially when the transmit function is distributed over all the radiating elements. Basic requirements for distributed active transmitters are: bandwidth 10 to 50%, 30–40 dB gain (multistage amplifier), high DC to RF power efficiency (>30%), 2 to 20 W power output, pulsed/CW mode of operation, low AM to PM conversion (<5°/dB), high spectral purity (<−120 dBc), and low input VSWR (1.5:1). There are additional phase and amplitude requirements relating to the transmit sidelobe performance of the antenna such as allowable pulse and amplitude stability over temperature. These must be matched to each application. GaAs power amplifiers are able to achieve uncompensated rms phase and amplitude errors on the order of 10° and 2 dB, respectively. Using feedback (or feedforward) stability compensation methods can readily reduce these errors to 1°–3° and 0.5–1 dB.

4.3.3 Antennas

The antenna in a PAR is a complex system of feeds, phase shifters, and radiating elements. The antenna in an APAR is even more complex since it also embodies the transmit/receive modules and all the associated electromechanical subsystems as shown in Fig. 18: the digital control network which distributes mode, beam pointing, timing, and amplitude information; the power distribution network which distributes power for circuit biasing; and a cooling network which removes heat from the active components.

The complexity of the networks in an APAR is dependent on the architecture and packing density of the T/R modules in the array. In a fully distributed array, these networks must provide resources to each T/R module element. In a partially distributed array, resources are provided to a subarray or a column or a row of elements. As the operating frequency or the number of elements increases, the packing density of the modules and the resource networks increase. Yet the same amount of heat must be swept away and the same number of control and bias lines must be interfaced with the modules. If the partially distributed approach is selected to simplify the networks then additional transmit/receive losses (e.g., 2 dB or more) are incurred and more power and gain are needed in the module. Also unwanted grating lobes may occur within the desired angle coverage envelope. Careful system analysis must be performed when deriving the power-aperture product of the APAR, since a significant price will be paid for the number and power of the elements and the associated architecture of the supporting networks.

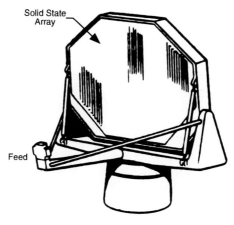

FIG. 18 An active phased-array antenna and its mechanical support.

The number of radiating elements required in an APAR depends primarily upon required power-aperture product and secondarily (except in tracking radars) on the number of elements. The power is that which is generated by the individual transmit power amplifier chains in the T/R module and which is actually radiated and integrated in space over all the elements. The transmit and receive aperture gain is determined by summing the isotropic gain of all the elements of the antenna, given by $G = 4\pi A/\lambda^2$ where A is the array area and λ is the operating wavelength. The beamwidth is inversely proportional to the aperture size (in either elevation of azimuth) [3]. Thus a phased-array radar might need thousands of radiators which can be in the form of dipoles, slots, or microstrip printed elements [29]. The type of radiating element, array size, and arrangement depends upon frequency of operation, the system requirements, and the platform.

4.3.4 Receivers

The primary functions of a receiver in a PAR are to amplify the received signal with minimum added noise power, to filter it from other interfering signals with maximum discrimination, and to frequency downconvert it from RF to IF to baseband for phase and amplitude detection. The interference signals might be generated in the receiver itself or come from other sources, such as reflected signals from moving and/or stationary objects, background clutter, neighboring radars, communication systems, jammers, and galactic sources. The receiver will usually consist of numerous channels to accommodate a single or double monopulse operation for angle measurement and to provide additional antenna beam channels for multibeam coverage in heavy clutter or jamming environments. The front-end of the receiver consists of a LNA which establishes the noise figure of the receiver unit. The receiver is also provided with a protection circuit which prevents LNA burn-out from the reflected transmitter signal and high-power interfering signals. In APARs the LNA and front-end limiter are contained in the receiver of the distributed T/R module. Individual spatial beams are formed in the receive mode by adjusting the phase of the distributed receivers. In this case the desired dynamic range (peak desired signal to noise ratio) must be established by the third-order intercept in the distributed low-noise receivers and preserved in the downconversion receiver. Typical requirements for an APAR low noise receiver are: 15 to 25 dB gain, 3 to 4 dB noise figure, 10 to 20 dBm third-order intercept point, amplitude control range 5 to 15 dB, and low VSWR (1.5:1).

4.3.5 Signal Processing/Control

The impact of GaAs technology will be felt both at the front end of the APAR (which uses analog-type circuits for microwave and intermediate frequency (IF) functions) and at the signal processors (which rely on ultra-high-speed digital circuits for analog-to-digital and digital-to-analog conversions, mathematical operations, timing, and filtering). The emerging GaAs digital IC technology (MESFETs as well as HBT based) will significantly improve the performance of signal processors in terms of speed, prime power dissipation, operating temperatures, and radiation hardness. Although most of the signal processing functions used in conventional radars and PARs can be used for APARs, complexity and speed requirements of the APAR processing subsystem is increased dramatically. This is due to the fact that most APAR applications operate in closed loop fashion for control purposes. The degrees of freedom are even greater in APAR since the phase, amplitude, timing, and frequency modulation of the transmitted signal can be varied. Thus, different waveforms for performing pulse compression and adaptive Doppler filtering and detection of the received signals can be generated. Further, this type of radar is usually environmentally adaptive and uses special search and track modes as scheduled by a resource manager to steer the beams by controlling the phase and amplitude of the hundreds and thousands of T/R modules. The high-speed pulse processing subsystems usually are implemented with array processors while the track and resource manager is implemented with microprocessors. The computations for beam steering are usually done by using high-speed microprocessors in real time while the control at the module is provided by a high-speed low-power VLSI ASIC (applications-specific integrated circuit). These integrated circuits play an important role because of the rapid response time required for the APAR.

4.4 TRANSCEIVER MODULE TECHNOLOGY

The previous sections have shown that future systems need wideband phased-array antennas that will provide high-speed electronic beam scanning and simultaneous operation of multiple functions. Such requirements are best met with an active array antenna. Active T/R modules are the basic elements of the active-array antenna for multifunction radar, EW, missile, and communication applications. The desirable features for such modules are low cost, small size, light weight, high reliability, and broadband performance.

A generic MMIC-based T/R module consists of single/multiple ICs, a circulator, a radiating element and a controller. The MMIC functions include switching, phase shifting, amplitude-controlled amplification, low-noise amplification, and power amplification. The circulator is a lightweight and high-power ferrite drop-in component. The radiating element could be a printed antenna (slot, patch, or dipole) or conventional dipole or slot. For broadband applications a flared notch or a parallel plate monopole or dipole may be more suitable. The controller is a digital circuit consisting of hybrid circuits or highly integrated random access memory (RAM) or digital ASIC currently implemented using Si devices. Figure 19 shows a block diagram of a T/R module which illustrates the complexity of the circuitry.

Two MMIC-based approaches for the T/R modules described in the literature are multichip [30–46] and single chip [47–52]. Small-size X-band multichip MMIC T/R modules with transmit power/receive noise figure of 1 W/4.7 dB, weighing less than 10 g, have been developed for airborne applications. A complete multifunction monolithic transceiver chip with transmit power/receive noise figure of 4 W/4 dB at C-band (discussed in Section 4.4.7) has been demonstrated [52]. This chip provides a major step toward a single chip T/R module. Another notable contribution in this area is wafer-level integration of 16 or more T/R modules on a single 3"-diameter wafer and combined as a subarray [51].

4.4.1 Transceiver Module Architectures

Selection of the APAR T/R module architecture is determined by the requirements cost, for weight, prime power dissipation, system cooling and support structure, bandwidth, transmitted power, effective radiated power, system noise figure, system input third-order intercept, scan range, sidelobe levels, antenna radar cross section, polarization, pulsewidth, duty cycle, and operating mode [53–56]. The design of a T/R module also depends upon the functions performed by the antenna and the platform on which the system will reside. Since this topic is vast, only salient features will be described in this section.

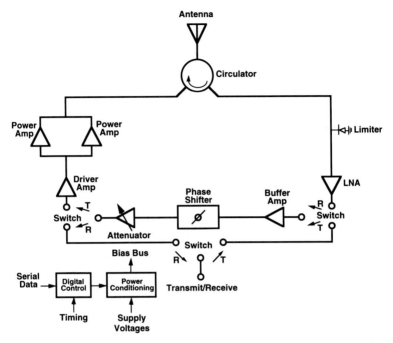

FIG. 19 Block diagram of a common leg T/R module, including control circuitry and power conditioning.

Figures 20–22 depict several possible T/R module architectures. For given aperture size and scan angle, the T/R modules' transmit channel power output, PAE, and gain required depend upon the target size and range. The required peak radiated power and PAE for each module also depends on the optimum trade-off in terms of cost, reliability, antenna array size, and other mechanical constraints such as support structure, transportability, and cooling requirements. For given input and output power levels, gain stages are selected depending upon the technology used and the operating bandwidth. The receiver channel's prime function is to condition the weakest possible signal and establish the noise floor for subsequent detection and processing. In order to process both strong and weak returns in a linear manner high dynamic range and gain are required.

The receiver parameters such as noise figure, bandwidth, and input third-order intercept are varied for each receiver component to establish the overall receivers gain and dynamic range. The low-noise receiver's key components are the LNA, attenuator, phase shifter, and usually a buffer amplifier. Different T/R module architectures shown in Figs. 20–22 are briefly described in the following subsections.

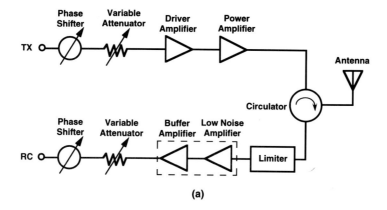

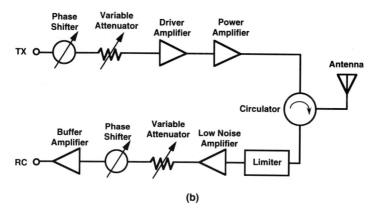

FIG. 20 T/R module with isolated transmit and receive paths. (a) Low noise figure configuration and (b) high third order intercept topology.

Isolated Transmit and Receive Channels

Figure 20 shows the T/R module configuration having completely isolated transmit and receive channels. The advantages of this configuration are (1) low noise figure due to elimination of a T/R switch before the LNA, (2) improved isolation between the transmit and receive channels, (3) the receive channels can be housed in one package sharing power supplies and controllers, and (4) improved radar efficiency due to elimination of switching between the transmitter and receiver channels. The major disadvantages of this scheme are higher component counts, larger T/R chip size, and greater cost.

The gain block in a T/R module can be optimized in terms of the system

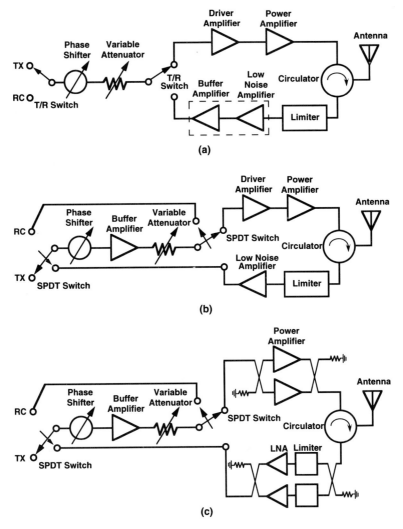

FIG. 21 T/R module architecture with shared components. (a) Shared phase shifter and attenuator, (b) shared phase shifter, amplifier, and attenuator, and (c) shared phase shifter, amplifier, and attenuator, and using balanced LNA and power amplifier.

application. For example, the lowest noise figure is obtained when all the gain stages (LNA and post/buffer amplifier) in the receive channel are lumped together near the radiating elements of the antenna. If the post/buffer amplifier is placed after the control components, the noise figure is slightly higher, but the third order intercept is improved. This arrangement does

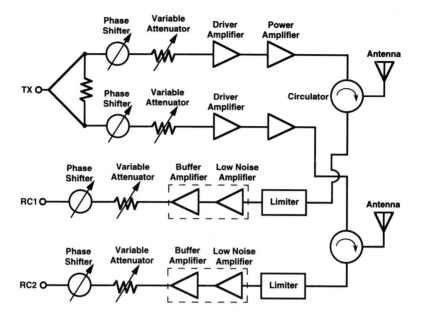

FIG. 22 T/R module configuration using two receive channels for monopulse radar.

provide a buffer between control components and the receiver's signal processing.

Shared Circuitry T/R Module

Figure 21 shows T/R module configurations with shared circuitry. The approach shown minimizes the number of components by sharing common circuitry between the transmit and receive channels. This approach reduces circuit complexity, chip size, and cost. This scheme also improves the receiver's dynamic range by distributing the gain before and after the "lossy" phase shifter, attenuator, and switches. Due to the high impedance mismatch present, the balanced amplifier near the radiating elements of the antenna as shown in Fig. 21c minimizes the effect of reflections from the antenna on the channel performance. The disadvantages of the shared circuitry approach are increased number of switches, possible signal losses, and the possibility of oscillations due to signal leakage or coupling caused by proximity effects. Figure 22 shows a scheme used for monopulse radar. In this case independent phase and amplitude control channels are used in the receive channel to satisfy complex beamforming networks requirements.

Several examples of T/R modules consisting of a digital controller and several MMIC chips are shown in Figs. 23 through 27. Figure 24 shows an S/C-band T/R module [45] which uses a dozen chips of six types. Module size, including the circulator and lid, is 4.4″ × 1.16″ × 0.5″. This module has achieved a maximum power of 21 W (at S-band) and minimum noise figure of 3.9 dB. A C-band T/R module using eight MMICs is shown in Fig. 24. The measured noise figure was about 4 dB and power output over 14 W with associated PAE greater than 35% [58]. T/R modules developed at X-band for airborne systems are shown in Figs. 25 and 26. Figure 27 shows a MMIC based C/X/Ku-Band T/R module used for broadband ESM and ECM applications. It has two T/R channels packaged wihin an aluminum silicon carbide housing measuring 4.7 × 0.825 × 0.25 inches.

The six subsections that follow describe the performance of several different types of MMICs developed over the past 10 years. Although these chips have been developed for multichip T/R modules, they can also be used for other applications.

4.4.2 Low-Noise Amplifiers

The LNA at the receiver front end sets the system noise figure or sensitivity. Thus, the LNA must have a low noise figure over its operational band. Both narrowband and broadband LNAs are required depending upon the system application. For most radar applications the LNA falls into the narrowband category—bandwidth less than an octave—with 20 to 30 dB of flat gain and noise figure (NF) on the order of 2 dB. If the LNA's NF is less than 2 dB, the T/R module NF will be less than 4 dB at room temperature; these levels are achievable up to X-band.

Desirable characteristics of an LNA are low noise figure and high gain, high dynamic range (in order to obtain a spurious-free signal), high third order intercept point, low input VSWR, wide bandwidth, compact size, and

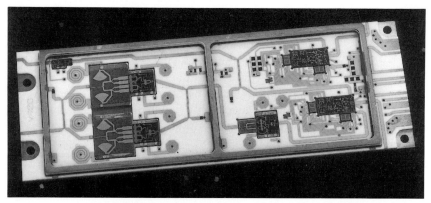

FIG. 23 Wideband S/C-band T/R module (photograph courtesy of Martin Marietta).

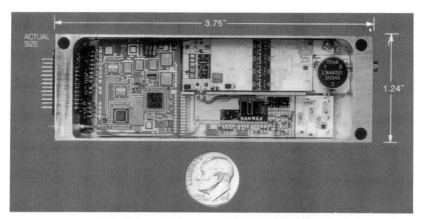

FIG. 24   C-band T/R module (courtesy of ITT).

low power dissipation. MESFET or HEMT technologies cannot meet all the requirements simultaneously. For a given technology and frequency range a circuit topology is selected to meet the most important system requirements. Among the possible LNA configurations, the common-source amplifier configuration is commonly used. In a multistage amplifier the input matching is designed for minimum noise figure, the interstage matching is designed for flat gain, and the output matching is designed for maximum gain and power

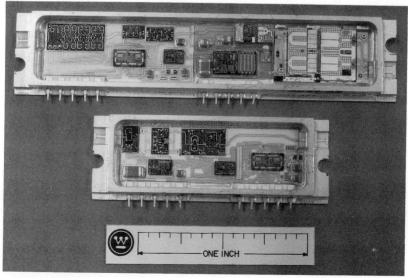

FIG. 25   Advanced transmit and receive modules (photograph courtesy of Westinghouse).

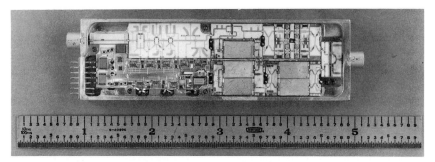

FIG. 26   Wideband T/R module (photograph courtesy of Texas Instruments).

output. A series-source feedback single-ended configuration provides good input match and minimum noise figure simultaneously over narrow bands, while a balanced topology is generally used for good input match and minimum noise figure over an octave bandwidth. The distributed amplifier configuration gives best noise figure over multioctave bandwidth.

In multistage amplifiers the FETs or HEMTs have gate peripheries (or gate width which is proportional to transistor size) increasing progressively in order to achieve higher dynamic range and higher intercept third order. The input stage is usually biased at low power levels for best noise figure and succeeding stages are biased for higher gain and higher power output. Shorter gate length (e.g., sub-half micrometer) devices provide lower noise figure and larger bandwidths as compared to longer gate length (e.g., half micrometer) devices, but the former are generally more expensive. HEMT LNAs have the lowest noise figure from microwave to millimeter-wave frequencies as compared with FETs, and FETs, in turn, are better than HBTs.

Figure 28 shows a photograph of a C-band LNA chip [39] used in the T/R module shown in Fig. 24. Typical measured performance includes 30 ± 1.0 dB, gain, 2.5 dB maximum noise figure, and 1.6:1 VSWR over the 5- to 6-GHz frequency range.

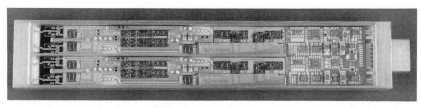

FIG. 27   Dual polarized broadband multifunction T/R module (photograph courtesy of Lockheed Sanders).

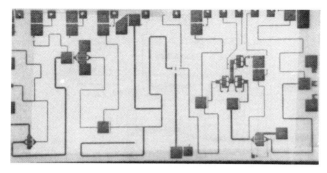

FIG. 28 Photograph of the variable-gain low-noise amplifier chip. (photograph courtesy of ITT).

Figure 29 shows the state-of-the-art in narrowband MMIC multistage LNAs developed using FETs and HEMTs. Noise figures of about 1 and 5 dB have been achieved at 10 and 94 GHz, respectively. The number in the parentheses represents the gate length of active devices. For broadband applications, distributed amplifiers have shown excellent performance in numerous applications covering multioctave to decade bandwidths. For example, a NF of 3.5 dB has been demonstrated over 2 to 20 GHz using 0.2-μm GaAs HEMT technology [59].

4.4.3 Power Amplifiers

Power amplifier requirements include high PAE and gain flatness over the operating frequency band, and relatively high pulsed-power output with phase and amplitude stability. Linearity of transmission phase with input power up to the 1-dB power compression point is another important parameter. Furthermore, heat sinking must be adequate if the channel temperature is to be maintained below 150°C for good reliability.

MMIC amplifiers are designed in Class A, AB, and B operation depending upon the frequency of operation and the type of active devices used. For high output power, a large gate periphery is required, decreasing the input and output impedances to low levels. However, with low impedance it is difficult to maintain a good match over broad bandwidths. In such cases the power amplifier is usually split into cells that can be conveniently prematched and then combined to produce the final output power. Reactively matched common-source amplifiers provide expected PAE up to about a two-octave bandwidth. A balanced configuration is usually used to meet good input and output match requirements, while the distributed amplifier topology is suitable for multioctave bandwidths and power levels up to 1W.

During the past decade there has been significant progress in monolithic power amplifiers operating over both narrow bands and broad bands [60,

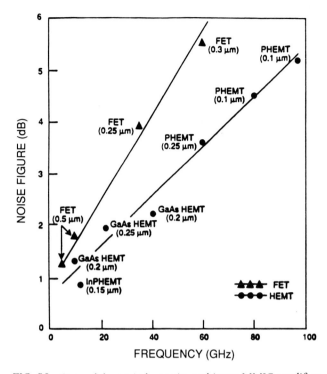

FIG. 29 State-of-the-art in low-noise multistage MMIC amplifiers.

61]. Power levels of 12 W with PAE of 36% at C-band and 6 W with PAE of 44% at X-band have been achieved. At 30 GHz, 2 W of power output has been obtained. For maximum efficiency, a C-band MMIC amplifier with 70% PAE, 8-dB gain, and 1.7-W power output has been demonstrated [61]. MMIC technology is very promising for broadband amplifiers having an octave or more bandwidth. Figure 30 depicts the current status for MMIC amplifiers working up to millimeter-wave frequencies. FETs used in the millimeter frequency spectrum have subhalf micrometer gate length. Some state-of-the-art results in high-efficiency and broadband power MMICs are summarized in Tables 6 and 7, respectively. The high-efficiency examples included in Table 6 have at least 36% PAE. Data presented in this section are for single-chip amplifiers. The photograph of a 12-W amplifier shown in Fig. 31 provides an example of a power MMIC.

4.4.4 Phase Shifters

One of the important elements in the T/R module is the programmable multibit phase shifter. The scanning of the beam in phased-array radars is

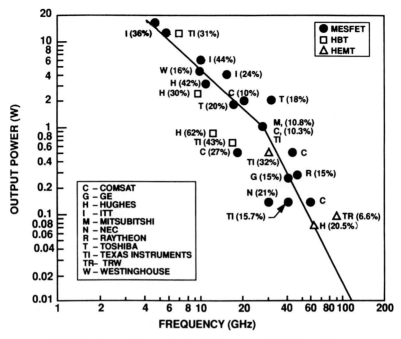

FIG. 30 State-of-the-art in single-chip power MMIC amplifiers.

achieved by changing the phase of the RF signal fed to or received from each radiating element. For beam steering, programmable bidirectional-phase shifters are required to adaptively adjust transceiver phase in both the transmit and the receive modes. To achieve maximum radar effectiveness, very low antenna sidelobes must be maintained, which imposes severe specifications on the programmable phase-shifter accuracy. For applications demanding very low sidelobe levels, up to 8-bit phase shifters have been used.

TABLE 6 State-of-the-Art Narrowband, High-Efficiency, Single Chip MESFET Power ICs

Freq. (GHz)	No. of stages	P_o (W)	PAE (%)	Gain (dB)	Company
1.3	1	9.0	52	16	ITT
5.0	1	5.0	60	9	ITT
5.5	1	1.7	70	8	ITT
8.5	2	3.2	52	14	Hughes
10.0	1	5.0	48	7	ITT
10.0	1	6.0	44	6	ITT
11.5	2	3.0	42	12	ITT
14.0	1	4.0	36	5	ITT

TABLE 7 State-of-the-Art Broadband, High Efficiency, Single Chip MESFET Power ICs

Freq. (GHz)	Configuration	No. of stages	Gain (dB)	P_o (W)	PAE (%)	Company
1.5–9.0	Reactive match	2	5	0.5	14	Teledyne
2–6	Reactive match	2	17	1.0	25	M/A-COM
2.0–8.0	Distributed	1	5	1.0	—	Raytheon
2.0–20.0	Distributed	1	4	0.8	15	TI
3.5–8.0	Reactive match	2	10	2.0	20	Raytheon
6–17	Dist./reactive	4	16	0.8	11	Raytheon
6–20	Distributed	1	11	0.25	—	HP
7–11	Reactive match	2	14	4.0	35	ITT
7–10.5	Reactive match	2	12.5	3.0	35	Hughes
7.7–12.2	Reactive match	2	8.0	3.0	14	Hughes
9–14	Reactive match	2	9.0	4.0	37	ITT
10–17	Reactive match	1	4.0	3.0	24	ITT
12–16	Reactive match	3	18	1.8	18	Hughes
14–33	Distributed	1	4	0.1	—	Raytheon

Alternatively, a 6-bit phase shifter (5 bits analog and 1 bit digital) satisfies most system requirements. The digital bits are 180°, 90°, 45°, 22.5°, and 11.25°, and the analog bit (0°–12°) is used to trim the transmission phase to the required accuracy. The size of a C-band 6-bit phase shifter MMIC chip can be as small as 8 to 9 mm².

There are four main types of solid-state digitally controlled phase shifters: switched line, reflection, loaded line, and low-pass/high-pass, as shown

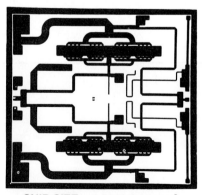

CHIP SIZE: 4.48 X 4.14 mm²

FIG. 31 Microphotograph of a 12 W C-band power MMIC (photograph courtesy of ITT).

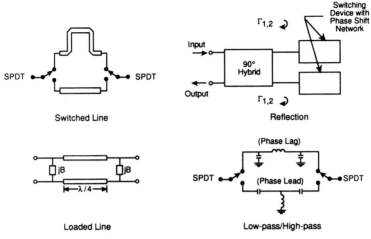

FIG. 32 Four basic phase shifter configurations.

in Fig. 32. The switched line and low-pass/high-pass configurations, which are most suitable for broadband applications and compact size, are not suitable for analog operation. Reflection and loaded-line phase shifters are inherently narrowband; however, the loaded-line small-bit phase shifters, 22.5° or less, can be designed to have up to an octave bandwidth. Phase shifters using the vector-modulator concept have also been developed in monolithic form. The progress on monolithic phase shifters over the past 10 years is summarized in Refs. [62, 63]. An analog bit is generally realized by using another lowest digital bit and varying the control voltage to obtain variable phase shift. As an example, a 6-bit phase-shifter circuit consisting of 5 digital bits (11.25°, 22.5°, 45°, 90°, and 180°) and an analog 0° to 12° bit cascaded in a linear arrangement is shown in Fig. 33. Table 8 summarizes the test results for phase shift, return loss, and insertion loss over the 5- to 6-GHz frequency range.

For broadband applications loaded-line small bits (less than 22.5°) can achieve up to an-octave bandwidth. For larger bits, switching between low-pass and high-pass lumped element networks is generally used. The Schiffman section [64] which employs a 90° tightly coupled line shorted at one end as shown in Fig. 34 is typically used to realize a 90° bit phase shifter when switched with regard to a 90° length of a transmission line. The bandwidth can be greater than an octave if an extremely tight coupling factor of approximately 0.7 dB is used.

An ultrabroadband 180° phase shift can be realized by using the phase reversal property of a tightly coupled (3 dB) four-port network [65]. When coupled and direct ports are switched from open circuited to short circuited, the transmission phase difference between the input and isolated ports

TABLE 8 Test Results for 6-Bit Phase Shifter Operating at 5 to 6 GHz

Bit	Phase shift (degrees)	Return loss (dB)	Insertion-loss difference (dB)
11.25°	12 ± 1	>19	<0.2
22.5°	23 ± 1	>18	<0.5
45.0°	44 ± 1	>15	<0.5
90.0°	94 ± 6	>17	<0.7
180.0°	173 ± 6	>17	<0.8
Analog	0 to 13°	>19	<0.3

changes from $-90°$ to $-270°$. When the open-circuited coupler is replaced by a pi or equivalent network of transmission lines, the 180° phase shift becomes independent of the electrical length of the two networks (i.e., pi and short-circuited coupler), resulting in wider bandwidth. Figure 35 shows the 180° phase difference sections. With this configuration a bandwidth of about two octaves can be achieved.

4.4.5 Switches

The transceiver requires several switches (low and high power). Low-power switches are used in phase shifters and attenuators. An excellent overview of these components is given in Refs. [62, 63]. Here only salient features are described.

FIG. 33 This 6-bit phase-shifter circuit consists of five digital bits and an analog bit cascaded in a linear arrangement. The analog, 11.25°, 22.5°, and 45° bits are of loaded line type and 90° and 180° bits are of reflection type.

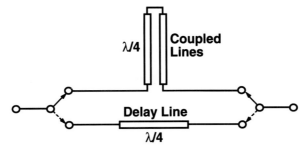

FIG. 34 Constant phase shift (e.g., 90°) with frequency, obtained with a Schiffman delay line section.

The MESFET can be used as an active switch (operated to have gain) or as a passive switch with insertion loss and negligible DC power dissipation. Both types of switches are compatible with MMIC technology. In the latter case, gate voltage controls the switching action by controlling the impedance between drain and source terminals. The integration of p–i–n diodes in monolithic switch circuits to achieve very low conversion loss has also been reported. The most commonly used configuration for microwave switches is the single-pole double throw (SPDT), as shown in Fig. 36, which requires a minimum of two switching devices (p–i–n diodes or transistors).

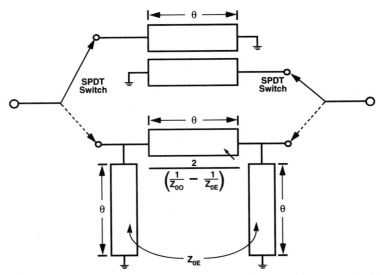

FIG. 35 180° Phase shifter bit configuration. Z_{OO} and Z_{OE} are the even and odd mode characteristic impedances of the coupled line.

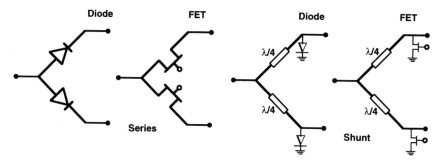

FIG. 36 SPDT switch configurations.

The SPDT switch can be designed using series, shunt, or series–shunt mounting of devices in a transmission line with proper impedance transformers.

Passive switches are bidirectional, provide wide bandwidth performance with great flexibility, and have small size, low cost, and light weight. They offer:

- virtually no DC-control power dissipation
- nanosecond switching speed,
- no DC blocking capacitors, and
- multiwatt power-handling capability.

Figure 37 shows a photograph of a high-power C-band T/R switch that provides an SPDT switching action. Here MESFETs in a shunt configuration have been used. For higher isolation, an inductor is connected between the drain and source terminals to resonate out the drain-source capacitance. The circuit exhibited 1-dB insertion loss and 20-dB isolation over the 5- to 6-GHz frequency range.

Single GaAs MESFETs can handle up to about 2 W in a 50-ohm system and up to 15 W in a properly designed SPDT switch. GaAs MMIC switches using shunt FETs in series, as shown in Fig. 38, can handle high power levels (10 to 100 W). The power handling capability increases approximately as the square of the number of FETs in series [66], because power handling in MESFET switches is proportional to voltage squared.

For broadband applications, low insertion and high isolation can be achieved if the FET's off-capacitance is incorporated into the transmission line by a distributed approach to match the circuit to 50 ohms as shown in Fig. 39. Typical performance for broadband SPDT switches developed using GaAs MESFET monolithic technology includes insertion loss less than 2 dB and isolation greater than 24 dB over dc to 20-GHz frequency range.

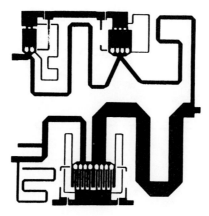

FIG. 37 Photograph of a monolithic C-band T/R switch (photograph courtesy of ITT).

4.4.6 Attenuators

In T/R modules, programmable attenuators are required to adjust the amplitude of the signal entering a particular radiating element for beam forming. They are also used to compensate for the change in gain caused by temperature variations. Several monolithic variable attenuators have been

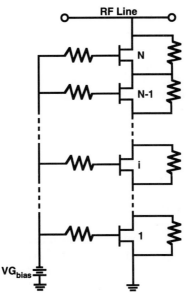

FIG. 38 Resistor bias network used to fabricate a high-power monolithic switch circuit.

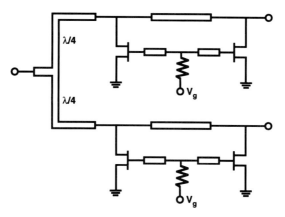

FIG. 39 A broadband SPDT switch incorporating the FET off-capacitance in a capacitively loaded line.

developed during the past several years. For APAR applications, the transmission phase characteristics are extremely important, and it is imperative that the phase remains constant as the gain is varied. Dual-gate FETs may be used to achieve variable gain/attenuation as well as programmable attenuation with constant phase. They possess two gates: an RF gate and a control gate. The gain/attenuation can be adjusted by varying the bias voltage on the control gate. The RF gate acts in the same manner as the gate on a standard single gate MESFET. Figure 40 shows a programmable attenuator using a dual-gate FET. Measured amplitude and phase variations vs frequency with various control voltages are shown in Fig. 41. This circuit offers a 10-dB attenuation range with ±3° phase change over the 5- to 6-GHz frequency range. Over this frequency band, transmission amplitude and phase remain constant for any attenuation value.

Variable attenuation can also be realized using switched attenuation sections. These attenuators are usually digitally controlled and consist of binary combination of attenuation. A broadband 0.05- to 14-GHz, 5-bit (8-, 4-, 2-, 1-, and 0.5-db) digital attenuator using a basic configuration as shown in Fig. 42 has been developed [67, 68]. The switched attenuator circuit consists of lower bits using T-networks made up of transistors and resistors in both reference and attenuation paths, while 4- and 8-dB bits are realized by switching T-configuration 4- and 8-dB attenuators with reference to phase compensated through paths using broadband SPDT switches. Additional transmission line lengths are provided to equalize phases in the two different paths. The circuit provides insertion loss of 5.6 dB at 50 MHz and 12 dB at 14 GHz in the reference state with 18 dB minimum return loss and 1°/GHz phase tracking. In Fig. 42(a), W_1 and W_2 denote FET's gate widths.

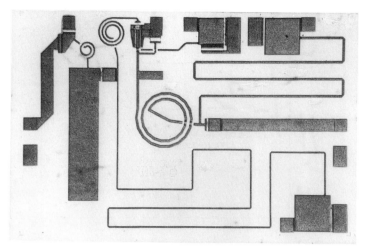

FIG. 40 This programmable attenuator uses a dual-gate FET (photograph courtesy of ITT).

4.4.7 T/R Module Chips

Reducing the cost of GaAs IC-based transmit/receive (T/R) modules is essential to the deployment of active phased-array antennas. GaAs MMIC

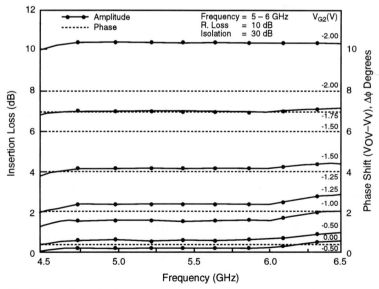

FIG. 41 Measured amplitude and phase variations for a programmable attenuator where V_{OV} corresponds to $V_{G2} = OV$ and V_V denotes V_{G2} as a variable (courtesy of ITT).

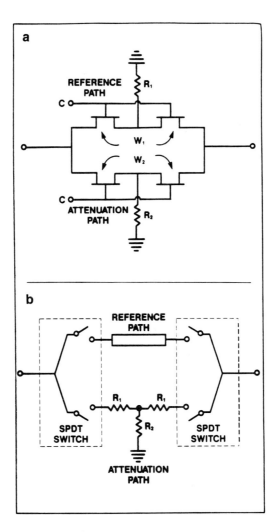
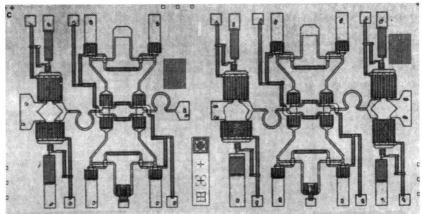

FIG. 42 Five-bit attenuator operating from 0.05 to 14 GHz: (a) 0.5, 1, and 2 dB bits; (b) 4 and 8 dB bits; and (c) fabricated chip (photograph courtesy of COMSAT).

technology offers part of the solution to the cost problem, since a high yield and a higher level of integration at the MMIC chip level reduces the number of chips required and results in lower module test and assembly costs. Several attempts have been made [50, 69, 70] to develop a single chip transmit/receive module on GaAs substrate. ITT has successfully demonstrated [71] what is believed to be the most complex GaAs monolithic C-band transceiver chip in the industry. The transceiver chip represents a major step forward in reducing the complexity and ultimate cost of T/R modules. A complete multifunction monolithic transceiver, containing 16 microwave circuits, has been developed on a GaAs substrate measuring 10.8 × 15.7 mm (170 mm^2). The chip includes a class-B 4 W power amplifier with 40% power added efficiency, a high-power T/R switch, several SPDT switches, buffer amplifiers, a 6-bit programmable phase shifter, digital and analog attenuators, and a low-noise amplifier. Excellent functional yield and performance have been obtained using the ITT multifunctional self-aligned gate (MSAG) fabrication process described in Refs. [72–75].

C-Band T/R Chip

The block diagram of the T/R chip is shown in Fig. 43, and the circuit breakouts along with signal flow charts are shown in Fig. 44. The chip is laid out so that the three blocks (the receiver section, the phase shifter with the buffer amplifier section, and the transmitter section) can be diced or tested on wafer individually for evaluation. A photograph of the chip is shown in Fig. 45 while the chip statistics are given in Table 9. The IC uses 58 separate FETs with a total gate periphery of about 63.5mm and has 87 via holes.

The receiver section is composed of five separate circuits: a high-power T/R switch, a LNA, an analog attenuator, a digital attenuator, and a buffer

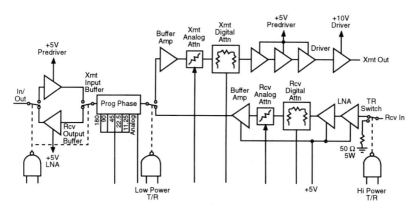

FIG. 43 MMIC transceiver architecture.

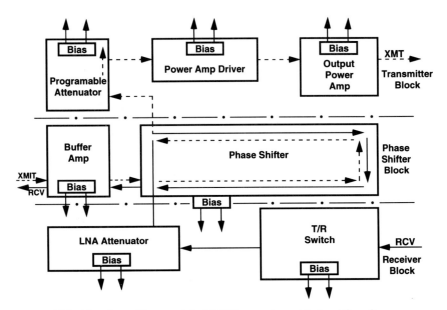

FIG. 44 Physical layout of the MMIC transceiver with signal flow diagram.

amplifier. A high-power SPDT switch serves as the T/R switch, with the common junction being the receive input. The high-power arm consists of two shunt 2.5-mm FETs in parallel with the output terminated in a 50-ohm resistor. The low-power arm consists of 1.25- and 2.5-mm FETs separated by a λ/4 line length. The LNA is a two-stage design, using 300 and 600 μm small signal FETs biased at 25 and 50% drain-source saturated current (I_{DSS}), respectively, and drain-source voltage (V_{DS}) of 5 V. The analog attenuator is based upon a 250-μm dual-gate FET. This stage provides roughly 25 dB of variable attenuation with an accompanying insertion phase shift less than 30°. The buffer amplifier uses a 600-μm FET biased for maximum dynamic range. Resistive feedback is used to obtain improved input/output VSWR and gain flatness. This circuit is biased at V_{DS} of 5 V and 50% I_{DSS}.

The phase shifter section is composed of a 6-bit phase shifter and a bidirectional buffer amplifier. A SPDT switch is included to direct the signal path from either the receiver or to the transmitter. Three types of phase shifting configurations are used in the phase shifter. For the analog 0°–11° bit, and the digital 11.25° and 22.5° bits, the loaded-line approach is utilized. The sizes of FETs used are 2.5, 2.5, and 1.8 mm, respectively. For the 45° and 90° bits, the switched-filter configuration is used. This particular circuit alternates between a high-pass and low-pass filter. The FETs used vary in size from 0.2 to 1.5 mm. The 180° bit is also realized by the switched-filter approach, whereby high-pass and low-pass filters are connected by a pair of SPDT switches. The high-pass/low-pass (HP/LP) filter is

4. Phased-Array Radar

FIG. 45 Photograph of the 16-microwave function MMIC T/R chip. Chip size = 170 mm² (photograph courtesy of ITT).

TABLE 9 MIMIC Transceiver Chip Component Population

Total circuits	16
FETs	58
Total gate periphery	63.5 mm
Capacitors	83
Resistors	153
Spiral inductors	6
Via holes	87
Air bridges	65
Chip size	170 mm² (10.8 × 15.7 mm)
Chip thickness	0.125 mm
RF interfaces	4
DC interfaces	37
Control interfaces	19

formed by using simple LC networks, while the SPDT switch employs a series/shunt configuration using 0.6-mm FETs. The bidirectional buffer amplifier consists of two single-stage amplifiers using resistive feedback networks. These amplifiers are sandwiched between a pair of SPDT switches. All together there are six microwave circuits in this part of the transceiver chip.

The transmitter section consists of a variable gain stage, a driver stage, and a 4-W output-power stage. The 4-W power stage can also be used to drive a higher power amplifier to create a higher power T/R module. The variable gain stage consists of a buffer amplifier, an analog attenuator based on a 250-μm dual-gate FET, and a digital 10-dB pad. The FETs in the buffer amplifier operate in a linear mode. The driver is a three stage medium power amplifier design that incorporates a 2.5-mm class-B FET in the third stage and is capable of delivering more than 1-W output power to the output power amplifier across the band when biased at drain voltage of 10 V. The output stage uses 10 mm of FET periphery to deliver greater than 3 W of output power across the band.

Limited on-wafer testing yielded 26% functional chips over several wafers. After processing, 10 transceiver chips were assembled on carriers and tested in fixtures and RF performance was evaluated. Five of 10 chips achieved the design goal. The measured performance is summarized in Table 10. Output power of 36.05 ± 0.65 dB (>3.4 watts) was obtained over the 5.2- to 5.8-GHz band with power added efficiency value between 30 and 40% as illustrated in Fig. 46. In the receive mode 23 dB of gain with ±0.75 dB flatness and maximum noise figure of 4.5 dB were achieved over the 5.2- to 5.8-GHz frequency range. Phase shifter performance of all digital bits from 4.5 to 6.5 GHz is given in Fig. 47. The 11°, 22°, 45°, and

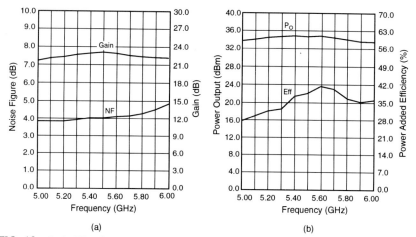

FIG. 46 GaAs T/R chip's measured performance (a) Typical receive channel and (b) minimum transmit channel.

TABLE 10 Transceiver Chip Measured Average Performance from Five Chips (Temperature = 25°C)

Parameter	Measured
Operating frequency (GHz)	5.2 to 5.8
Receive mode	
Noise figure (dB)	4.5 Maximum
Gain (dB)	23
Gain flatness (dB)	±.75 Over band
VSWR input/output	≤2.1 : 1, 50 Ohms
Transmit mode	
Gain (dB)	40 dB
Output power (1 dB compression)	3.5 to 4.6 W (35.4 to 36.7 dBm)
VSWR input/output	<2.0 : 1
Power efficiency (%)	30–40
Maximum duty cycle (%)	100
Harmonic output	−30 dBc
Phase shifter	
Bits	5 Digital, 1 analog
Phase accuracy (degrees)	
(11°, 22°, 45°, 180° bits)	±5°
(90° bit)	75–89°
Amplitude variation (dB)	0.5 dB RMS
Analog bit range (degrees)	±10°
Programmable attenuator (receive/transmit)	
Analog attenuation range	20 dB
Phase change	<3°/dB
Digit attenuation range	10 dB

180° bits were within 5° of nominal value, while the 90° bit varied from 75° to 89° over the 5.2- to 5.9-GHz band. Over the 0 to 85°C temperature range, the amplitude and phase variations were within 1 dB and 10°, respectively.

The next generation of this C-band T/R chip has been completed in which the chip area is reduced from 170 to 65 mm^2 and the number of on-chip functions increased by 35% [58]. An even smaller (55 mm^2) version is under development.

Broadband T/R Chip

A single 2- to 20-GHz T/R chip on GaAs has also been developed [50]. The chip's block diagram and photograph are shown in Fig. 48. The chip, which measures 17.6 mm^2, consists of T/R switches, driver amplifier, power amplifier and LNA. A summary of the T/R chip is given in Table 11. A basic traveling wave design approach was used for amplifiers. The measured performance includes maximum noise figure of 10 dB with associated gain of 18 dB and minimum power output of 200 mW with associated gain of 12 dB.

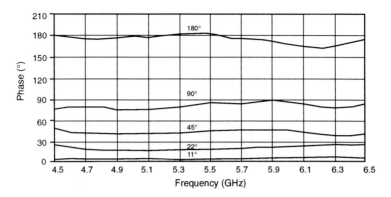

FIG. 47 Measured performance of the phase shifter in the T/R chip.

4.4.8 Circulators

A T/R module circulator is an important component since it isolates the RF circuitry from the antenna and separates the transmit and receive paths. A circulator used as a duplexer provides good input VSWR toward the power amplifier. It minimizes the interaction between the transmitter and the antenna even when the antenna impedance changes during beam scanning. Most circulators are of drop-in type and are connected to other circuits in the module by soldering.

At S-band or lower frequencies, a lumped element design approach for circulators reduces the size and weight drastically compared to conventional microstrip design techniques. Both lumped and microstrip drop-in circulators can handle several hundred watts of pulsed power (short pulsewidth) and up to 20 W of continuous wave (CW) power with no measurable change in the transmitter performance.

4.4.9 Housing

Advanced APAR technology mandates new materials for T/R module housings and antenna elements. Desirable characteristics are: lightweight, good thermal conductivity, thermal coefficient of expansion (TCE) to match GaAs, hermetically sealable, and low cost in large quantities. Aluminum has been extensively used for module housings because of its good thermal properties, light weight, low cost, and ease with which it is molded or machined. However, the TCE of aluminum is significantly different from both that of GaAs and that of power IC carriers and substrates, which leads

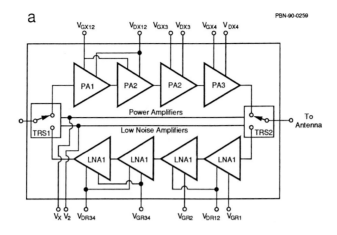

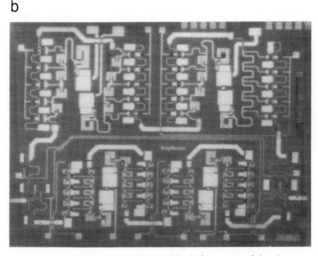

FIG. 48 Broadband (2–20 GHz) T/R chip (a) block diagram and (b) photograph. (Reprinted with permission from M. J. Schindler [50], and IEEE. © 1990.)

to degradation of the overall module performance including module reliability. T/R module housing materials also depend upon the radiated power/module, polarization, and antenna array configuration.

Several different new materials meeting some of the above requirements are now available. These are silicon carbide-embedded aluminum (SiC/Al) alloys also known as metal matrix, Be/Al alloys, graphite composites, and aluminum alloys. By controlling SiC content in aluminum, the material's properties such as electrical, mechanical, and thermal can be closely

TABLE 11 Summary of Broadband T/R Chip

Chip size	0.143" × 0.194" (3.6 × 4.9 mm)
Total gate periphery	11.5 mm
No. of FETs	44
Total capacitance	190 pF
No. of capacitors	44
No. of resistors	69
No. of spiral inductors	60
No. of via holes	60
No. of bias terminals	13

matched to aluminum, GaAs and the chip carrier. The weight is lower than that of aluminum. Metal matrix and graphite composite materials look promising. Metal matrix materials are rigid and lightweight but are difficult to machine, weld, and braze. Conversely, SiC/Al alloy can be laser sealed for hermeticity. The technical challenge in the use of these materials is developing manufacturing processes that permit taking advantage of their superior structural properties.

4.5 SUMMARY AND FUTURE TRENDS

Reduced cost, enhanced functionality, increased performance, improved reliability, and easy maintainability are the important features of currently developed active phased-array radars (APARs). Each of these multifunction radars can replace several conventional radars to reduce weight and to reduce the total operating and maintenance costs. These radars are heavily dependent on MIC and MMIC manufacturing technologies to meet performance and cost goals for affordable systems.

Future high-performance APARs will employ fully monolithic T/R chips to reduce system cost, to achieve increased scan flexibility, and to reduce prime power consumption. Using T/R chips with precise phase and amplitude control, APARs will be able to transmit beams with very low spatial sidelobes.

The developments in the future T/R modules must support low cost, light weight, improved performance and more functionality in terms of bandwidth and antenna polarization. Specific improvements include reduced noise figure and increased power output and PAE, low-loss high-power T/R switch and limiter, higher level of integration (including power, digital interface and control, and built-in-test and calibration capability), reduced size and weight, multiple antenna polarizations, and low-cost packaging and testing.

Future evolution of the active array radar into a high-performance, multifunction, low-cost and highly reliable system for civilian and military applications depends upon realizing the full benefit of gallium arsenide monolithic integrated circuit manufacturing technology. This technology holds the future of many new systems that require thousands of low-cost microwave circuits.

Optoelectronic feed for radiating elements of the APAR is progressing but at a slow pace. Optoelectronic control of rapid beam steering will also have the advantage of compactness, light weight, increased scan flexibility, low attenuation, large bandwidth, and immunity to both cross-talk and electromagnetic interference. HBTs and HEMTs are making rapid progress toward integration of optical devices such as lasers, LEDs, modulators and photodetectors, leading to fully integrated optoelectronic ICs. This technology will be crucial in developing optical feed for very large and millimeter-wave APARs.

REFERENCES

[1] M. I. Skolnik, *Introduction Radar Systems*. McGraw–Hill, New York, 1962.
[2] A. W. Rihaczek, *Principles in High-Resolution Radar*. McGraw–Hill, New York, 1969.
[3] M. I. Skolnik (Ed.), *Radar Handbook*. McGraw–Hill, New York, 1970.
[4] S. A. Hovanessian, *Radar Detection and Tracking Systems*. Artech House, Norwood, MA, 1973.
[5] D. K. Barton, *Radars*, Vols. I–VII. Artech House, Norwood, MA, 1976.
[6] J. J. Kovaly, *Synthetic Aperture Radar*. Artech House, Norwood, MA, 1976.
[7] E. Brookner, *Radar Technology*. Artech House, Norwood, MA, 1977.
[8] D. C. Schleher, *MTI Radar*. Artech House, Norwood, MA, 1978.
[9] J. V. Difranco, and W. L. Rubin, *Radar Detection*. Artech House, Norwood, MA, 1980.
[10] D. Rhodes, *Introduction Monopulse*. Artech House, Norwood, MA, 1980.
[11] S. L. Johnston (Ed.), *Millimeter Wave Radar*. Artech House, Norwood, MA, 1980.
[12] M. I. Skolnik, *Introduction to Radar Systems*, 2nd ed. McGraw–Hill, New York, 1980.
[13] E. K. Reedy and G. W. Ewell, Millimeter radar. In *Infrared and Millimeter Waves* (K. J. Button and J. C. Wiltse, Eds.), Vol. 4, pp. 23–94. Academic Press, New York, 1981.
[14] C. R. Seashore, Missile guidance. In *Infrared and Millimeter Waves* (K. J. Button and J. C. Wiltse, Eds.), Vol. 4, pp. 95–150. Academic Press, New York, 1981.
[15] D. K. Barton, A half century of radar. IEEE *Trans. Microwave Theory Tech.* **MTT-32**, 1161–1170, Sept. 1984.
[16] P. Bhartia and I. J. Bahl, *Millimeter Wave Engineering and Applications*. Wiley, New York, 1984.
[17] E. A. Wolff and R. Kaul, *Microwave Engineering and Systems Applications*. Wiley, New York, 1988.
[18] H. Sauvageot, *Radar Meteorology*. Artech House, Norwood, MA, 1992.
[19] N. C. Currie, R. D. Harps, and R. N. Trebils, *Millimeter-Wave Radar Clutter*. Artech House, Norwood, MA, 1992.
[20] W. W. Goj, *Synthetic-Aperture Radar and Electronic Warfare*. Artech House, Norwood, MA, 1992.

[21] A. V. Jelalian, *Laser Radar Systems*. Artech House, Norwood, MA, 1992.
[22] W. C. Morchin, *Radar Engineers' Source Book*. Artech House, Norwood, MA, 1992.
[23] T. M. Miller and H. A. Corriher, Jr., *Basic Radar Principles, Workshop Notes*. Georgia Institute of Technology, Atlanta, GA, 1987.
[24] B. Edde, *Radar—Principles, Technology, Applications*. PTR Prentice Hall, Englewood Cliffs, NJ, 1993.
[25] D. N. McQuiddy et al., Transmit/receive module technology for X-band active array radar. *Proc. IEEE* **79**, 308–341, 1991.
[26] P. Gwynne, Sharper eyes. *High Technology* 46–52, Nov. 1986.
[27] D. R. Carey and W. Evans, The Patriot radar in tactical air defense. *Microwave J.* **31**, 325–332, May 1988.
[28] S. B. Moghe and T. J. Holden, High performance GaAs MMIC oscillators. *IEEE Trans. Microwave Theory Tech.* **MTT-35**, 1283–1287, Dec. 1987.
[29] I. J. Bahl and P. Bhartia, *Microstrip Antennas*. Artech House, Norwood, MA, 1980.
[30] J. P. Sasonoff et al., L-band GaAs MMIC transceiver modules. *GOMAC*, 335–338, 1986.
[31] D. N. Jessen et al., Ten-watt monolithic S-band transceiver modules. *GOMAC*, 399–402, 1987.
[32] M. T. Borkowski et al., L-band transceiver modules for space-based radar. *GOMAC*, 403–406, 1987.
[33] R. Ali et al., A C-band low noise MMIC phased array receive module. *IEEE MTT-S Microwave Symp. Dig.* 951–954, 1988.
[34] J. R. Potukuchi et al., MMIC modules for active phased array applications in communications satellites. *MSN & CT* **18**, 20–28, Nov. 1988.
[35] A. M. Pavio, J. S. Pavio, and J. E. Chapman, Jr., The high volume production of advanced radar modules. *Microwave J.* **32**, 139–148, Jan. 1989.
[36] D. Willems et al., Multifunction chip set for T/R module receive path. *Microwave Millimeter-Wave Monolithic Circuits Symp. Dig.* 95–98, 1990.
[37] A. M. Pavio et al., Transmit/receive modules for 6 to 18 GHz multifunction arrays. *IEE MTT-S Int. Microwave Symp. Dig.* 127–1230, 1990.
[38] D. E. Meharry et al., 6–18 GHz phased array active module. *Appl. Microwave*, **2**, 83–93, Spring 1990.
[39] D. A. Willems et al., Multifunction small-signal chip set for transmit/receive modules. *IEEE Trans. Microwave Theory Tech.* **38**, Dec. 1990.
[40] J. A. Windyka, Methodology for MMIC insertion into systems. *GOMAC* 5–8, 1991.
[41] M. Alastair et al., Application of MMIC technology to C-band radar modules. *GOMAC* 9–12, 1991.
[42] P. Maloney and J. Sasonoff, Lightweight L-band T/R modules for SBR applications. *GOMAC* 191–194, 1991.
[43] M. Borkowski et al., GaAs MMIC module technology for L-band spaceborne radar applications. *GOMAC* 199–203, 1991.
[44] D. M. McPheron et al., Advanced L-band module for spaced based phased array applications: Electrical design. *GOMAC* 313–316, 1991.
[45] J. J. Komiak and A. K. Agrawel, Design and performance of octave S/C-band MMIC T/R modules for multifunction phased array. *IEEE Trans. Microwave Theory Tech.* **39**, 1955–1963, 1992.
[46] T. Sakai et al., Ultra small size X-band MMIC T/R modules for active phased array. *IEEE MTT-S Int. Microwave Symp. Dig.* 1531–1534, 1992.
[47] G. Jerinic et al., X-band single chip T/R module development. *GOMAC* 343–345, 1986.
[48] W. R. Wisseman et al., X-band GaAs single-chip transmit/receive radar module. *Microwave J.* 167–173, Sept. 1987.

[49] A. A. Lane, S- and C-band GaAs multifunction MMICs for phased array radar. *IEEE GaAs Ic Symp. Dig.* 259–262, 1989.
[50] M. J. Schindler et al., A single chip 2–20 GHz T/R module. *IEEE Microwave Millimeter-Wave Monolithic Circuits Symp. Dig.* 99–102, 1990.
[51] L. R. Whicker et al., A new approach to active phased arrays through RF wafer scale integration. *IEEE MTT-S Int. Microwave Symp. Dig.* 1223–1226, 1990.
[52] C. Andricos et al., 4-Watt monolithic GaAs C-band transceiver chip. *GOMAC* 195–198, 1991.
[53] C. Andricos and I. Bahl, "GaAs Monolithic ICs Applied to Military Radar Design, MSN." *The Microwave System Designer's Handbook,* pp. 335–346, July 1986.
[54] S. S. Bharj et al., Manufacturing cost analysis of monolithic X-band phased-array-radar T/R module. *MSN & CT* 32–43, Aug. 1987.
[55] R. H. Chilton, MMIC T/R modules and applications. *Microwave J.* 30, 131–146, Sept. 1987.
[56] T. R. Turlington, F. E. Sacks, and J. W. Gipprich, T/R module architectural consideration for active electronically steerable arrays. *IEEE MTT-S Int. Microwave Symp. Dig.* 1523–1526, 1992.
[57] J. L. B. Walker (Ed.), *High-Power GaAs FET Amplifiers.* Artech House, Norwood, MA, 1993.
[58] *Active Array Technology for Radars of the Future.* ITT Defense & Electronics, Van Nuys, CA.
[59] R. Dixit et al., A family of 2–20 Ghz broadband low noise AlGaAs HEMT MMIC amplifiers. *IEEE Microwave Millimeter-Wave Monolithic Circuits Symp. Dig.* 15–19, 1989.
[60] H. Q. Tserng and P. Saunier, Advances in power MMIC amplifier technology in space communications. *Proc. SPIE-Monolithic Microwave Integrated Circuits Sensors, Radar, Commun. Syst.* 74–85, 1991.
[61] D. Willems and I. Bahl, Advances in monolithic microwave and millimeter wave integrated circuits. *IEEE Int. Circuits Syst. Symp. Dig.* 783–786, 1992.
[62] A. K. Sharma, Solid-state control devices: State of the art. *Microwave J.* 95–112, Sept. 1989.
[63] V. Sokolov, Phase shifters technology assessment: Prospects and applications. *Proc. SPIE-Monolithic Microwave Integrated Circuits Sensors, Radar Commun. Syst.* 1475, 228–332, 1991.
[64] B. M. Schiffman, A new class of broad-band microwave 90-degree phase shifters. *IRE Trans. Microwave Theory Tech.* **MTT-3**, 232–237, April 1958.
[65] D. C. Boire, J. E. Degenford and M. Cohn, A 4.5 to 18 GHz Phase Shifter. *IEEE Int. Microwave Symp. Dig.* 601–604, 1985.
[66] M. Shifrin, P. Katzin, and Y. Ayasli, Monolithic FET structures for high-power control component applications. *IEEE T-MTT* 37, 2134–2141, Dec. 1989.
[67] R. Gupta, L. Holdeman, J. Potukuchi, B. Geller, and F. Assal, A 0.05 to 14 GHz MMIC 5-Bit Digital Attenuator, GaAs Integrated Circuits Symposium, pp. 231–234, 1987.
[68] R. Gupta, J. H. Reynolds, and P. J. McNally, Modeling and CAD of an ultra broadband monolithic 5-bit digital attenuator. *Eur. Microwave Conf.* 151–155, 1988.
[69] L. C. Witkowski et al., A GaAs single chip transmit/receive radar module. *GOMAC Dig.* 339–342, 1986.
[70] W. R. Wisseman et al., X-band GaAs single-chip T/R radar module. *Microwave J.* 30, 167–173, Sept. 1987.
[71] C. Andricos et al., 4 Watt monolithic GaAs C-band transceiver chip. *GOMAC Dig.* 195–198, 1991.
[72] A. E. Geissberger, R. A. Sadler, E. L. Griffin, I. J. Bahl, and M. L. Balzan, Refractory self-

aligned gate technology for GaAs microwave FETs and MMICs. *Electron. Lett.* **23**, 1073–1075, Sept. 1987.
[73] A. E. Geissberger, I. J. Bahl, E. L. Griffin, and R. A. Sadler, A refractory self-aligned gate technology for GaAs microwave power FETs and MMICs. *IEEE Trans. Electron Devices* **35**, 615–622, Dec. 1988.
[74] I. J. Bahl *et al.*, Multifunction SAG process for high-yield, low-cost GaAs microwave integrated circuits. *IEEE Trans. Microwave Theory Tech.* **38**, 1175–1182, Sept. 1990.
[75] I. J. Bahl *et al.*, GaAs ICs fabricated with the high-performance, high-yield MSAG process for radar and EW applications. *IEEE Trans. Microwave Theory Tech.* **38**, 1232–1241, Sept. 1990.
[76] D. H. Reep *et al.*, Transportability of MMIC technology. *GOMAC Dig.* 215–217, 1991.

5

Electronic Warfare I
Transmitters

E. Ronald Schineller
ITT Avionics, Clifton, New Jersey

5.1 Introduction
    5.1.1 Historical Perspective
    5.1.2 Technology Requirements/Challenges
    5.1.3 Chapter Outline/Overview
5.2 EW Subsystems
    5.2.1 Search Lock Oscillator/IFM System
    5.2.2 Frequency Synthesizer/Digital RF Memory
    5.2.3 Phased-Array Antenna
    5.2.4 Adaptive Polarization Jamming
5.3 Generic EW Chips
    5.3.1 Small Signal Amplifier
    5.3.2 Medium-Power Amplifier
    5.3.3 Single-Pole, Two-Throw Switch
    5.3.4 Digital Attenuators
    5.3.5 Direct Digital Synthesizer
5.4 Summary
5.5 Future Trends
    References

5.1 INTRODUCTION

5.1.1 Historical Perspective

The term electronic warfare, often abbreviated EW, has come to be used to designate those systems generally defensive in nature, used to thwart offensive weapons under the guidance and control of an electronic radar system. Perhaps the most common example is a radar-guided surface-to-air missile, used to attack and destroy aircraft. The function of the EW system, located on the aircraft, is to detect the radar signal, identify it as hostile, and to then radiate a signal which will jam or misguide the radar so as to cause the attacking missile to miss the aircraft. It is an electronic countermeasure; hence the name electronic countermeasure (ECM) system is often used interchangeably with the name EW system.

In addition to this role as a radar jammer, electronic warfare systems have also been developed to jam or disrupt communication systems. However, discussions in this chapter will treat primarily radar jammers.

Systems/Circuit Types

The basic elements of an EW system include antennas for receive and transmit, a receiver, signal sorting and processing circuitry, and a transmitter with appropriate waveform modulators.

Since EW systems are inherently defensive in nature and must react in response to other systems, their operating frequency is dictated by the operating frequency of the radar or communication system being addressed.

Since their inception in World War II, radars have been developed which operate over a wide portion of the frequency spectrum, starting in the VHF/UHF bands (0.1 to 1 GHz), extending through the microwave region (1 to 20 GHz), and more recently up into the millimeter-wave region (20 to 100 GHz). As modern radars have evolved, particular bands have been selected for different types of radar.

Long-range acquisition and surveillance radars now generally operate in the lower portion of the microwave region (L-band and S-band), while terminal guidance radars for use with surface-to-air and air-to-air missiles tend to operate in the higher portion of the microwave band (X-band and Ku-band).

In order to combat this range of radars, modern EW systems are required to operate over a broad frequency band, typically 2 to 18 GHz. In fact, many component manufacturers now advertize that their parts are operating in all or some portion of the EW band, i.e., 2 to 18 GHz.

Until the advent and maturing of monolithic microwave integrated circuit (MMIC) technology, it was very difficult to design and produce components operating over such a wide band. Consequently, several subbands have evolved with a range of components covering each. The most typical bands are 2 to 6 GHz and 6 to 18 GHz. With the advent of gallium arsenide (GaAs) integrated circuit (IC) technology, certain types of microwave circuits have been made practical, which are capable of very broadband performance. The most notable is the broadband distributed amplifier which has been demonstrated to operate routinely over the range of 2 to 20 GHz and in some cases from 1 to 40 GHz.

EW systems also make extensive use of digital circuits, and here, too, GaAs is having an impact. Digital frequency dividers operating in the GHz range are a prime example of new circuits made possible by GaAs IC technology. As discussed later, the integration of microwave and digital circuitry on the same chip can provide major benefits in reducing the complexity of EW components and subassemblies.

Communications systems have historically operated in the VHF, UHF, and lower microwave region below 2 GHz. Thus, the EW systems designed for comm-jam applications will generally operate in this lower frequency range, as compared to radar jammer EW systems.

Characteristics of EW Transmitters   As stated previously, the function of an EW system is to receive a radar signal from a hostile source and to retransmit a signal very similar to the received signal, but with some altered characteristics so as to confuse or jam the radar. The *transmitter subsystem* is required to perform several functions.

1. It must either receive a signal from the receiver subsystem and amplify it or generate a signal internally from an oscillator or frequency synthesizer.
2. It must amplify the signal and impress some type of modulation, either amplitude or frequency.
3. It must radiate the signal back to the radar.

The type of modulator required is determined by the ECM technique being used, which in turn will depend on the type of radar. Radar waveforms include pulsed, pulsed-Doppler, and continuous wave (CW), and each requires a different jamming technique. Radars can also be conical scan or monopulse, depending on the type of antenna used to acquire angle information on the target.

Details of deception or "jamming" techniques are generally classified and are beyond the scope of this book. However, in general terms, the objective is to induce errors in either the range information or the angle information of the radar being used to guide the attacking missile to its target. In the case of pulsed radars, jamming may involve some sort of modulation of the pulsewidth or pulse repetition frequency (PRF) to change the apparent range of the target. For pulse-Doppler and CW radars, jamming may require a slight change in the frequency of the signal retransmitted by the EW system to affect the apparent target range.

Angle information is derived from the radar antenna, which may be conical scan or monopulse, and changing the apparent angular direction of the target is dependent on the type of antenna system employed. Monopulse radars are more difficult to jam than conical scan, and new techniques are under development. One of these, discussed later, involves changing the polarization of the EW signal compared to that of the radar signal to induce angle errors.

Since the EW system functions by transmitting a signal back to the radar which must obscure the normal radar return, a key parameter for success is the amplitude of the radiated signal. For an EW system to be effective, it must radiate a signal whose amplitude exceeds that of the radar signal reflected from the target by at least 10 dB. As discussed above, the EW signal must be similar to the radar in its frequency and modulation characteristics in order to be received by it, but must have subtle differences which yield false target information. These subtle differences must be carefully controlled; otherwise, the stronger EW signal could actually help guide the radar to its target. The power level of the reflected signal depends primarily on three parameters:

- the effective radiated power (ERP) of the radar
- the range (R) between the radar and the reflector (aircraft)
- the effective radar cross section (Ac) of the aircraft.

The reflected power (P_r) from a target aircraft with an effective radar cross section Ac is given by

$$P_r = \frac{ERP \times Ac}{4 \pi R^2},$$

A graph showing the reflected power vs range for several values of radar cross section is given in Fig. 1.

The radar cross section can vary widely depending on the type of aircraft and its size. Large strategic aircraft such as the B-52 can have a cross section of 1000 m², conversely, the latest stealth aircraft such as the F-117 and F-22, designed specifically to have a small radar cross section, may have a cross section of only 1 m².

From the graph, it can be seen that a 1000-m² target, illuminated by a +100-dBm ERP radar at a range of 2 nautical miles reflects a power of +50

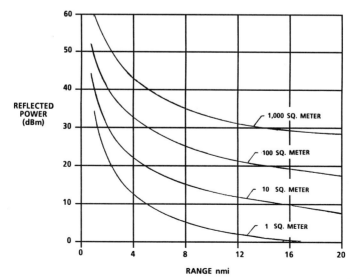

FIG. 1 Reflected power vs range for threat radar (ERP of 10 MW).

dBm. In this case, the EW transmitter is required to transmit an effective radiated power of +60 dBm (1 kW) in order to provide the needed 10-dB jammer-to-signal ratio (J/S).

The ERP of the EW jammer is the actual transmitted power (in dBm) plus the gain of the antenna in dB. For a simple horn antenna with a gain of 3 dB, the required EW transmitter power in the above example is 57 dBm or 500 W.

Until the advent of stealth aircraft the transmitter power required of EW systems has typically ranged from 100 to over 1000 W. Broadband solid-state amplifiers operating in the 2- to 20-GHz frequency band are not capable of these power levels. Consequently, EW transmitters have generally required traveling wave tube amplifiers (TWTAs) to generate the required power, and solid-state amplifiers operating in the $1/4$ W range are used as driver amplifiers for the traveling wave tube (TWT). For a target of 10 m^2 cross section, the reflected signal is reduced to +30 dBm, and the required ERP is only +40 dBm (10 W). In this case, the required EW transmitter power is only 5 W, which is approaching the range of today's solid state amplifiers.

The use of phased-array antennas in EW systems can dramatically change the required transmitter power in two ways. First, the phased array can generate a narrow beam, with appreciable antenna gain. The required transmitter power is thereby reduced below the ERP by the antenna gain, typically 20 to 30 dB. Second, the use of active aperture phased arrays, made possible by the development of low-cost transmit/receive (T/R) modules using MMIC technology, distributes the power among many individual amplifiers. By the combination of these two factors, the required ERP can be achieved with a phased array of less than 100 elements, using solid-state amplifiers having an output power of as little as a few watts. Thus, the need for the TWT can be eliminated and an all solid-state transmitter antenna can be used. Solid-state amplifiers have the advantages of compact size, ruggedness, low operating bias voltages, and reliability. A more detailed discussion of active aperture phased-array antennas for EW systems is presented later in this chapter. A discussion of phased-array antennas for radar applications is presented in Chapter 4.

5.1.2 Technology Requirements/Challenges

EW systems of the type discussed here form a class of avionics equipment intended for deployment on aircraft ranging in size from helicopters such as the Apache and Cobra, to tactical fighters such as the F-14, F-15, F-16, and F-18, to large strategic aircraft such as the B-52 and B-1 bombers. In most cases, and certainly for helicopters and tactical fighters, the available space and the acceptable weight is very limited. Consequently, there is great emphasis on the traditional requirements for small size, light weight, and low cost. However, as the sophistication of the radar threats has increased and improved over the years, the demands on the EW system have increased accordingly. The total number of threats encountered and the increase in the diversity of these threats has required a tremendous increase in the complexity of the EW system to maintain its effectivity. Unfortunately, the space allocation and the weight allocation for the EW system has not increased—in fact it is generally decreasing.

The challenge to the EW system designer then is to "cram" more and more functionality into less and less space and at lower cost. One answer of course is the use of higher and higher levels of integration in the electronic circuitry. Since a major portion of the EW system involves microwave circuitry, GaAs IC technology provides such a solution.

This was a primary motivation for, and objective of, the MIMIC program sponsored by ARPA—i.e., to develop monolithic microwave integrated circuits (MMICs) with high performance and which are affordable, available, and reliable.

Considerable emphasis on the MIMIC Phase I program was directed to EW systems, with several contractors developing MMIC chips specifically aimed at broadband EW systems. For example, ITT developed over 20 broadband MIMIC chips for application to EW systems, including small signal amplifiers, medium-power amplifiers, limiting amplifiers, multithrow switches, analog and digital attenuators, power dividers, power combiners, and several types of mixers. These chips, many of which are described later in this chapter, provide the capability for greatly reduced size, weight, and cost in EW systems.

5.1.3 Chapter Outline/Overview

The following sections in this chapter will describe some of the circuit and component requirements in the transmitter portion of EW systems and will discuss the use of GaAs IC technology to meet these requirements. A discussion of several system architectures is presented first; this is followed by a description of the individual subsystems used in each architecture, including the various design considerations involved in applying MMIC technology. A description of several common MMIC chips with their performance characteristics is then presented. The chapter will conclude with a brief discussion of the future trends in EW systems and how these impact the technology required.

5.2 EW SUBSYSTEMS

A large number of EW systems have been developed and fielded over the years since World War II, when modern radar systems emerged. These systems have been developed by many different companies for application to a variety of different types of aircraft, and consequently many different circuit approaches have evolved. These varying approaches appear more in the front-end or receiver part of the system. Such receiver types as superheterodyne receivers, channelized receivers, instantaneous frequency measurement (IFM) receivers, and compressive receivers are discussed in Chapter 6.

In the transmitter portion of the system, there are fewer fundamentally different approaches, although many alternative types of circuitry are employed. Also, there is not a clear distinction as to whether some components should be classified as in the receiver or transmitter part of the system—in fact some components such as the frequency synthesizer serve both receive and transmit functions.

For purposes of discussion, several EW system architectures have been selected here, and a description of their operation and the major components utilized in them is presented in the following sections.

One approach uses a search lock oscillator (SLO) and an IFM circuit to generate the EW signal; another uses a frequency synthesizer and a digital RF memory (DRFM). With regard to the actual power generation, the alternatives are either a single or more TWT amplifiers which radiate through a few simple antennas, or a solid-state phased array using distributed transmitters. Finally a discussion of the use of polarization control as an ECM technique is presented, together with circuit approaches for implementing it.

5.2.1 Search Lock Oscillator/IFM System

A block diagram of an EW system using an SLO in conjunction with an IFM to generate the EW signal is shown in Fig. 2. The incoming signal from the antenna is first amplified and then distributed to different parts of the receiver for analysis and classification. Details of these circuits are presented in Chapter 6. One of these functions, namely to determine the frequency of the incoming radar signal, is performed by the IFM.

The IFM measures the frequency of the received signal and generates a voltage used to control the frequency of the SLO. Coarse control of the SLO frequency can be either analog or digital: fine tuning is accomplished by mixing the output signal of the SLO with the received radar signal in a quadrature IF mixer (QIFM). The output of the QIFM provides an error signal proportional to the difference in frequency between the SLO and received signal, which is used to drive the difference to zero.

A block diagram of a SLO used in a particular EW system is shown in Fig. 3. In this case, the SLO consists of two voltage-tuned oscillators (VTOs) and a quadrature IF mixer. Either of the VTOs can be locked to a signal supplied to the QIFM from the receiver subsystem. The output of the VTO to be locked is connected to the radio frequency (RF) input port of the QIFM, and the receiver signal is connected to the local oscillator (LO) port.

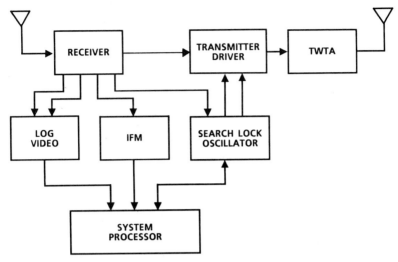

FIG. 2 EW system utilizing search lock oscillator and instantaneous frequency measurement.

The QIFM generates two intermediate frequency (IF) signals in phase quadrature at the difference frequency. These I and Q signals drive an error counter circuit which generates an error signal proportional to the difference in frequency between the VTO and received (radar) signals. By driving this error signal to zero (and hence the frequency difference to zero) the

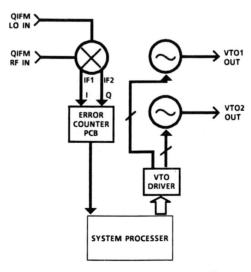

FIG. 3 Basic elements of search lock oscillator.

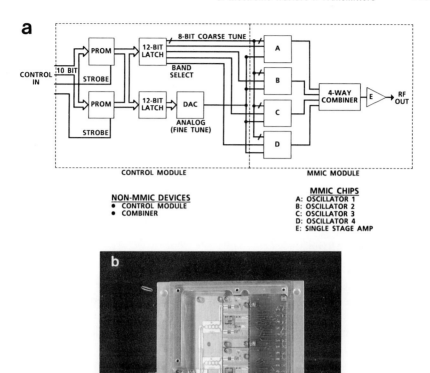

FIG. 4 Broadband VTO using four varactor-tuned MMIC oscillators. (a) Circuit diagram. (b) Photo of module.

VTO is locked to the received signal. MMIC implementations of the VTO and QIFM are described below.

Voltage-Tuned Oscillator

A block diagram of the voltage tuned oscillator is shown in Fig. 4a and a photo is shown in Fig. 4b. The VTO consists of four individual oscillators (A, B, C and D), each driven by an 8-bit coarse tuning word and a 10-bit fine tuning word. The individual oscillators are varactor-tuned MMIC chips, each of which covers approximately one-fourth of the total operating band. The outputs of the four oscillators are combined in a four-way combiner followed by a broadband MMIC amplifier; the combiner can be a passive MIC combiner or an active MMIC combiner. The VTO control circuitry consists of programmable read-only memory (PROM) frequency look-up table circuitry, programmable array logic (PAL), and D to A converters.

TABLE 1 Performance of Broadband VTO Implemented with Five MMIC Chips

Output VSWR	1.5 : 1
Power output	+12 TO +19 dBm
Power output variation	6dB maximum
Tuning command	10-Bit digital
Frequency accuracy	40 MHz

A listing of typical performance specifications for such a VTO is given in Table 1. The frequency band is generally some portion of the 2- to 18-GHz EW band.

Quadrature IF Mixer

The quadrature IF mixer illustrated in Fig. 5 consists of three MMIC chip types. The mixer provides an in-phase and a qudrature output at the IF (difference) frequency. The amplifier at the LO port is used to provide the required drive level for the mixer, while variable attenuators are used to adjust the signal level at the RF input port. A listing of the key performance specifications for the QIFM is given in Table 2.

Transmitter Amplifier/Modulator

As discussed previously, a function of the EW transmitter is to provide amplification and modulation of a signal to be radiated from the transmit antenna. This function may also require switching of a variety of signal paths.

A diagram of a transmitter subassembly designed for use with the SLO

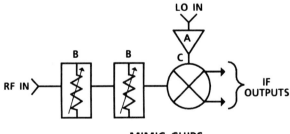

MIMIC CHIPS

A: HIGH BAND LIMITING AMPLIFIER
B: VARIABLE ATTENUATOR
C: QUADRATURE IF MIXER

FIG. 5 Quadrature IF mixer MMIC module.

TABLE 2 Performance of Quadrature IF Mixer (QIFM) Implemented with Four MMIC Chips

IF frequency	DC TO 45 MHz
RF input level	+5 TO +12 dBm
Conversion loss	3 dB maximum
IF amplitude tracking	1.0 dB
IF differential phase	90 ± 10°

described above and implemented with MMIC technology is illustrated in Fig. 6a, and a photo is shown in Fig. 6b. The MMIC module utilizes five chip types and a total of 13 MMIC chips.

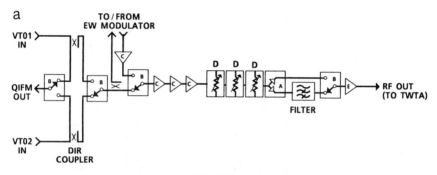

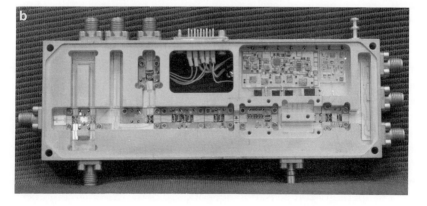

FIG. 6 Transmitter amplifier/modulator module. (a) Circuit diagram. (b) Photo of MMIC module.

TABLE 3 Performance of the Transmitter Amplifier MMIC Module

VSWR	
VTO inputs	1.8 : 1
All others I/O ports	2.0 : 1
Gain (VTO in to QIFM out)	−6.5 +/− 1 dB
Power output (TWT out)	7 dBm min
Power output flatness (TWT out)	4 dB
Second harmonic output (TWT out)	−30 dBc
Variable attentuation dynamic range	0 TO 60 dB

The transmitter module accepts two VTO signals, selects a sample of one of them (the signal being search locked), sends it back to the SLO QIFM, and generates an output signal which is used to drive the output TWTA. This output signal must be low in harmonic content and may require amplitude modulation. Table 3 lists the key transmitter module specifications.

The input section of the module contains two MIC directional couplers serving as power dividers and two MMIC SP2T switches denoted (B) in Fig. 6a. These circuits are configured such that either VTO input can be independently selected as either the signal fed back to the QIFM or as the signal transmitted to the output TWTA. This permits locking one VTO to a received signal while the other is being used for transmit, which may be necessary in handling multiple radar threats.

The input section is followed by another directional coupler, a limiting amplifier denoted (C) in Fig. 6a, and a SP2T switch (B). The coupler provides an output signal that is applied to an external modulator. The amplifier (C) raises the signal level returned from the modulator to the required level; the SP2T switch (B) is used to select either the straight-through path or the external modulator path.

Following this switch is a cascade of amplifier stages (C) that amplifies and then limits the power level to approximately a +20-dBm level. This signal is then passed through a switched filter for harmonic suppression before it is applied to the amplitude modulation section (3 chips designated D), which is capable of a 60-dB modulation range. The complete module includes 13 MMIC chips with several MIC components, namely the filter and directional couplers.

The peripheral dc bias and dc control circuitry (non-MMIC) include two voltage regulators (+3.5 V and −5.0 V), a modulation control circuit, and a temperature compensation network. All of the driver circuitry needed to control the SP2T switches is included on chip, so it can be driven directly from externally applied transistor–transistor logic (TTL) control signals.

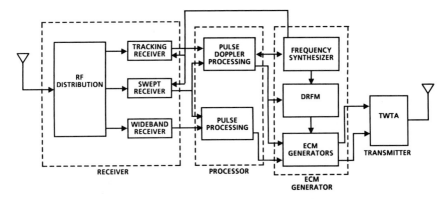

FIG. 7 EW system using frequency synthesizer and digital RF memory.

5.2.2 Frequency Synthesizer/Digital RF Memory

Another method of generating the transmitted signal in an EW system is through the use of a fast hopping frequency synthesizer, operating in conjunction with a DRFM. A block diagram of an EW system using a frequency synthesizer and a DRFM is shown in Fig. 7. A description of the receiver portion of systems of this type is presented in Chapter 6. The ECM Generator section includes the frequency synthesizer, the DRFM, and modulation circuitry for providing specialized waveforms for ECM techniques. The frequency synthesizer actually serves as a local oscillator to convert the received signal to a fixed, narrow, IF band which can be stored in the DRFM and subsequently upconverted for transmission.

The primary requirements on the synthesizer are that it be tunable over the full EW band of interest (e.g., 2 to 18 GHz) and that it be capable of fast switching from one frequency to another (in a few microseconds). Several types of frequency synthesizers have been applied in EW systems using direct synthesis, mix and divide, and phase lock loop techniques. In all cases, GaAs IC technology (analog and digital) has been utilized for improved performance and reduced size, weight, and cost. High-speed digital GaAs ICs have increased the operating frequency range of direct digital synthesizers (DDS) and form a key element (the programmable frequency divider) in high-speed phase lock loops in the phase-locked loop (PLL) type. A description of a single-chip DDS is given in Section 5.3.

In the mix and divide type of synthesizer, the individual output frequencies are derived from several oscillators whose outputs are multiplied, translated, and divided in various combinations through a set of RF switches and mixers. In order to provide a large number of closely spaced frequencies over a wide EW band, a complex set of circuits is required. The resultant size, weight, and cost can be relatively high even with the application of

MMIC technology (switches, mixers, and amplifiers). However, such synthesizers have the added benefit of very fast switching time, for example a few hundred nanoseconds.

The indirect synthesis PLL, synthesizer, has been applied in EW systems and appears especially well suited for meeting the requirements of small size and weight with moderately fast switching speed. Through the use of GaAs digital IC technology which has become available in recent years, the performance of such synthesizers has improved steadily (faster switching with lower spurious output) while the size, weight, and cost have been reduced. The use of a technique called fractional division enables PLL synthesizers to achieve very fine channel spacing with fast switching speed, i.e., 1 µs. A description of a PLL synthesizer using GaAs digital ICs is presented below. For a more detailed discussion, see Refs. [1–4].

Phase-Locked Loop Synthesizer Design

In order to cover the full EW band with a PLL type synthesizer, a combination of VCOs, downconverter mixers, and phase locked loop circuitry is generally used. A block diagram showing the major elements of such a synthesizer is given in Fig. 8. The loop circuitry including the VCO is shown on the left. The circuitry on the right (the RF converter) is used to

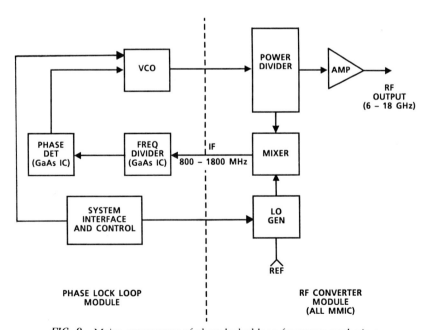

FIG. 8 Major components of phase locked loop frequency synthesizer.

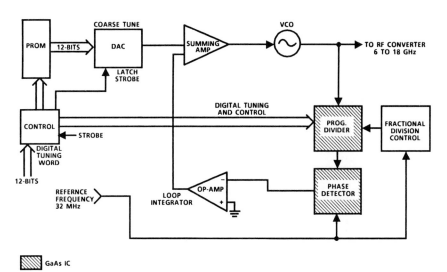

FIG. 9 Diagram of phase-locked loop circuitry with GaAs chips highlighted.

sample the VCO output and, by a set of switched local oscillators, to downconvert the full 6- to 18-GHz band into an IF band of 800 to 1800 MHz, which is in the operating range of the digital frequency divider in the phase locked loop. A description of the PLL circuitry is given below, followed by a description of an RF converter implemented with GaAs MMIC technology.

Figure 9 shows a circuit diagram of the phase locked loop portion of the fast-settling frequency synthesizer with the digital and analog GaAs ICs highlighted. The GaAs ICs include a programmable divider (digital IC), a phase detector (analog IC) and a D-to-A converter (analog and digital). The output of the voltage-controlled oscillator (VCO) is connected to the GaAs divider, which in turn provides the input to the GaAs phase detector. The phase detector compares the phase relationship of the divided VCO signal to the reference and generates an error voltage to tune the VCO. The VCO is initially tuned open loop to the approximate frequency to minimize the settling time. This is accomplished by a digital tuning network consisting of a PROM and a 12-bit digital-to-analog converter (DAC). The DAC can be implemented as a single GaAs chip. Fractional division techniques are used in the synthesizer to obtain resolutions as fine as 10 Hz with a 32-MHz reference frequency. The fractional division controller enables the programmable divider to, in effect, divide by fractional (i.e., noninteger) numbers.

IC Circuit Design

The programmable divider is a GaAs digital chip that is capable of dividing inputs from 10 MHz to 2.0 GHz by any integer from 12 to 2047. A

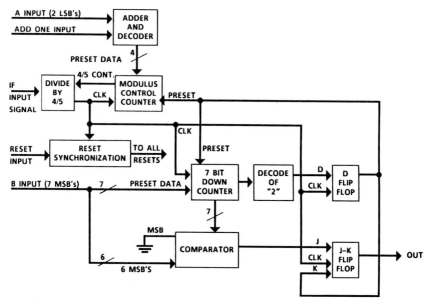

FIG. 10 GaAs digital IC high-speed programmable divider (400-gate complexity).

diagram of the divider chip is shown in Fig. 10. The divider has a total of 11 bits of digital control; 2 bits are used to control the modulus control counter and 9 bits for the downcounter. Direct-coupled FET logic (DCFL) is used throughout the design based on a well characterized standard-cell library. DCFL offers very high speed at very low power. The divider interfaces with TTL or ECL control levels. The GaAs divider replaces a thick film hybrid containing 35 ECL chips in an existing synthesizer.

The GaAs IC implementation provides several advantages. First, the higher operating frequency (2 GHz) eliminates the need for a fixed frequency divider (×4) in the ECL version. This results in a 12-dB reduction in spurious output level. Second, the lower power dissipation, together with higher operating temperature capability of the GaAs chip, enables operation over the full military temperature range of $-55°C$ to $+125°C$.

A diagram of the sample and hold circuits (S/H) in the phase detector is shown in Fig. 11. A S/H phase comparator is used instead of the more common digital phase comparator because it provides an order of magnitude increase in gain with reduced phase noise and sampling spikes. Furthermore, the reduced phase noise and spikes allows a wider loop bandwidth for faster response. The S/H uses a FET switch design that consists of over 50 active devices.

The S/H uses ±5-V power supplies and can accept signals as large 3 V peak to peak without saturating. The most significant benefit of the GaAs

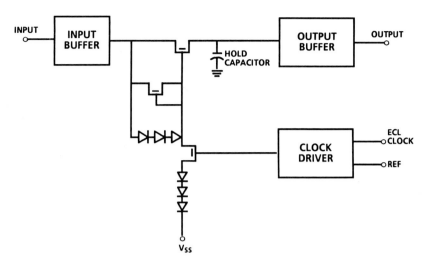

FIG. 11 GaAs analog IC sample and hold phase detector.

S/H chip is the reduction in power and size. A silicon hybrid S/H measuring 1.5" × 0.75" and dissipating 1.5 W is replaced with a single GaAs S/H chip measuring 1 × 1 mm and dissipating only 200 mW of power.

RF Converter Module A diagram of the RF converter section of the synthesizer is shown in Fig. 12. The output of the VCO is received from the PLL section and divided into two paths in a MMIC power divider. One path is amplified and the signal is then available for use in the system. The other signal is connected to a downconverter mixer. One of six local oscillators is selected by the switching circuitry as shown in order to downconvert the VCO output frequency into the IF band (800 to 1800 MHz). This IF signal is amplified and then sent back to the PLL module for frequency division and locking to the reference oscillator. By this combination of the switched LOs and the PLL circuitry, the VCO is locked to the stable reference over the full output band, i.e., 6 to 18 GHz.

Although the RF converter shown here could have been implemented with conventional MIC technology, MMIC technology results in a major reduction in the size, weight, and cost. An illustration of the size reduction can be seen by reference to Figs. 13 and 14. The unit in Fig. 13 was built in 1984 using individually packaged components interconnected with coaxial cables; it measures 8" by 10" by 1.5" in size (120 inch^3) and weighs about 2 pounds. The unit shown in Fig. 14 is a MMIC version utilizing 13 chips including single-pole two-throw switches, single-pole three-throw switches, small signal amplifiers, variable gain amplifiers, a mixer, and an intermediate frequency amplifier. The unit is configured in a standard electronic

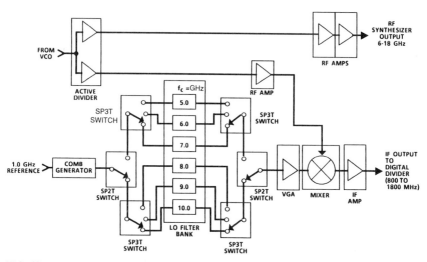

FIG. 12 Frequency synthesizer RF converter module implemented with 13 MMIC chips (6 switches, 3 RF amplifiers, 1 variable gain amplifier, 1 mixer, 1 IF amplifier, 1 active divider).

FIG. 13 Photo of RF converter module implemented with discrete (Pre-MMIC) technology; size is 120 in^3.

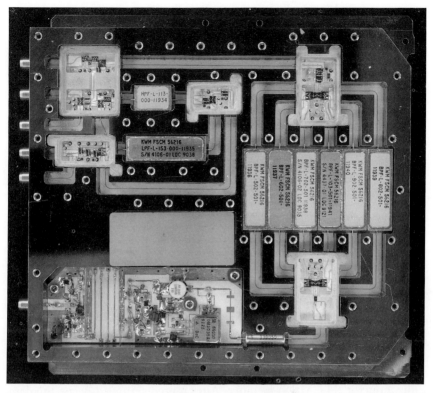

FIG. 14 Photo of RF converter module implemented with MMIC technology; size is only 18 in^3, a factor of 7 reduction from using pre-MMIC technology.

module (SEM-E) format measuring 5″ by 6″ by 0.6″ thick (18 inch^3) and weighing less than one-half pound. Thus a size reduction by a factor of about seven has been achieved. Furthermore, a significant reliability improvement and power consumption reduction is realized owing to the elimination of many RF cables and connectors. A listing of key performance parameters for the complete frequency synthesizer is given in Table 4.

Digital RF Memory The function of a DRFM is to receive a portion of an RF waveform, to store it in memory for some period of time, and then to retransmit it. The signal retransmitted must be a very faithful reproduction of the received signal, so the DRFM must maintain very precise control of the frequency and phase of the stored signal. Since the signal is actually stored in digital format using random access memory (RAM) circuitry, the RF signal must first be converted to a lower frequency and then digitized. Some of the frequency downconversion is typically performed in circuitry external to the DRFM, but some may be within the DRFM. The use of a

TABLE 4 Performance of Broadband, Fast-Hopping Frequency Synthesizer

Parameter	Value
Output frequency	6 to 18 GHz
Channel spacing	16 KHz
Settling time	1 μs
Spurious output	
within 10 MHz	−30 dBc
beyond 10 MHz	−60 dBc
SSB phase noise	
@ 1 KHz	−64 dBc
@ 10 KHz	−69 dBc
@ 100 KHz	−84 dBc
Operating temperature	−55°C to +90°C

stable frequency synthesizer can enable conversion of the RF signal to be stored to a relatively narrow band.

A diagram of a DRFM designed to operate at an input frequency of 50 MHz is shown in Fig. 15. The signal is compressed in an RF logarithmic amplifier and digitized in a 1-bit quantizer. A key element in the DRFM is the GaAs demux/mux chip, which transforms serial input data from the quantizer into 16-bit parallel words for the H-complementary metal oxide semiconductor (H-CMOS) memory. After a predetermined delay, the data are recalled from memory and converted back to serial data for retransmission. The regenerated signal is filtered for removal of harmonics and out-of-band spurious signals. The demux/mux circuit is clocked at a 224-MHz rate, made possible by the GaAs implementation. Data are written into and read from the CMOS memory at a 14-MHz rate.

Figure 16 is a detailed block diagram of the GaAs demux/mux/delay IC. To illustrate the improvement afforded by GaAs technology compared to silicon ECL technology, the solid blocks denote the functions of an earlier ECL discrete component demux/mux/delay circuit used in DRFM applications. The dotted blocks show the additional capability and capacity of the approximately 2500-gate GaAs chip. As can be seen, two delay channels have been added and the registers have been lengthened by 8 bits. The shifting rate has been doubled to 224 MHz, resulting in lower spurious output levels, and the memory capacity has been doubled to accommodate the 224-MHz shifting rate. The GaAs demux/mux and H-CMOS memory dissipate 50 W less power than the previous ECL components which enables high-density packaging without major thermal problems. Additional information on the DRFM chip is presented in Ref. [5].

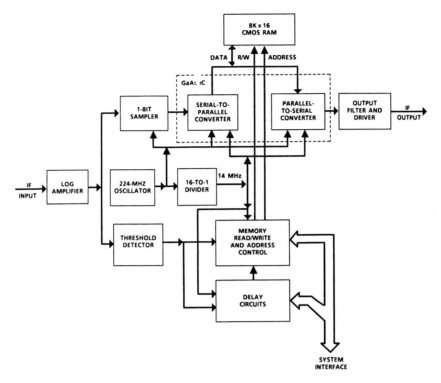

FIG. 15 Diagram of digital RF memory utilizing a GaAs demux/mux IC.

5.2.3 Phased-Array Antenna

As mentioned earlier in this chapter, phased-array antenna technology can be used to produce a completely solid-state implementation of an EW transmitter.

During the 1980s, a great deal of effort by private companies and government agencies was directed to the development of T/R modules for active phased-array antennas. A detailed discussion of this technology is presented in Chapter 4, which treats the application of phased arrays to radar systems. Although not studied nearly as intensively, the application to EW systems is also being addressed. The required number of elements in an EW phased array will generally be less than that for radar, but the basic configuration of the T/R module is similar.

Trade-off studies have been carried out to determine the required number of elements to meet the effective radiated power (ERP) and coverage requirements for several EW systems. These studies indicate that for an output power of approximately 1 W from each element (achievable with

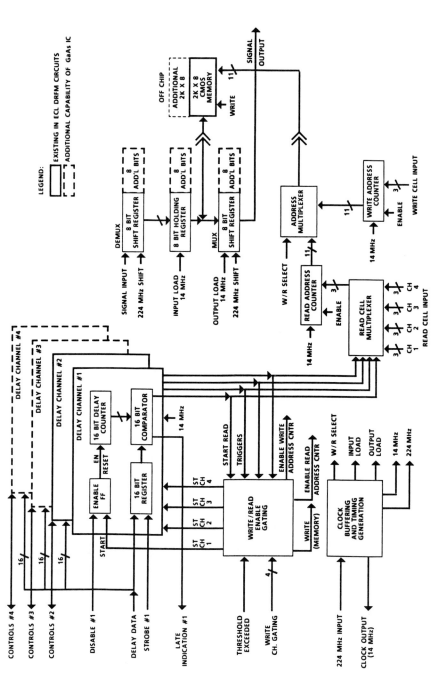

FIG. 16 GaAs digital demux/mux IC (8000-gate complexity).

GaAs MMIC amplifiers) the array is required to have about 100 elements, for a relatively low radar cross section (RCS) aircraft. For ship defense, an array of approximately 1000 elements with about 4 watts per element is required.

Since radar T/R modules are discussed in detail in Chapter 4, a detailed discussion of EW T/R modules is not presented here. The primary difference between the radar and the EW application is the operating bandwidth—the bandwidth is much wider for EW. Otherwise the basic configuration and circuit elements are the same, namely amplifiers, switches, phase shifters, and attenuators.

5.2.4 Adaptive Polarization Jamming

One type of radar which is very difficult to jam effectively is monopulse. In simple terms, the monopulse antenna derives angle (or directional) information of an incoming signal by a comparison of the relative amplitude of signals received from its sum and difference radiation patterns. Since the signals are received and processed simultaneously, the angle information can be derived instantaneously from a single pulse, hence the name monopulse. This property also makes electronic countermeasures (such as angle deception) very difficult, since amplitude or frequency modulation techniques are not effective.

One EW technique which has been developed to counter a monopulse radar, is to radiate a signal back to the radar which is similar in all characteristics to the reflected signal, but of a different polarization. The radiation pattern of most monopulse antennas has a different shape when measured in a polarization orthogonal to the nominal polarization, (called Condon lobes; see Refs. 5, 6) so the angle information received in the orthogonal polarization is different.

In order to utilize this characteristic for angle deception, the EW system must detect the polarization of the radar signal and then radiate a signal with a different polarization. An illustration of this technique, using conventional technology and hardware is shown in Fig. 17. In this case, a polarimeter network is inserted between the high-power (TWT) transmitter and the antenna. The orthogonal ports of a dual-polarized antenna provide sufficient information to determine the polarization of the received (radar) signal. The polarimeter network is then used to radiate an arbitrary polarization (typically orthogonal to the radar) for the EW signal. The polarimeter network consists of a combination of phase shifters and hybrid junctions which can be adjusted to vary the relative amplitude and phase of the signals from the two ports so as to generate any arbitrary polarization from two orthogonal antennas.

With this approach, the polarimeter must be capable of handling the high power of the transmitter. A disadvantage is that the insertion loss of the polarimeter will lower the effective output power radiated from the antenna, so this must be made up by increased power from the TWT.

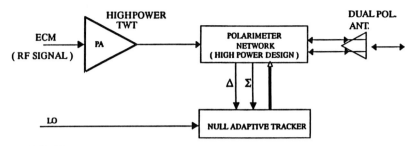

FIG. 17 Adaptive polarization jammer utilizing a single high-power polarimeter.

An alternative approach, illustrated in Fig. 18, is to locate the polarimeter on the input side of the transmitter. In this case, the polarimeter need not handle the high transmitter power level, and any insertion loss can be made up with gain on the input (low level) side of the transmitter. Since the polarimeter operates at small signal levels (below 23 dBm), monolithic (MMIC) technology can be used. The polarimeter requires a considerable amount of microwave circuitry; hence the size, weight, and cost savings derived from MMIC technology can be significant. A disadvantage is that the power amplifiers (shown here as Mini-TWTs) will contribute phase and amplitude errors which will change the effective polarization radiated. This can be compensated for by the addition of a second polarimeter which is used with an adaptive, self-calibration scheme to remove the effects of the phase and amplitude variations of the TWT.

A diagram of a polarimeter network is shown in Fig. 19a. The amplitude and phase of the signals connected to the dual antenna ports are controlled by a combination of phase shifters and hybrid/magic-tee power dividers. The key circuit element is the phaser, which must provide 0° to 360° phase shift over the full EW frequency band with very fine resolution, i.e., a fraction of 1°. To accomplish this, a combination of digital and analog

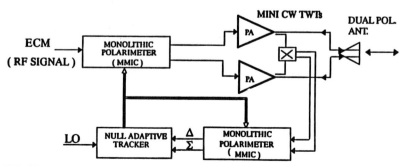

FIG. 18 Adaptive polarization jammer utilizing two low-power (MMIC) polarimeters.

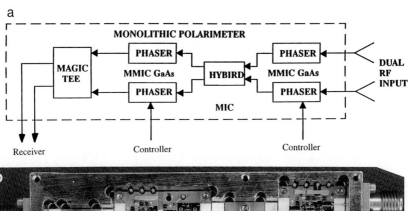

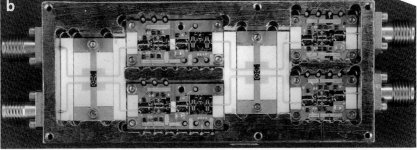

FIG. 19 Monolithic polarimeter module. (a) Circuit diagram. (b) Photo of MMIC (18 chips) module.

phase shifters are typically used. The analog phase shifter, used in conjunction with the adaptive control circuitry, enables very precise control of the polarization which is required for effective nulling. A photo of an early version of a polarimeter module using MMIC chips is shown in Fig. 19b. This polarimeter utilizes a total of 18 MMIC chips including amplifiers, digital phase shifters, analog phase shifters, and digital attenuators. This module was tested over the 6- to 18-GHz band and demonstrated the ability to provide the very precise insertion phase and amplitude required for precision polarization control.

The capability to implement the polarimeter in a very small size using monolithic GaAs technology makes possible an all solid-state phased-array antenna with polarization jamming capability. As discussed previously, a solid-state T/R module of this type might be used in a phased-array for both airborne and shipboard applications.

5.3. GENERIC EW CHIPS

A description of several key components used in EW systems was presented in the previous sections of this chapter. GaAs ICs, both microwave

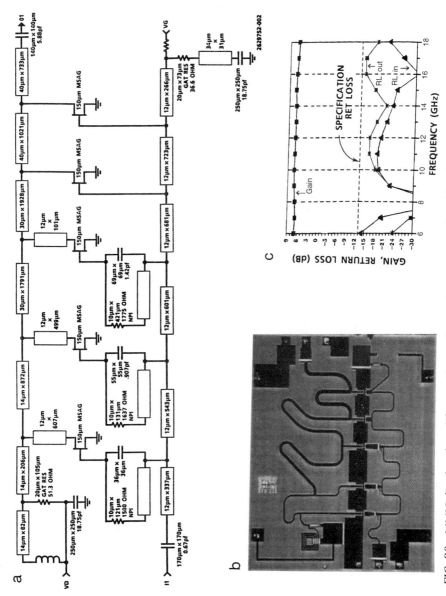

FIG. 20 MMIC distributed amplifier provides excellent performance over a broad bandwidth. (a) Schematic. (b) Photo of MMIC chip. (c) Measured performance.

and digital, provide major benefits in terms of improved performance, small size, reduced cost, and increased reliability. Many MMIC chips have been developed for use in these systems and many others are still in development. Some chips tend to be specialized, while others are quite generic in nature with potential application in many different systems.

Presented below are descriptions of several chips used in the subassemblies described previously and which are believed to have widespread applicability in EW systems. These chips are available from one or more MMIC foundries. Limited design information that can be used as a starting point for design of chips with slightly different requirements or for combining several chips to form a more highly integrated chip is provided.

5.3.1 Small Signal Amplifier

Broadband amplifier chips used as gain blocks are a basic element in every EW system. Because of its inherent broadband capability, the most widely used circuit for such amplifiers is the distributed architecture, typically utilizing four, five, or six FETs (Refs. 7, 8). A schematic of a typical five-stage distributed amplifier is shown in Fig. 20a, and a photo is shown in Fig. 20b. The primary design objectives are constant gain and good input/output match to permit cascading of several chips. Measured performance is illustrated in Fig. 20c and a listing of typical performance parameters is given in Table 5.

5.3.2 Medium-Power Amplifier

Medium-power amplifiers capable of power sufficient to drive a TWTA are also a key element in most EW systems. Such amplifiers utilize designs similar to the small signal amplifier, but with power FETs replacing the low-noise FETs. Depending on the system requirements, the key design objective may be either gain flatness or power output and efficiency. Designing for high power and efficiency involves nonlinear modeling and simulation technique. Typical performance characteristics of a medium power amplifier are given in Table 6.

TABLE 5 Measured Performance of Broadband Single-Stage Amplifier

Frequency	6 to 18 GHz
Gain	6.5 dB
Gain flatness	±0.3 dB
Input/output return loss	14 dB
Noise figure	7 dB
Power out (1 dBc)	12 dBm

TABLE 6 Performance of MMIC Medium Power Amplifier

Parameters	Value
Frequency	6 TO 18 GHz
Gain	5.5 dB
Gain flatness	±0.5 dB
Pout, 1 DBC	+23 dBm
VSWR	2.0 : 1
Drain voltage	8 V

5.3.3 Single-Pole, Two-Throw Switch

Together with the small signal amplifier, the single-pole two-throw (SP2T) switch is probably the most widely used MMIC chip in EW systems (Refs. 9, 10). A schematic of a broadband SP2T switch is given in Fig. 21a, and a photo is shown in Fig. 21b. Key features of MMIC switches are good input/output match, high port-to-port isolation, and ease of driving the switch from digital control signals. The chip illustrated in the photo includes on-chip TTL translators, enabling direct control by digital signals. Measured performance is illustrated in Fig. 21c and a list of key performance characteristics of a SP2T broadband switch is given in Table 7.

5.3.4 Digital Attenuators

Variable attenuators are employed in EW systems to adjust signal levels within the system and to provide amplitude modulation of the output signal. Both analog and digital attenuators have been developed and both are widely used. A four-bit digital type which has been shown to possess very desirable properties is illustrated in Fig. 22. Each bit utilizes a set of FET switches to direct the signal to either a through path (low attenuation) or a resistive path (high attenuation). The attenuation values of the individual bits can be tailored to meet the system requirements. The chip in Fig. 22b has bit attenuation values of 1.5, 3, 6, and 12 dB. The measured performance from 2 to 20 GHz is shown in Fig. 22c. A desirable feature of such digital attenuators is that this performance is nearly constant over the full military temperature range, without any compensation. The chip size is quite small—the 4-bit device measures only 2 × 2.5 mm. This single chip can replace a complex assembly of p–i–n diode switches and MIC attenuators with a major reduction in size and cost. A summary of key performance parameters for the four-bit attenuator is given in Table 8. A detailed description of MMIC digital attenuators is given in Refs. [11–14].

TABLE 7 Measured Performance of Single-Pole Two-Throw Nonreflective Switch

Frequency	2 to 18 GHz
Insertion loss	2.5 dB
IL flatness	±0.3 dB
Isolation	58 dB
Input/output VSWR	1.5 : 1
Switching speed	5 ns
Control signals	TTL

5.3.5 Direct Digital Synthesizer

DDS have existed for some time, but until the advent of high-speed GaAs digital technology, their use was limited to relatively low frequencies (less than 50 MHz). With GaAs digital ICs capable of 1 to 2 GHz clocking rates, DDSs can generate output frequencies up to 500 MHz. A diagram of a DDS is shown in Fig. 23. The key elements are an accumulator, a read-only memory (ROM), and a D to A converter. This DDS was fabricated as a single GaAs IC with about 10,000 gate complexity. A summary of key performance parameters is given in Table 9.

5.4 SUMMARY

Throughout the 1980s, despite a great deal of development, a very large expenditure of money and human resources, and tremendous progress in the development of GaAs integrated circuits, very few chips were actually deployed in operational systems. This was due in part to the very long development cycle for military systems, where the technology to be used

TABLE 8 Measured Performance of Broadband 4-Bit Attentuator

Frequency	2 to 20 GHz
Insertion loss	4.3 dB
IL flatness	±0.5 dB
Bit 0	1.6 ± 0.1 dB
Bit 1	3.0 ± 0.1 dB
Bit 2	6.9 ± 0.1 dB
Bit 3	12.4 ± 0.1 dB
VSWR	1.6 : 1
Control signals	TTL

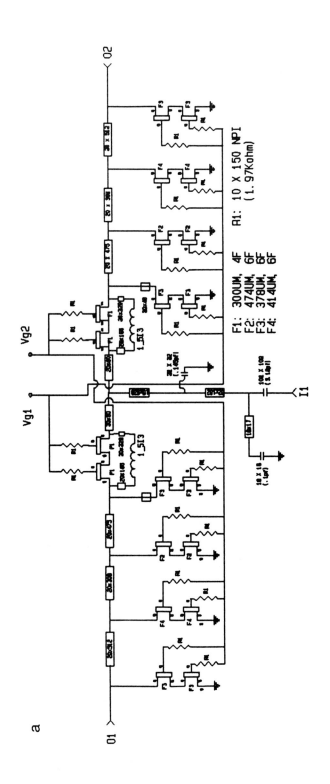

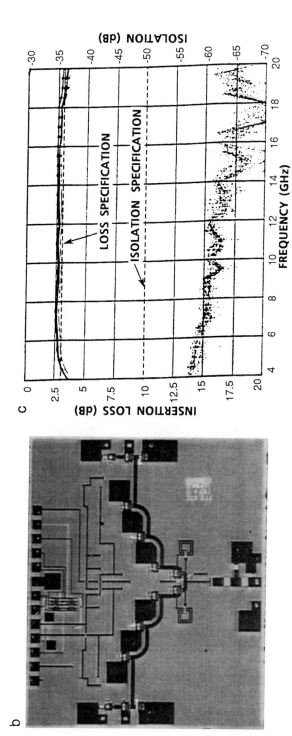

FIG. 21 Broadband MMIC switch. (a) Schematic. (b) Photo of MMIC chip. (c) Measured performance.

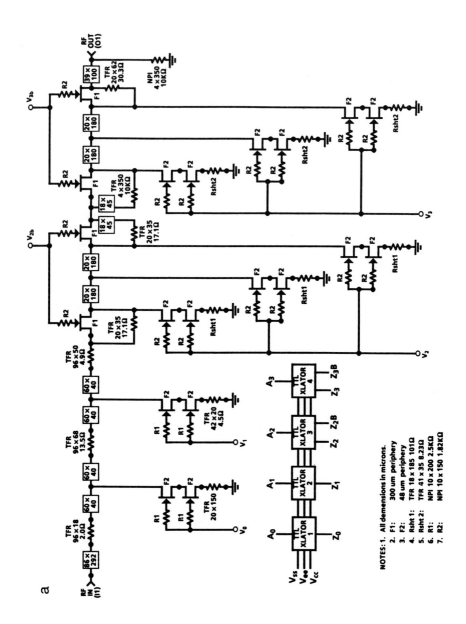

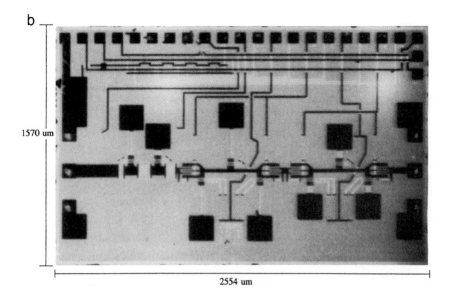

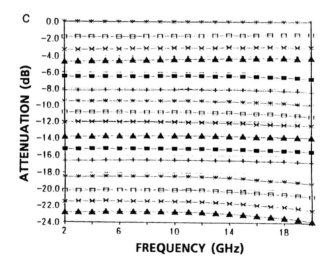

FIG. 22 Broadband, 4-bit digital attenuator. (a) Schematic. (b) Photo of MMIC chip. (c) Measured performance.

must be selected many years prior to production release, and in part to the reluctance of program managers to insert new and potentially "risky" technology into their ongoing programs. However, this situation started to

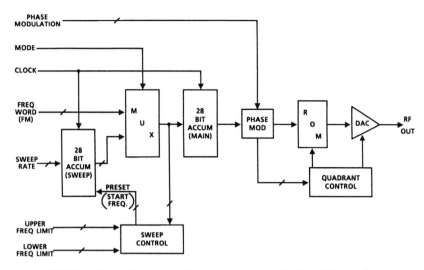

FIG. 23 Direct digital synthesizer implemented with GaAs IC technology.

change dramatically around 1990, due in large measure to the ARPA-sponsored MIMIC program. Due both to the technological achievements of the MIMIC program in making GaAs chips more producible and more readily available, and to the demonstration of MMIC brassboard hardware in actual systems, GaAs MMICs became widely accepted by both the technologists and the program managers. The promise of MMIC technology, namely to provide system hardware which was high performance, low-cost, small-size, lightweight, and more reliable was demonstrated. As a result, starting around 1990, the application of GaAs IC technology in military systems increased dramatically, and in fact it has now become the rule rather than the exception.

In EW systems, MMIC technology was first retrofitted into existing systems by building MMIC-based hardware configured as exact form/fit/function replacements of the existing hardware. In this case, the primary

TABLE 9 Performance of GaAs Direct Digit Synthesizer

Clocking frequency	1200 MHz
Output frequency	0 TO 300 MHz
Frequency resolution	1 Hz (30 bit accumulator)
Spurious levels	−45 DBC
Switching speed	100 ns
Input signal	TTL

benefits are lower cost and improved reliability. Typical benefits resulting from such MMIC insertion are listed below, based on a direct comparison of the MMIC hardware and non-MMIC hardware being replaced.

TABLE 10 Typical Benefits of MMIC Insertion

Cost savings	50%
Weight reduction	30%
Reliability improvement	4 : 1

5.5 FUTURE TRENDS

Starting around 1992, all new EW systems have been designed with extensive use of GaAs IC technology. In the microwave portion of the system, the MMIC chips described previously, including oscillators, amplifiers, switches, mixers, and variable attenuators, are being applied. In addition, there is a trend to integrate multiple functions on a single chip in order to achieve still further reductions in size and greater cost savings. A complete transmit/receive chip has been demonstrated which includes all the functions in a phased-array T/R module. This use of higher and higher levels of integration is the answer to meeting the demand for providing more and more capability in a smaller size and at lower cost.

The trend to include both microwave and high-speed digital functions on a single chip will continue. The simplification of MMIC modules resulting from the use of digital translators on MMIC switches and digital attenuators has been demonstrated. For digital phase shifters the use of digital circuitry on chip to do serial-to-parallel conversion can greatly simplify the external control wiring.

For the output power amplifier in EW systems, the exclusive use of the TWTA will likely be challenged by solid-state power amplifiers. Although the power output of individual broadband MMIC chips will probably remain in the 1- to 10-W range for the near future, the use of active aperture phased-arrays and the reduced ERP requirement of stealth aircraft make their use practical and desirable.

In the digital area, the greatest use of GaAs ICs in EW systems appears to be for frequency synthesizers and DRFMs. The ever increasing clocking rate of digital GaAs ICs will result in frequency synthesizers with more and more digital circuitry. The useful output frequency of direct digital synthesizers, currently limited to about 500 MHz (one-third the clocking rate) will increase to the GHz range. Likewise as the clocking frequency of programmable dividers increases, the operating frequency of phase locked loops will increase in the PLL type of frequency synthesizer. This in turn will enable broader portions of the VCO output band to be phase locked, thus requiring less microwave (MMIC) circuitry. The goal of producing a complete frequency synthesizer with just one (or a few) GaAs ICs appears reachable.

In the DRFM area, GaAs ICs are presently used primarily for serial-to-parallel conversion of data to permit the RF or IF input signal to be stored in RAM implemented with silicon technology. The availability of GaAs ICs with ever increasing gate counts (100,000 gates or more) may eventually result in DRFMs made entirely with GaAs technology.

In conclusion, the future of GaAs ICs is bright. Next-generation EW transmitters will make widespread use of GaAs IC technology in both the microwave and digital area. The trend to higher and higher levels of integration with both microwave and digital functions on the same chip will continue.

REFERENCES

[1] W. Tanis, Frequency synthesizer having fractional frequency divider in phase-locked loop. U.S. Patent 3,959,737, May 1976.
[2] H. Singh, R. Sadler, W. Tanis, and A. Schenberg, GaAs prescalers and counters for fast-settling frequency synthesizer. *IEEE J. Solid-State Circuits* 25 (1), 239–245, Feb. 1990.
[3] H. Singh, R. Sadler, A. Geissberger, W. Tanis, and E. R. Schineller, High speed, low power GaAs programmable counters for synthesizer applications. *Proc. GaAs IC Symposium*, Oct. 1987.
[4] J. Naber, H. Singh, W. Tanis, A. Koshar, and G. Segalla, A fast-settling GaAs-enhanced frequency synthesizer. *IEEE J. Solid-State Circuits* 27 (10), 1327–1331, Oct. 1992.
[5] P. Gamand, A. Deswarte, M. Wolny, J. Meunier, and P. Chambergy, 2 to 42 GHz flat gain monolithic HEMT distributed amplifiers. *Proc. GaAs IC Symp. Tech. Dig.*, 109, 1988.
[6] E. Condon, *Theory of Radiation from Paraboloid Reflectors*. Westinghouse Report 15, Sept. 1941.
[7] A. P. Chang, K. B. Niclas, B. D. Cantos, and W. A. Strifler, Design and performance of a 2-18 GHz monolithic matrix amplifier. *IEEE Micro. Millimeter-wave Mono. Circ. Symp. Tech. Dig.* 139, 1989.
[8] S. Houng and T. Tsukii, 60–70 dB Isolation 2-19 GHz MMIC Switches. p. 173, 1989. (Raytheon Electromagnetic Systems Division, Goleta, CA, and M. Schindler, Raytheon Research Division, *Proc. GaAs IC Symp. Tech. Dig.*
[9] B. Khabbaz, S. Morais, and S. Powell, *A GaAs DC-20 GHz SPDT Absorptive Switch*, pp. 165–167. Eighth Biennial UGIM Symposium, 1989.
[10] D. Payne, D. Bartle, S. Bandla, R. Tayrani, and L. Raffaelli, A GaAs monolithic PIN SPDT switch for 2-18 GHz applications. Alpha Industries, Inc., Woburn, MA.
[11] R. Gupta, L. Holdeman, J. Potukuchi, B. Geller, and F. Assal. A 0.05 To GHZ MMIC 5-Bit digital attenuator. *Proc. GAAS IC Symp. Dig.* 231–234, 1987.
[12] McGrath and Pratt, An ultra-broadband DC-12 GHz 4-bit GaAs monolithic digital attenuator. *Proc. GaAs IC Symp. Dig.* 247–250, 1991.
[13] F. Ali, S. Mitchell, and A. Podell, Low-loss, high-power, broadband GaAs MMIC multibit digital attenuators with on-chip TTL drivers. *Proc. GaAs IC Symp. Dig.* 243–246, 1991.
[14] B. Khabbaz, and H. P. Singh, DC-20 MMIC multi-bit digital attenuators with on chip TTL control. *IEEE J. Solid-State Circuits* 27 (10), 1457–1462, Oct. 1992.

6

Electronic Warfare II
Receivers

Sanjay B. Moghe, S. Consolazio, and H. Fudem

Advanced Microwave Technology, Electronics Systems and Integration Division, Northrop Grumman Corporation, Rolling Meadows, Illinois

6.1 Introduction
    6.1.1 Basic Receiver System Description
    6.1.2 Classification of EW Receivers
6.2 Crystal Video Receiver Structures
    6.2.1 Introduction
    6.2.2 Crystal Video Receiver Sensitivity
6.3 Superheterodyne Receiver Structures
    6.3.1 Principle of Operation
    6.3.2 MMIC Mixers for Receivers
    6.3.3 MMIC Oscillators for Receivers
6.4 Channelized Receiver Structures
    6.4.1 Introduction
    6.4.2 Principle of Operation
    6.4.3 System Requirements
    6.4.4 MMIC Downconverter
    6.4.5 Example of a Channelized Receiver
6.5 IFM Receiver Structures
    6.5.1 Introduction
    6.5.2 Phase Discriminators (Correlators)
    6.5.3 Limiting Amplifiers
    6.5.4 Passive Components
6.6 Integrated EW Transmit/Receive Modules
    6.6.1 Introduction
    6.6.2 Agile Receive/Transmit Module (ARTM) System Overview
    6.6.3 ARTM Architecture
    6.6.4 ARTM Construction—Dual ARTM
    6.6.5 ARTM Measured Performance
    6.6.6 Wideband Agile Sources
    6.6.7 Wideband Millimeter-Wave VCO
    6.6.8 VCO Circuit Design
    6.6.9 Millimeter-Wave MMIC Mixer Design
    6.6.10 DRO Circuit Design
    6.6.11 Wideband Agile Millimeter-Wave Source Performance
6.7 Other EW Receiver Types
6.8 Summary and Future Trends
    References

6.1 INTRODUCTION

Receivers are a very important part of modern electronic warfare (EW) systems. The study and development of modern electronically controlled, directed, and commanded weapons has caused great expansion in the field of electronic warfare systems. The basic concept of EW is to have the capability to explore the electromagnetic emissions from enemy aircraft in specified parts of the electronic spectrum. The information derived from these observations provide strategic intelligence on the enemy, revealing their intentions and capabilities. This information is used to invoke countermeasures that deny the effective use of enemy communications and weapon systems, thus protecting one's own use of the same spectrum. The most common signals detected are those from radars designed to guide modern electronic weapons [1,2]. The receivers described in this chapter are used to detect these radar signals and gather information about the enemy. Radars also are used today in commercial applications such as the guidance systems for airplanes and ships to detect impending weather conditions when there are stormy weather and poor visibility conditions. Thus, there are many commercial applications for military EW receiver technology.

Detection of radar signals by EW receivers is relatively easy since the signal received is much stronger than after it is received back at the radar. While radar signal detection is straightforward, the electronic signal environment has become very complicated as radar technology has improved. Today's receivers have to handle millions of processes per second. This has imposed serious constraints on receiver technology, forcing significant improvements over the past few years. The range in the cost of receivers can be as low as $2000 up to as high as $300,000 for very sophisticated receivers. Since modern receivers cover very wide bandwidths, from UHF frequencies to K-band frequencies (up to 20 GHz) and even up to 40 GHz, the microwave component cost of these receivers is significant. This cost can be substantially reduced through the use of monolithic microwave integrated circuit (MMIC) technology [3] for typical receiver components such as amplifiers, mixers, oscillators, low-noise and power amplifiers, and phase-locked loops. MMICs have an additional advantage in terms of significant size reduction over microwave integrated circuit (MIC) components. Typically, there is a 5 to 1 reduction in size of receiver subassemblies by use of MMIC technology over MIC technology. This is particularly advantageous in airborne application where space is at a premium. Another important reason to use MMICs in receivers is the significant improvement in system reliability. This comes about because microwave components designed using MMICs have typically one-tenth the number of parts and bond wires as compared with MIC technology. With MMIC technology, most of the components are realized monolithically on a single chip. For example, all the RF amplifiers, mixers, oscillators, and IF amplifier components can be realized monolithically on individual chips, reducing the number of parts and bond wires. This typically provides a 10 to 1 improvement in system reliability. In some cases, some or all of these functions can be further integrated onto a single chip. The specific reliability improvement and size reduction are dependent on the complexity of MMIC chips, system architecture, and other system considerations. However, in general, MMIC technology offers a significant benefit over conventional MIC technology in electronic warfare receivers, in terms of size, cost, performance, power consumption, and reliability.

In this chapter, we will describe a number of different types of EW receivers and applications including superheterodyne receivers, channelized receivers, and instantaneous frequency measurement receivers (IFM). We will discuss MMIC components required to build these receivers and give examples of actual receivers built using MMIC technology. We also discuss performance requirements that can be uniquely realized with MMIC technology. The later part of this chapter includes a discussion of other types of EW receivers, such as compressive and acouto-optic, and future trends in electronic warfare receivers.

6.1.1 Basic Receiver System Description

A basic EW receiver system consists of an antenna, a receiver, a digitizer, a processor, and a display unit. In concept, this is no different than a receiver system that would be found in common communications devices such as cellular phones. The antenna captures electromagnetic energy from free space and feeds it to the receiver. The receiver downconverts the signal to an IF frequency. The signal is further fed to the digital processor and then converted into data or voice. A basic EW receiver block diagram is shown in Fig. 1.

The receiver downconverts the radio frequency signal to a lower frequency ("video") output. The digitizer converts the video output to a digital signal and the processor (computer) analyzes the digital data to extract the necessary information. The display unit presents this information in a visual or audio form. (Sometimes, the outputs from the signal can control some special function that can automatically take the necessary action such as turning on jammers.) The receiver diagram shown in Fig. 1 can be further expanded into various types of receivers as shown in Fig. 2. This figure shows four different types of receivers: (a) crystal video receiver (CVR), (b) IFM system, (c) wideband super-het, and (d) yttrium–iron–garnet (YIG)-tuned narrowband superheterodyne (superhet).

There are three ways to classify EW receivers: by their operating frequency, by their applications, and by their structures. In this chapter, the receivers will be discussed according to their structures. However, since frequency is such an important attribute of EW systems we will first give a brief overview of the frequency nomenclature.

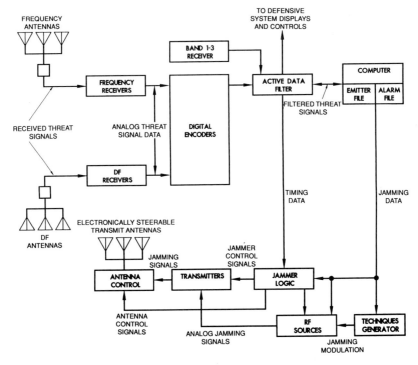

FIG. 1 EW receiver block diagram.

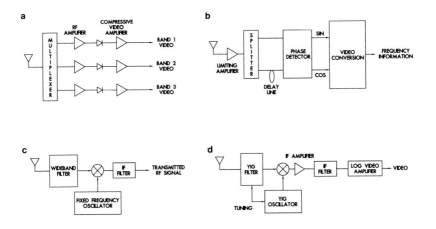

FIG. 2 (a) Crystal video receiver block diagram. (b) Instantaneous frequency measurement receiver block diagram. (c) Wideband superheterodyne receiver block diagram. (d) YIG-tune narrowband superheterodyne receiver block diagram.

TABLE 1 Frequency Band Designations Below 40 GHz (in GHz)

(Previous frequency designations)								
UHF	L	S	C	X	Ku	K	Ka	
C	D	E	F	G	H	I	J	K
				(Current frequency designations)				
0.5	1.0	2.0	3.0	4.0 6.0 8.0 10.0	12.4	18.0	20.0 26.5	40.0

According to their operating frequency, EW receivers can be classified as very high frequency (VHF), ultrahigh frequency (UHF), microwave, and extremely high frequency (EHF) receivers. Sometimes they are simply divided into two groups: communication intercept receivers and radar intercept receivers. There is no clear separation in frequency range between the communication and the radar operations. In general, the communication frequency range is the lower spectrum range and the radar operating frequency is in the higher frequency range. Of course, there is an overlap of the frequency range in which both communication receivers and radar operate.

The microwave frequency range has been divided into bands, with each band designated by a letter or letters. The original purpose was to ensure military secrecy; subsequently the letter designations were adapted for peacetime use. Since these band designations are difficult to remember, new designations have been assigned to frequencies below 40 GHz. Both the new and the old designations are presently used. Table 1 lists both the designations below 40 GHz.

The frequency bands above 40 GHz are listed in Table 2. It should be noted that under the current band designations some of the same letters are used for bands below as well as above 40 GHz. For example, the letter D represents the 1- to 2-GHz band in the low-frequency designation and the 110- to 170-GHz band in the high-frequency designation. The application context or the actual frequencies can always be used to resolve the ambiguity in frequency band designations.

TABLE 2 Frequency Band Designations above 40 GHz (in GHz)

Q	U	V	E	W	F	D	G
33–50	40–60	50–75	60–90	75–110	90–140	110–170	140–220

TABLE 3 EW Receiver Structures

	Receiver structure type					
	Crystal video	Superheterodyne	IFM	Channelized	Compressive	Bragg
EW receivers						
RWR	√	√	√			
RWHR	√	√	√	(Coarse)		
ECM		√	√	(Coarse)		
ESM	√	√	√	√	√	√
ELINT		√	√	√	√	√
Relative maturity	Production	Production	Developed	In development	In development	In development

6.1.2 Classification of EW Receivers

According to their applications, EW receivers can be classified as radar warning receivers (RWRs), radar homing and warning receivers (RHWRs), electronic countermeasure (ECM) receivers, electronic support measure (ESM) receivers, and electronic intelligence (ELINT) receivers (see Table 3). Each receiver type will be briefly discussed in the following paragraphs [4–8].

RWR

An RWR is used to detect weapon radars and provide warning to the pilot or navigator. A weapon control radar becomes a threat only when the main beam of the radar is directed toward the aircraft or ship. Since there is substantial energy in the main beam of a radar, a receiver with moderate sensitivity is sufficient. Knowing the frequency of the hostile radar will provide useful information for sorting and identification. However, it is not essential to have fine-frequency information for a warning application. In order to cover all the threats, the receiver must have wide frequency and spatial coverage.

RHWR

The function of an RHWR is almost identical to that of an RWR. However, in addition to the warning function, the receiver includes a homing function that will provide passive guidance toward the hostile radar. Passive ranging or location information is often required for standoff operation and weapon release. In order to provide the homing function, the receiver should generate accurate angle of arrival (AOA) information. The instantaneous angle and frequency coverage of the homing function need not be very wide; however, for the warning function, both the frequency and the angle coverage need to be wide.

ECM Receivers

An ECM receiver is used to obtain all the information from an electronic order battle (EOB). In other words, the receiver should collect information on all radars in the battlefield. Therefore, the receiver needs to cover wide frequency ranges and wide spatial angles. In order to detect radars that are not pointed toward the receiver, the sidelobes of the radar should also be detected. Therefore, the sensitivity of the receiver should be very high. In order to provide enough information for a signal processor to sort the environment, the receiver must measure all the parameters of the input signals with high resolution. In particular, the receiver should generate fine-frequency and AOA resolutions. Typical ECM receiver performance requirements are given in Table 4.

TABLE 4 Typical Modern ECM Receiver [4] Performance Requirement

Frequency range	0.5–40 GHz
Signal type	50 ns pulse to
Sensitivity	>-60 dBm
Frequency resolution	<5 MHz
Dynamic range	>50 dB
Amplitude accuracy	1 dB
Bearing accuracy	Better than 5°
Pulsewidth resolution	25 ns
TOA resolution	50 ns
Probability of intercept	100%
Pulse rate	10^6 Pulses/s

ELINT Receivers

An ELINT receiver is used to receive special signals, isolate them, and perform a high-resolution analysis on them. The input frequency range of the receiver should be wide enough to cover the signals of interest. However, the instantaneous bandwidth of the receiver need only be wide enough to cover one or a few signals of interest. The receiver should provide very fine measurements on all the parameters of the hostile radar. If the data cannot be analyzed at the collection station, they can be stored and analyzed at a later time. Thus the receiver does not have to process signals in near real time, as do the other receivers discussed.

The discussion in this chapter will concentrate on the operating principles of receivers and their designs. Therefore, the receivers are discussed according to their structures: crystal video, superhet, IFM, channelized, compressive (microscan), and Bragg cell. Each type of receiver structure is discussed in a separate chapter section.

Crystal video receiver structures can be used in EW RWRs and RHWRs because of their small size and wide frequency coverage. They can also be three types, channelized, compressive, and Bragg cell, are still under development at various stages. Each receiver structure has its own figure of merit. However, only the anticipated performances of the latter three types of receivers will be discussed here. It should also be emphasized that these anticipated performances are subject to change if new devices or technologies are discovered.

In the following paragraphs, the potential applications of each type of receiver will be discussed briefly. Table 3 summarizes these applications to the various EW receivers.

Crystal video receiver structures can be used in EW RWRs and RHWRs because of their small size and wide frequency coverage. They can also be used in ESM receivers. The wide-open crystal video receiver consists of a frequency multiplexer which splits the input signal spectrum into broad contiguous bands, each with its own detector, an envelope detector (generally square-law over the signal range of interest), and a log-video amplifier. A low-noise radio frequency (RF) amplifier might precede the video detector if enhanced sensitivity is desired.

Superhet receiver structures can be used in almost any application because of their high selectivity, sensitivity, and dynamic range. However, because of their narrow-input bandwidth, this type of receiver is usually used in conjunction with other types of receivers in EW applications. A scanning superhet receiver structure is used in ESM applications. It commonly consists of a YIG preselector, a mixer driven by a swept-frequency local oscillator, and a narrowband IF amplifier and envelope detector. A variation is the wideband superheterodyne type which uses a fixed tuned local oscillator to translate various RF bands to a common IF band for processing.

IFM receiver structures can be used in RWRs, RHWRs, and ECM receivers because of their wide input frequency range and small size. They may be used in conjunction with other receivers including ESM and ELINT receivers. This type of receiver structure consists of a limiting RF amplifier followed by a frequency discriminator. Table 5 gives IFM receiver performance characteristics.

Channelized receiver structures can be used in ESM and ELINT receivers because of their superior performance. Receivers with coarse channelization may be used as RWRs and RHWRs because of their potentially small size.

TABLE 5   Digital IFM Receiver Characteristics [4]

	L-band	S-band	C-band	X-band	Ku-band
Frequency range (GHz)	1–2	2–4	4–8	8–12	12–18
Unambiguous bandwidth (MHz)	1060	2120	4240	4240	6360
Sensitivity (threshold) (dBm)	−65	−65	−65	−65	−60
Dynamic range (dB)	70	70	70	70	65
Input impedance (nom.) (Ω)	50	50	50	50	50
VSWR (max)	2.1	2.1	2.1	2.1	2.1
Capture ratio (at discriminator input) (dB)	10	10	10	10	10
Resolution (11 bits) (MHz)	.52	1.04	2.08	2.08	3.12
Accuracy (RMS) (MHz)	1.25	2.5	5.0	6.5	12
Throughput delay (ns)	185	150	135	135	130
Shadow time (ns)	70	50	50	50	50
Pulsewidth (min. for full accuracy) (ns)	95	60	45	45	40

TABLE 6 ELINT Type Compressive Receivers [4]

f_o	2250 MHz
Analysis bandwidth	250 MHz
Resolution[a]	
3 dB	6 MHz
30 dB	10 MHz
50 dB	60 MHz
Sensitivity/processing gain	
50-ns pulse	−78 dBm/+4 dB
100-ns pulse	−84 dBm/+10 dB
250-ns pulse or longer	−92 dBm/+18 dB
Dynamic range	60 dB min
Processing delay time	750 ns max
Scan revisit time	1 Ls max
Sweep rate	1000 MHz/Ls
Resolvable frequency cells	41
Output data rate[b]	164 MHz
Probability of intercept	50 to 100%

[a] Two tones 250 ns or longer, 1 Ls = 0.5 μs.
[b] Resolution cells/s.

Compressive receiver structures can be used in ESM and ELINT receivers because of their high performance. Table 6 gives ELINT type compressive receiver performance requirements.

Bragg cell receiver structures can be used in ECM, ESM, and ELINT receivers because of their potentially high performance and small size.

6.2 CRYSTAL VIDEO RECEIVER STRUCTURES

6.2.1 Introduction

The basic crystal video receiver [9,10] is the simplest type of microwave receiver, low in cost, lightweight, and small in size. It is used primarily for wideband (e.g., 2 to 18 GHz) detection of low-duty cycle pulse signals, although it can be adapted to the detection of CW signals by the addition of appropriate switching or chopping of the CW carrier. The detection technique removes both frequency and phase information, and hence the basic crystal video receiver has no capability to measure either of these quantities. Rapid degradation occurs in a dense signal environment due to amplitude distortion resulting from overlapping signals within the RF bandwidth. The primary disadvantage of this type of receiver is that key threats can be masked by other signals, such as jamming or pulses from other emitters which are coincident in time with the desired signal. Fig. 2a shows the block diagram of the basic crystal video receiver.

6.2.2 Crystal Video Receiver Sensitivity

The sensitivity of a crystal video receiver is generally limited by the sensitivity of the detector itself [11–13]. The best detectors available have a tangential sensitivity signal-to-noise ratio (SNR) equal to 8 dB in video (or 4 dB in RF) between -50 and -60 dBm at 10 MHz video bandwidth. Inherent RF and filtering losses combine to give a system sensitivity that is reduced further by typically 10 dB. This sensitivity is generally adequate for detecting main-beam radiations from most radars. Also, the wide-open nature of the receiver provides a high probability of intercept.

Addition of a low-noise, wideband amplifier ahead of the crystal video detector can increase sensitivity by 20 dB or more [14]. The effective noise bandwidth of this configuration for a high-gain RF amplifier can be expressed as $B_n = B_{RF} - (2B_v)^\alpha$, where B_{RF} is the wideband predetection bandwidth, B_v is the relatively narrow postdetection video bandwidth, and α is an exponent whose value is between 0.5 and 1. The value of α primarily depends on the ratio $B_{RF}/2B_v$, but is also a weak function of the receiver's probabilities of detection and false alarm. As the ratio $B_{RF}/2B_v$ becomes very large, the statistics at the output of the video amplifier becomes Gaussian distributed. In this case, the value of α approaches 0.5 which results in an effective noise bandwidth equal to $B_n = \sqrt{2B_{RF}B_v}$. This value of effective noise bandwidth can also be obtained using a two-moment analysis, which implicitly assumes Gaussian statistics. The receiver sensitivity is generally maximized by selecting a video bandwidth, B_v, which just supports the narrowest pulsewidth ($B_v = 0.35/\tau$ to $0.45/\tau$), where τ is the narrowest pulsewidth.

Another variation of crystal video receivers applies a YIG-tuned RF filter in the predetection portion of the receiver that serves as a narrowband bandpass filter. Typically, the filter is switched in or out depending upon the degree of frequency sorting desired. An alternative approach is to use a multiple filter bank with each filter followed by a crystal video receiver.

A primary component of communication intelligence (COMINT) ESM is the double- or triple-conversion superheterodyne receiver. Table 7 gives performance requirements for a COMINT type receiver. These receivers are normally designed for operation over the entire HF band, or part of the VHF or UHF bands. A high performance HF receiver [14] uses a 1-Hz step synthesizer which has the memory capability to hold the 100 most significant threat channels. Frequency stability is ±1 ppm over the temperature range and the single sideband sensitivity is 1 μV for 10-dB output SNR. Dynamic range is 80 to 100 dB for signals spaced at least 20 kHz apart. Single-sideband (SSB), AM, FM, and CW signals can be individually identified, and frequency shift keying (FSK) can be decoded with a modem.

TABLE 7 Comint Type Compressive Receivers [4]

Multiplier configuration	Long	Short
Bandwidth (1 dB)	25 MHz	6 MHz
Resolution (3 dB)	50 kHz	50 kHz
Number of transform points	500	120
Input signal		
Duration	40 μs	25 μs
Max. level	−16 dBm	+5 dBm
Dynamic range		
Sidelobe limited	35 dB	40 dB
Noise limited	60 dB	60 dB
Power consumption	16 W	40 W
Weight	5.4 kg	5 kg

6.3 SUPERHETERODYNE RECEIVER STRUCTURES

6.3.1 Principle of Operation

A Superheterodyne receiver converts an RF signal to an intermediate frequency (IF) by mixing the RF frequency with a local oscillator (LO) frequency in a nonlinear device such as a diode or FET. Further processing of the lower frequency IF is accomplished more easily than at the high-frequency RF signal [15–17]. A basic superhet receiver block diagram is shown in Fig. 2c. The desired IF signal, $F_{IF} = F_{RF} \pm F_{LO}$, is present at the mixer output along with additional signals of the form $mF_{RF} \pm nF_{LO}$, where m and n are integers. The higher order signals generated by the mixer are referred to as single-tone intermodulation products or spurs and are generally reduced by using an IF filter at the mixer output. The IF section of a superhet receiver is typically narrowband which reduces the RF noise bandwidth to a much narrower IF bandwidth and enhances the sensitivity of the receiver.

IF amplifiers are used after the mixer to increase the IF signal strength prior to detection which also improves the receiver sensitivity. The narrow IF bandwidth also rejects spurious responses generated in the mixer circuit which would otherwise limit the receiver dynamic range. The detector output contains the desired information which was carried by the RF signal due to the linear translation from RF frequency to IF frequency using the heterodyne process. The primary reason for the development of the superheterodyne receiver was to improve the sensitivity of the crystal video receiver and add the capability to measure frequency. Since linear frequency translation occurs in the heterodyne process, frequency measurement capability and increased receiver sensitivity is accomplished.

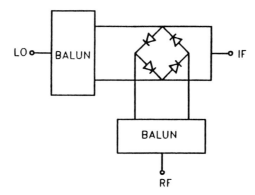

FIG. 3 Mixer schematic.

6.3.2 MMIC Mixers for Receivers

A microwave mixer is a three-port device shown schematically in Fig. 3. The RF input port and LO port are generally the high-frequency ports while the IF output port is a low-frequency port. Matching transformers (baluns) are indicated for the RF and LO ports.

The RF and LO frequencies are mixed in a nonlinear device such as a diode or FET. The conversion loss (gain) of a mixer is the ratio of IF output power to RF input power. The noise figure of a diode mixer is approximately equal to its conversion loss in dB. Mixer isolation is an indicator of the mixer circuit balance. Isolation is the ratio of input power at one port to the leakage at any other port. Mixer dynamic range is bounded at the lower limit by the spurious signal level and at the upper limit by the third-order intercept point.

Receivers required for most EW systems must meet very tight performance requirements. LO leakage and spurious signal levels in mixers must be kept to a minimum since they strongly influence overall system performance [18–24]. The MMIC mixer in Fig. 4 is a high dynamic range (60 dB), eight-diode double-balanced mixer designed for high LO to IF isolation (35 dB) and high single-tone intermodulation suppression (−50 dBc) [25].

Careful design of the on-chip RF and LO port baluns is essential in meeting these goals. A unique passive balun design was implemented to ensure the circuit balance and eliminate the dynamic range limitations of active baluns. The passive wideband baluns have typical insertion loss of about 1 dB at 20 GHz. The diodes used in the mixer are 0.5-μm-gate-length FETs with drain and source tied together. Figure 5 shows conversion loss and RF port return loss of 10 MMIC mixer subassemblies used in a Ku-band receiver; repeatability of the subassemblies is quite respectable.

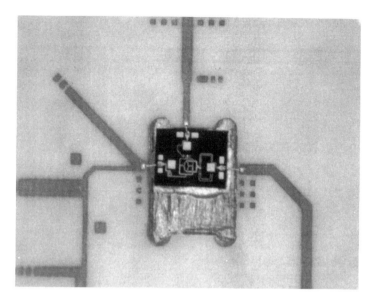

FIG. 4 MMIC mixer chip in subassembly.

The measured 2R–L single-tone spur suppression is < −50 dBc at 0 dBm RF signal input and +20 dBm LO drive. The repeatable performance of the mixer is a result of its MMIC implementation which ensures that

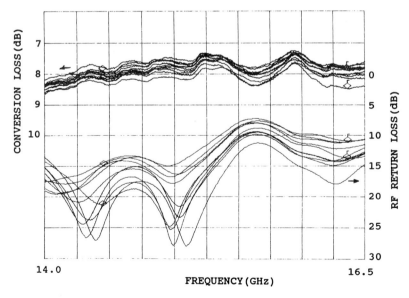

FIG. 5 Mixer conversion loss/RF return loss (10 pcs).

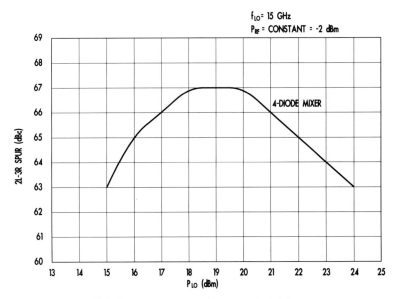

FIG. 6 Mixer spur suppression vs LO drive.

circuit balance and diode-quad symmetry is identical on each chip across the MMIC wafer. The counterpart MIC diode mixer is fabricated with manual assembly techniques which lead to errors in circuit symmetry and performance degradation. A 10- to 15-dB degradation in spur suppression performance has been measured in a MIC diode mixer relative to its MMIC counterpart. MMIC mixer single-tone spur suppression vs LO drive is shown in Fig. 6. The 2L–3R spur suppression of < -65 dBc with $> +16$ dBm LO drive is superior to any MIC style mixer in this frequency band.

6.3.3 MMIC Oscillators for Receivers

Previous work on MMIC oscillators has shown good voltage controlled oscillator (VCO) and dielectric resonator oscillator (DRO) performance at X- and Ku-bands [26,27]. For many switched oscillator applications, the device is either shut off by switching the gate or drain bias or by using a separate switch structure at the RF output. Switching the device current is a slow process as it takes time for the current to stabilize, typically on the order of milliseconds. The variation in drain current, after the device is turned on, results in small changes in device Cgs and consequently affects the oscillator frequency. One solution to this problem is to keep the device current constant and turn the oscillator off by suppressing the negative resistance. A method of quenching the negative resistance which uses a

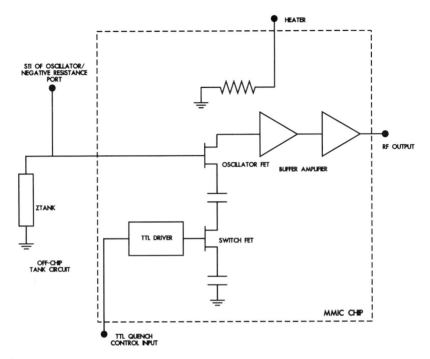

FIG. 7 Quench oscillator schematic.

control diode placed in the feedback path of the oscillator has been recently reported [28]. This oscillator circuit makes use of a passive FET switch as the negative resistance control device, allowing for easy integration of the quench circuit in the MMIC.

A common-source oscillator configuration was selected to achieve wide negative resistance frequency coverage while exhibiting a minimum of S_{11} phase angle excursion. This configuration allows the widest tuning range performance to be achieved while allowing the quench function to be added by inserting a switch FET at the source of the oscillator FET as shown in Fig. 7.

The FET oscillator with quench circuit was analyzed using small signal equivalent circuit FET models to predict and optimize circuit performance. The size of the switch FET used to quench the negative resistance needs to be chosen carefully. The FET must be large enough so that when quenched ($V_{gs} = 0$), the oscillator S_{11} falls below unity. In the other condition ($V_{gs} = V_p$), the drain-source capacitance should be small enough to achieve the negative resistance required for high-frequency operation while not becoming a significant portion of the feedback for temperature stability. A DC block is utilized at both terminals of the switch FET to isolate it from the

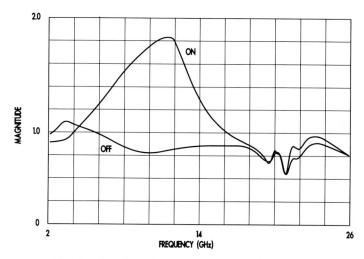

FIG. 8 Quench oscillator S_{11} for "ON" and "OFF" states.

oscillator bias and to DC float the switch FET above ground. This eliminates the need for any negative supply voltages to control the switch FET. The oscillator feedback consists of the parasitic capacitance of the quench circuit plus additional bondpads which were included to allow for feedback adjustment. Using the gate to turn on the FET switch, the drain-source resistance of the switch quenches the negative resistance and hence the oscillator is turned off. When the source FET is pinched off, the negative resistance of the oscillator is restored and the oscillator turns on. The variation of small signal negative resistance as a function of the source FET state is shown in Fig. 8. In either of the two states, the oscillator current remains relatively constant, avoiding problems with frequency changes due to current.

In the common-source FET oscillator configuration, the impedance at the source terminal is quite high due to the low value of feedback capacitance needed to optimize the negative resistance. This requires the switch FET impedance to be held very constant during the oscillator "ON" state; otherwise, any impedance changes in the switch FET would affect the oscillator frequency. The problem is minimized by biasing the switch FET well into its pinch-off state. To further minimize any possible influence of the switch FET on the oscillator, a FET inverter buffer was added to the quench control input line. This allows very accurate and controlled switching using a conventional transistor–transistor logic (TTL) control signal.

These ideas have been used to integrate a negative resistance oscillator with a FET switch quench circuit and a buffer amplifier into a single 1.8 × 1.2-mm chip. Figure 9 shows a photograph of the MMIC oscillator chip with the TTL compatible quench circuit visible in the lower left corner.

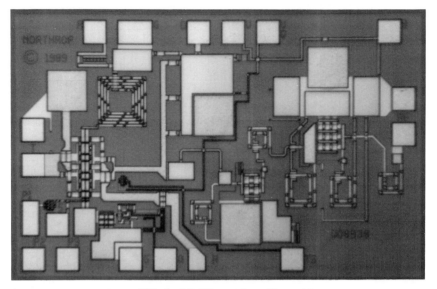

FIG. 9 MMIC quench oscillator chip.

The chip was fabricated using an ion implantation process. The FET devices utilized 0.5-μm gate geometry which was defined using a lift-off technique. The MMIC contains 1.4 mm of FET periphery, 56 pF of MIM capacitance, 20 nichrome and implant resistors, 6 rectangular inductors, and 2 n+ implant diodes. Wet etch backside vias are used for low inductance grounds, and the chip is passivated to improve reliability and yield. An on-chip heater circuit, used to control temperature, utilizes nichrome resistors for adequate current handling and temperature stability.

Two versions of the quenchable oscillator were built using a single MMIC chip design, including a stabilized DRO and a wideband VCO. The operation of the quench circuit was verified and no spurious oscillations were detected for either type of oscillator in the "OFF" state.

A DRO was assembled using a 10-mil thick alumina substrate with 50-ohm match-terminated transmission line to interface with the MMIC chip. A 5.2 × 2.1-mm dielectric resonator was coupled to the 50-ohm transmission line using the TE_{01} mode. The output frequency of the quench DRO was measured to be 11.583 GHz, with an output power of 11.8 dBm.

The settling time of the DRO was measured using a high-speed frequency acquisition test setup. The DRO output frequency was measured to be within 0.6 MHz of the final frequency only 0.5 μs after the quench control was switched to turn the DRO on. The measured frequency versus time response is shown in Fig. 10.

A wideband VCO was also built using a silicon hyperabrupt chip varac-

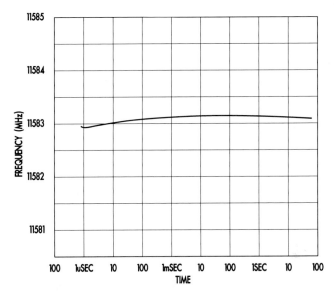

FIG. 10 Quench DRO "turn on" frequency response.

tor diode bonded to the negative resistance port of the quench oscillator MMIC circuit. This VCO was tested over a varactor tuning range of 0 to 20 V. The measured MMIC VCO tuning response is shown in Fig. 11. The tuning range was 9.7 to 14.4 GHz, and the RF power was -10.2 ± 2.4 dBm. The bias requirements for the quench VCO are 55 and 30 mA at 8 V for the oscillator and buffer amp sections, respectively. The quench circuit allows the VCO to be switched on and off in less than 0.5 μs. This result is significant in terms of the VCO's usefulness in agile source subsystem applications.

An important feature of a VCO is good frequency stability over wide temperature variations. Since a wideband VCO has a relatively low Q, its temperature stability is not very good. A heater can be used to maintain a constant chip temperature, thereby achieving frequency stability. A resistive heater in close proximity to the oscillator active device allows the device temperature to be maintained using a minimum of DC power in the heater circuit. The quench VCO was tested over the temperature range of -55 to $+95°C$, and the current through the heater was adjusted to maintain a constant frequency over the full temperature range. The heater current required as a function of temperature is shown in Fig. 12. The relationship is nearly linear from -55 to $\sim+85°C$.

The maximum DC dissipation required was 2.4 W at $-55°C$. This compares favorably against a typical MIC VCO where the complete assembly on its carrier was heated and required up to 10 W of heater power,

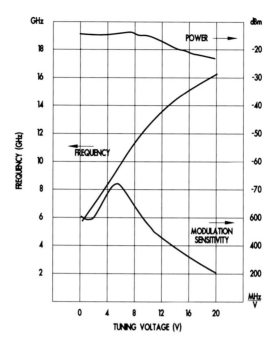

FIG. 11 Quench VCO frequency and power.

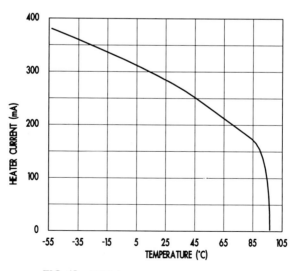

FIG. 12 VCO heater current vs temperature.

representing a 4 to 1 reduction in maximum heater power. Because the MMIC heater is in such close physical proximity to the active device, the time required to stabilize the device temperature is significantly reduced. This heater circuit achieved its final temperature in less than 1 s, compared to 30 s for a VCO on a heated carrier. This would imply that the temperature may be controlled more accurately, resulting in less temperature deviation, and quicker initial "warmup" time.

6.4 CHANNELIZED RECEIVER STRUCTURES

6.4.1 Introduction

Many EW systems use channelized receivers for sorting different threats at different frequencies over a wide radio frequency (RF) band [29–34]. Figure 13 shows a block diagram of a typical microwave channelized receiver.

Signals in the 2- to 20-GHz RF frequency band cover the typical EW frequency bands. Any signal within this range can be input to the receiver. The task of an EW receiver is to determine the frequency of the incoming signal. If the input signal accuracy is to be within 10 MHz, for example, then the full 18-GHz input would need to be divided into 1800 channels. To

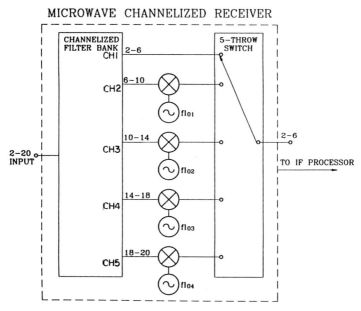

FIG. 13 Channelized receiver block diagram.

avoid this unrealistic number of channels, the front-end of the receiver typically has a bank of filters which divide the input RF signal into smaller bands. These smaller bands are then downconverted to a common IF band and the bandwidth of the smaller RF bands corresponds to the bandwidth of the IF band. The IF signal is then sent to an IF signal processor where the frequency can be determined to the necessary accuracy.

6.4.2 Principle of Operation

The example in Fig. 13 has an IF bandwidth of about 4 GHz. Typical channels in the receiver have mixers which convert the input signal to a 2- to 6-GHz IF. The one exception to this is the first channel which covers the 2- to 6-GHz band. Input signals within this 2- to 6-GHz band are output from the receiver without any frequency conversion. The output of the channelized receiver then is a signal which falls within a 2- to 6-GHz range. A five-pole switch selects the single channel where the signal is located and this signal goes to an IF processor for further processing. If the switch in Fig. 13 is in the position of the third channel, and the local oscillator frequency flo2 is 12 GHz, a 4-GHz signal at the receiver output corresponds to an 8-GHz input signal. If flo3 is 16 GHz and the switch is in the position of the fourth channel, a 4-GHz IF signal corresponds to an input at 12 GHz.

6.4.3 System Requirements

The exact filter bands and local oscillator frequencies must be carefully selected to have the lowest level of spurious signals at the output of mixers. A mixer is a nonlinear device which takes two signals, one at the input RF band (R) and one at the local oscillator frequency (L) and mixes them together resulting in the IF signal of R−L. Unfortunately, this is not the only signal present at the mixer IF port. There is the image signal at R + L as well as all single-tone intermodulation products (commonly referred to as spurious signals) of the form $mR \pm nL$ where $m \geq 0$ and $n \geq 0$. When these signals fall inside the IF band they cannot be filtered and therefore limit the dynamic range of the receiver. Typically, if $m \geq 2$ (for high-quality double-balanced mixers), the spurious signal levels will be low enough to allow the receiver sufficient dynamic range for optimum performance.

6.4.4 MMIC Downconverter

The downconverter sections of the receiver are all basically the same. Each downconverter has a mixer, local oscillator, and an IF amplifier. If the mixer and the local oscillator are made sufficiently wideband, then each downconverter can use the same wideband mixer, oscillator, and IF ampli-

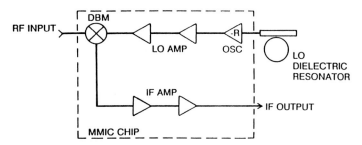

FIG. 14 MMIC downconverter block diagram.

fier. Figure 14 shows an example of a channelized receiver where a single MMIC chip can be used for all the channels where a frequency conversion is necessary.

The photograph in Fig. 15 is a MMIC converter which can be used in an EW channelized receiver (chip size = 2.5 × 2.5-mm). The block diagram of this MMIC is shown in Fig. 16.

The MMIC has a wideband double-balanced mixer, a wideband oscillator, and a three-stage LO buffer amplifier for driving the mixer with sufficiently high LO power for high dynamic range operation. The chip also has a two-stage IF amplifier. Figure 17 shows the conversion gain of the con-

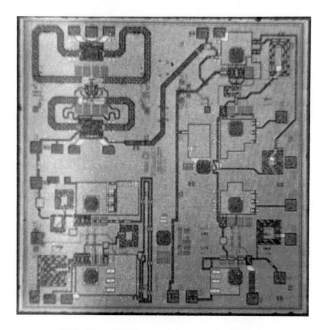

FIG. 15 MMIC converter chip photograph.

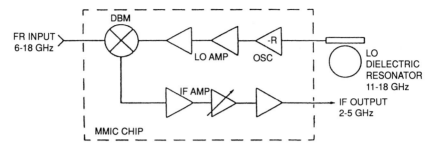

FIG. 16 Converter block diagram.

verter with a local oscillator frequency set at about 12.5 GHz and a swept RF input frequency from 6 to 18 GHz. This MMIC can be used in a channelized receiver like the one shown in Fig. 13 by tuning the oscillator to the desired LO frequency for each of the individual channels.

Because a single chip can be used for different channels by tuning with different off-chip tuning elements, the chip quantities increase, which help drive the cost down. In order to keep the chip yield as high as possible and keep chip cost low the chip size needs to be as small as possible. Figure 18 shows that the "RF good" yield is 65% for a typical chip. For a chip of this complexity this is a good yield, resulting in a chip cost under $50 and a significant cost savings over a discrete component solution.

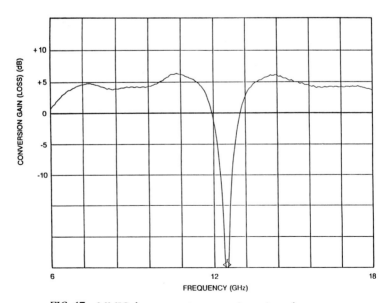

FIG. 17 MMIC downconverter conversion gain vs frequency.

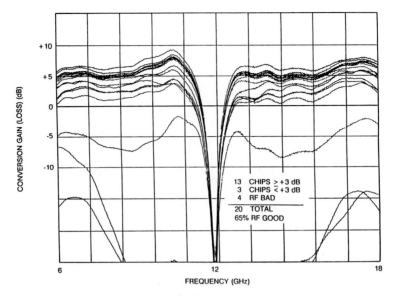

FIG. 18 MMIC converter yield.

Figure 19 shows an entire converter assembly with a single MMIC converter chip. The size of the converter assembly is about 12.7 × 20.3 mm.

By comparison, a discrete component solution would be about 5 times larger. Also because everything is integrated on a single chip the reliability (mean time between failure, MTBF) is increased by a factor of 10 because the majority of bond wires and ribbons are eliminated. In addition, the MMIC converter dissipates ¼ of the power of a discrete design.

6.4.5 Example of a Channelized Receiver

A 5- to 20-GHz microwave channelized receiver used in an EW system currently in production is shown in Fig. 20. One set of the modules pictured achieves the entire 5- to 20-GHz frequency band coverage.

The bandpass filters and couplers realized with thin-film alumina substrates make up the front-end multiplexing section of the receiver. Multiplexing in this fashion establishes excellent receiver selectivity. The 5- to 20-GHz input signal is separated into six channels and then downconverted to a common 2- to 6-GHz IF band. The downconverter is a single MMIC chip featuring a DRO, buffer amplifier, double-balanced mixer, and IF gain stages. Single-tone spurs falling in the IF band are at a level < -55 dBc with -10-dBm RF input. The IF signal is filtered and then processed by successive detection log-video amplifiers shown in Fig. 21.

FIG. 19 MMIC converter assembly photograph.

The function of the log amplifier section is to detect the incoming threat signal strength in any of the specific RF frequency bands. This information, along with the frequency information obtained in the IFM section of the EW system, is essential in performing the radar jamming operation.

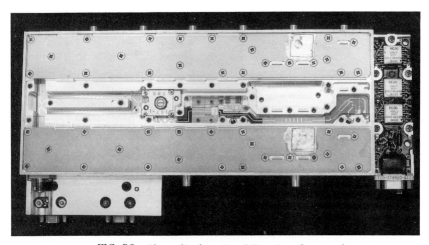

FIG. 20 Channelized receiver RF section photograph.

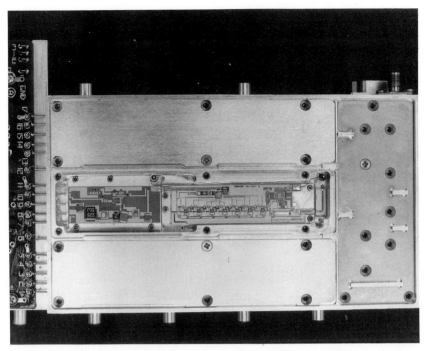

FIG. 21 Channelized receiver video section photograph.

6.5 IFM RECEIVER STRUCTURES

6.5.1 Introduction

An IFM receiver uses delay line phase discriminators (correlators) to measure the frequency of a received signal. IFM receivers feature wide RF bandwidth and the ability to measure the frequency of both pulsed and CW signals. The main attractions of using IFM in EW systems are high-frequency measurement accuracy and resolution with 100% pulse intercept probability instantaneous bandwidths over greater than an octave. Figure 22 shows a block diagram of an IFM receiver used in a typical EW system.

The wideband RF input signal is separated into several RF bands which are converted to a common IF. Further band limitation is achieved by employing IF preselection which reduces the system noise bandwidth and widens system dynamic range. The IF signals are then fed through a high-speed switch into the IFM section of the receiver to determine the exact frequency of the received signal. The frequency information obtained from the IFM section along with amplitude information obtained using detector log video amplifiers (DLVA) is used to synthesize the transmitted output signal of the EW system.

Up through the early 1980s [35–39], classical IFM receiver types had

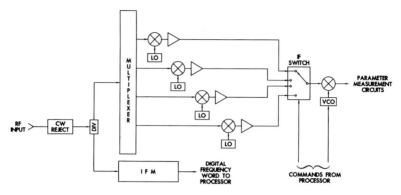

FIG. 22 IFM receiver in EW system.

been used in EW systems to monitor relatively low-density radar environments. Improved performance in modern-day IFM receivers has been achieved by advances in the design and fabrication of MMICs as well as the application of digital signal processing (DSP) techniques to multiple discriminators. In the case of IFM receivers, many identical phase discriminators are required to make up a high-performance system, making the use of MMIC circuits extremely attractive. MMIC technology is inherently repeatable because the symmetry required to achieve desired performance in a balanced circuit is fixed in the wafer mask. The absence of assembly interconnects eliminates unwanted and nonreproducible bond wire parasitics that limit circuit performance.

6.5.2 Phase Discriminators (Correlators)

Figure 23 shows a photograph of a basic frequency measurement circuit prototype assembly.

The received signal is divided into two paths: a delayed path and an undelayed (reference) path. Referring to Fig. 24, assume that the RF input signal is $\sin(\omega t + \theta)$, where ω is the angular frequency and θ is its phase angle. At the phase discriminator input, the two signals are $\sin(\omega t + \theta)$ and $\sin(\omega t - \omega \tau + \theta)$, where τ is the delay time introduced with a delay line in one arm of the power divider. Each signal is fed into a phase correlator which has two voltage outputs, I and Q, representing the in-phase and quadrature signals, respectively. The comparison of I and Q voltages is accomplished using high-speed differential comparators followed by digital logic circuits which decode the comparator outputs (see Fig. 25). The logic network combines the digital code and provides an accurate unambiguous binary code representing the received signal frequency.

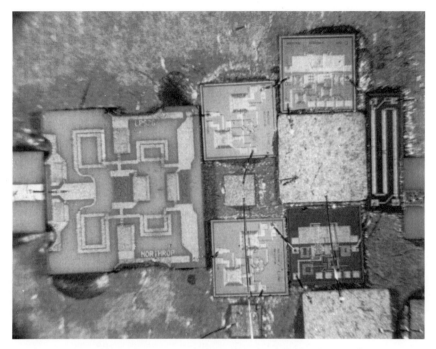

FIG. 23 Frequency measurement circuit assembly photograph.

Many phase correlator approaches can be used in IFM design. The IFM element shown in Fig. 25, using mixers as phase detectors, can be easily implemented using MMIC techniques. Since circuit mismatches introduce phase errors, reducing VSWR effects in the phase correlator circuit results in improved performance. This makes MMIC technology very suitable for phase correlator designs. Near perfect layout symmetry in monolithic mixers can be achieved and this leads to high-performance double-balanced mixers applicable to phase correlator designs.

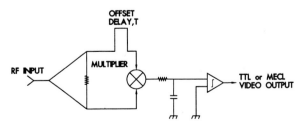

FIG. 24 IFM element—mixer as Ø-detector.

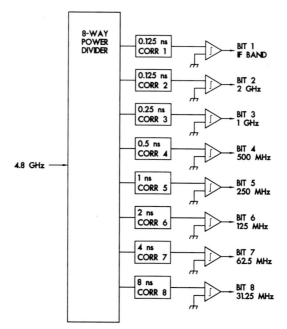

FIG. 25 Digital logic and decoder circuitry.

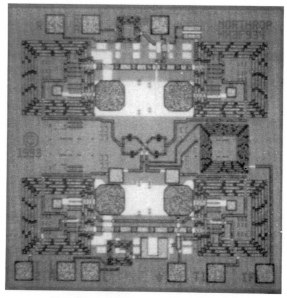

FIG. 26 MMIC double-balanced mixer.

Figure 26 shows a photograph of a MMIC dual double-balanced mixer chip which is ideally suited for this application. The circuit is fabricated on a 100-μm-thick semi-insulating GaAs substrate. It features Schottky diode mixer quads and on-chip, wideband, passive baluns on the LO and RF ports, as well as an IF bandwidth covering DC to 6 GHz. The mixers have low conversion loss (~7 dB) and high spurious signal suppression which is useful for other wideband IF applications. The LO and RF ports of each mixer are driven with the same 2.5- to 5.0-GHz signal, resulting in a DC IF output, eliminating the need for separate detectors (which would be required in a classical IFM element).

6.5.3 Limiting Amplifiers

Amplitude mismatch prior to the phase detection circuitry in an IFM receiver will cause errors in the frequency measurement [40–42]. The amplitude mismatch is reduced by using an RF limiting preamplifier at the front-end of the IFM section. The limiting amplifier provides a constant amplitude signal to the input of the frequency measurement circuit. The gain required of the limiting amplifier should be high enough to boost the minimum detectable signal level into saturation. In achieving limiting, the RF waveform shape is clipped, which produces an output having a strong harmonic content. If clipping of the output signal is symmetrical, the waveform would contain odd harmonics only and a bandwith greater than an octave could be covered. In practical limiting amplifiers, achieving symmetrical clipping increases amplifier alignment and tuning time considerably. Therefore, the bandwidth of most IFM receivers is limited to slightly less than an octave so that filtering can be used to suppress second harmonics. An important advantage of using a limiting amplifier in an IFM receiver is the inherent suppression of weak signals under limiting conditions when two or more signals arrive at the input simultaneously. The limiting amplifier reduces the input signal-to-interference ratio due to the balancing effect of other intermodulation components generated by the limiting process. Figure 27 shows a MMIC limiting amplifier sub-assembly currently used in production.

Since the phase correlators sum the component amplitudes vectorially, the resultant phase error is reduced. A typical IFM system is tolerant of about a $\pm 5°$ phase error while still producing accurate frequency measurement.

GaAs MMIC technology is suitable for implementing a limiting amplifier design because of overall small size, reduced power consumption, high reliability, and repeatable performance. Reduced VSWR effects due to elimination of bond wire interconnects is always an advantage of MMIC circuits which leads to performance higher than that of the discrete hybrid circuit approach. Figure 28 shows a block diagram of a MMIC limiting amplifier which is currently in high-volume production.

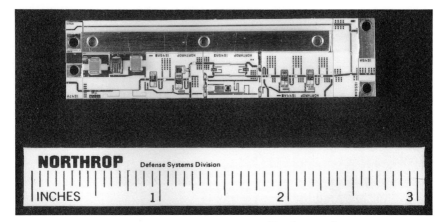

FIG. 27 Limiting amplifier assembly photograph.

The amplifier has 30 ± 0.5 dB small signal gain in the 2.5- to 5.0-GHz frequency range. The small signal gain drift versus temperature is compensated to $<\pm 0.5$ dB from -55 to $+95°C$.

Figure 29 shows a graph of the small signal gain of 25 production assemblies in the "as built" condition, emphasizing excellent circuit performance and repeatability. Figure 30 shows a graph of the limiting characteristics of the amplifier versus frequency under high drive conditions. The saturated output power versus frequency falls within a 2-dB window over a wide input power range. The limiting amplifier described can easily be tailored for use in many IFM applications.

6.5.4 Passive Components

Lange couplers and Wilkinson power dividers can also be realized on a GaAs MMIC substrate. Figures 31 and 32 show these passive components

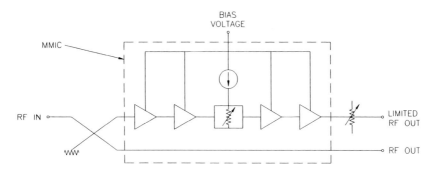

FIG. 28 Limiting amplifier block diagram.

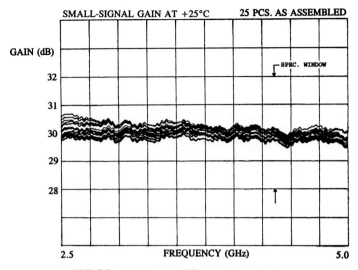

FIG. 29 Limiting amplifier small signal gain.

which can be incorporated into a phase correlator design. Advantages are small size (0.64 × 1.27 mm) and excellent symmetrical repeatability, with the drawback being slightly greater insertion loss than their MIC counterparts.

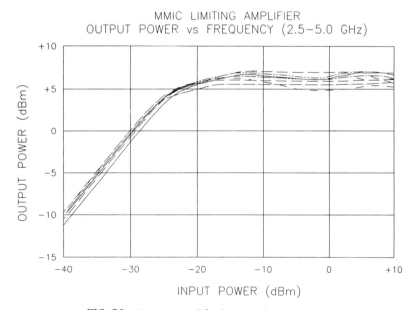

FIG. 30 Limiting amplifier limiting characteristics.

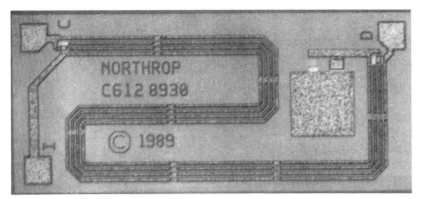

FIG. 31 MMIC Lange coupler photograph.

6.6 INTEGRATED EW TRANSMIT/RECEIVE MODULES

6.6.1 Introduction

Many EW systems require a high performance front/back-end covering all or part of the 2- to 18-GHz band. ECM systems, radar warning receivers, unmanned vehicles, decoys, expendables, and a variety of other EW systems and platforms are placing increasingly stringent requirements on the size, weight, power consumption, and reliability of the front-end. An effective method of meeting these requirements is to make extensive use of MMICs to allow packaging of major sections of the front/back-end in each of a small number of MIC modules. This can provide a significant savings in

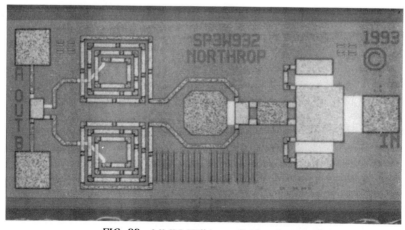

FIG. 32 MMIC Wilkinson divider photograph.

space otherwise required for connectorization and mounting hardware, in addition to the space saved by using the inherently smaller MMIC chips themselves. In addition to reducing size, simplifying assembly, enhancing reliability, and reducing power consumption, the use of MMICs results in improved RF performance. In mixers, the excellent balance and reduced parasitics of the MMIC implementation reduce the level of single tone spurious products and the level of LO leakages [43]. In the development of high-frequency VCOs, wider tuning bands can be achieved using MMIC technology due to the elimination of critical bond wire parasitics. Several state-of-the-art MIC modules, with up to 100 MMICs per module, have been designed and fabricated using this approach. These modules can be combined to meet various EW system front/back-end requirements.

The MIC/MMIC module discussed in this section was designed to form a high-performance core component for these systems.

6.6.2 Agile Receive/Transmit Module (ARTM) System Overview

Figure 33 presents the block diagram of one potential EW system front/back-end. Any input signal in the 2- to 18-GHz band can be converted to a 700-MHz IF where a variety of signal processing techniques may be efficiently applied. The signal can then be reconverted up in frequency for transmission. In the implementation shown, blocks of frequencies in the 2- to 18-GHz input band are first converted to a common 6.8- to 10.7-GHz band in a block converter module. The ARTM then provides conversion between the 6.8- to 10.7-GHz band and the 700-MHz IF. The only variable frequency LO required in this scheme is the first LO in the ARTM module. Fast tuning is provided by the use of a high-speed synthesizer for this LO. The MMIC-based ARTM module, pictured in Fig. 34, is discussed here.

6.6.3 ARTM Architecture

The ARTM includes dual converted receive and transmit chains, LO distribution chains, and the microwave portion of the frequency synthesizer which is used as the tunable first LO. The 19.1- to 23-GHz output from this synthesizer tunes the desired frequency from the 6.8- to 10.7-GHz RF input band into the 12.3-GHz first IF. The 12.3-GHz first IF is converted down to the 700-MHz second IF with the use of a fixed 13-GHz second LO. This conversion sequence is reversed in the transmit chain. High-speed phase-locking circuits allow the synthesizer to tune for the desired signal in under 5 μs.

As shown in Fig. 33, the 19.1- to 23-GHz synthesizer output is generated from the doubled output of a VCO which can be phase locked to a 48- to 78-MHz DDS or other suitable reference. To translate the doubled VCO

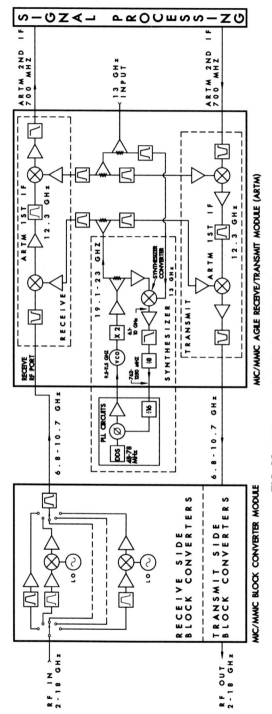

FIG. 33 EW system front/back-end block diagram.

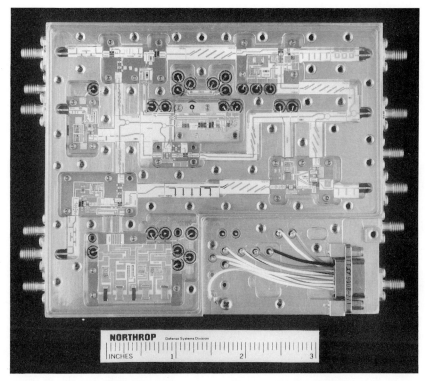

FIG. 34 Agile receive/transmit module photograph.

output down in frequency for phase locking, the 19.1- to 23-GHz band is first converted to a 6.1- to 10-GHz band by mixing with the same 13-GHz input used for the second LO in the receive and transmit chains. The 6.1- to 10-GHz band is then divided down to a 762- to 1250-MHz band using a divide-by-8 MMIC frequency divider. This 762- to 1250-MHz synthesizer IF is further divided down to the reference frequency using an external divider. Differing external components and reference sources can be used as required to provide the desired phase noise and close-in spurious characteristics.

The architecture of the ARTM was selected to allow module size to be kept to a minimum while providing excellent performance in all key areas. The selected frequency translation scheme uses relatively high LO and IF frequencies to keep the filters small, while avoiding requirements for a tunable preselector or switched filter bank at the input. High isolations between the receive and transmit chains are achieved using simple filters in the LO paths due to the separation of the RF and IF frequencies from the LO frequencies. Selection of the conversion scheme in the receive and transmit chains was heavily based on keeping the higher level mixer single tone

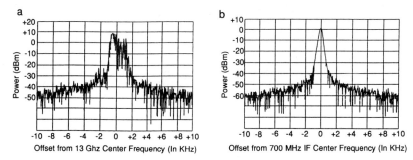

FIG. 35 (a) "Noisy" signal used as 13-GHz Source. (b) Received signal frequency stability when signal shown in Fig. 35a is used as the 13-GHz Input.

spurious products out of the IF bands and on minimizing LO leakages. Amplifiers and other components were selected and positioned in the RF chains to maximize dynamic range while minimizing power consumption.

An important feature of the ARTM architecture results from the use of the 13-GHz input as both the second LO and as a source in the synthesizer downconverter. Injection of the 13-GHz signal into the synthesizer converter causes the frequency stability characteristics of this signal to be superimposed on the 19.1- to 23-GHz first LO under phase-locked conditions. These characteristics are therefore injected into both mixers of the dual conversion receive (or transmit) chain, but with phasing that results largely in cancellation. This scheme allows the frequency stability of the received and transmitted signals to be much better than that of the 13-GHz source itself. Figures 35a and 35b show a comparison of the frequency stability of the received signal versus the frequency stability of the source used as the 13-GHz input. For this test a high-stability 7.1-GHz signal was input at the receive side RF port of the ARTM and the converted output was viewed at the 700-MHz second IF.

6.6.4 ARTM Construction—Dual ARTM

The MIC/MMIC circuits comprising one ARTM are all located on the visible face of the housing shown in Fig. 34. The back face of the housing is identical to the front face so that two ARTMs are available in the 4.6" × 3.9" × 0.9" package. The housing is machined from aluminum and incorporates soldered hermetic feedthrough pins and laser-welded covers to provide hermeticity. Field replaceable SSMA connectors are used on the RF ports to reduce connectorization area.

There are a total of 72 MMICs and 28 filters in the two-sided module. MMICs and other active components are mounted on screwed downcarrier plates for repairability, while filters and interconnects are attached directly to the housing with flexible epoxy. Voltage regulators are included in the

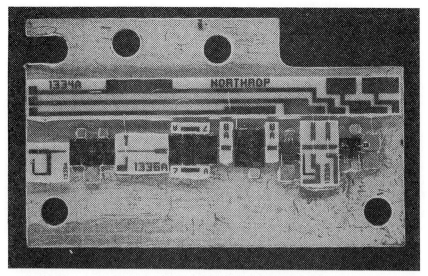

FIG. 36 VCO/doubler assembly photograph.

module to protect the MMICs from supply transients and to keep VCO pushing low. Filters were fabricated on alumina except for the narrowband 12.3-GHz IF filter. This filter was fabricated on a temperature-stable ceramic Al_2O_3 and exhibited under 3 MHz drift in center frequency between +25 and +95°C versus the 53-MHz drift measured for the same filter on 99.6% pure alumina.

A typical MMIC carrier assembly used in the module is shown in Fig. 36. This 7-MMIC assembly is the 19.1- to 23-GHz source used in the synthesizer. The 19.1- to 23-GHz output is obtained by doubling the output from a 9.5- to 11.5-GHz MMIC VCO. Doubling is accomplished by mixing the VCO output with itself as shown in Fig. 37.

A number of MMICs were designed for use in the ARTM including the VCO, mixer, 19- to 23-GHz amplifier, and power splitter chips used on the

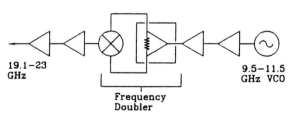

FIG. 37 VCO/doubler block diagram.

assembly shown in Fig. 36. The VCO MMIC provides negative resistance over a 9- to 16-GHz band. An external varactor/inductor tank circuit is used to tune the 9.5- to 11.5-GHz band required for this application. As mentioned earlier, wider tuning bands can be achieved using MMIC technology due to reduced parasitics. This was exhibited in the parallel development of MMIC and discrete fundamental 19- to 23-GHz VCOs to replace the VCO/doubler assembly in the ARTM. The MMIC VCO was able to tune 18.6–25.5 GHz, about twice the band of the discrete VCO, when both used the same varactor/inductor tank circuit. The 19- to 23-GHz amplifier provides 9 dB of gain and over 16 dBm of saturated output power. The mixer is a double-balanced passive design with an 8- to 23-GHz RF band and a 0- to 11-GHz IF band. Mixer conversion loss averages about 9 dB. This mixer is also used for conversion between the 6.8- to 10.7-GHz RF band and the 12.3-GHz IF in the receive and transmit chains. The two-way passive power splitter MMIC exhibits 1.5- to 2-dB insertion loss, 14-dB output port isolation, 0.3-dB amplitude tracking, and 7.5° phase tracking from 6 to 20 GHz.

6.6.5 ARTM Measured Performance

The receive side response of the ARTM is shown in Fig. 38, with the first LO tuned to several discrete frequencies. The bandpass response exhibited in this figure is determined by the 12.3-GHz IF filter which has a 3-dB bandwidth of 320 MHz. If desired, a narrowband filter may be added at the 700-MHz second IF port to provide greater selectivity or reduce total noise power.

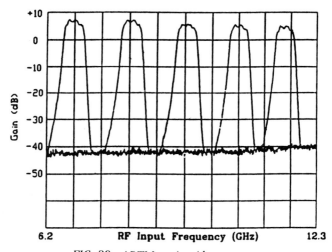

FIG. 38 ARTM receive side response.

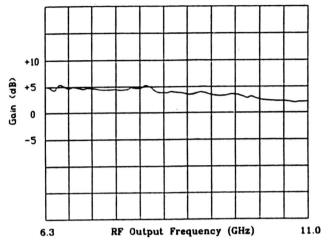

FIG. 39 ARTM transmit side response.

The transmit side response of the ARTM is shown in Fig. 39. A 700-MHz input signal is tuned across the 6.8- to 10.7-GHz band at the transmit side output by sweeping the first LO across its band. Table 8 shows additional measured ARTM performance.

6.6.6 Wideband Agile Sources

Today's EW systems require wideband VCOs to handle threats in multiple frequency bands. Current 5- to 20-GHz MIC VCO subsystems typically use four narrowband VCOs to cover the full band. An alternate

TABLE 8 ARTM Measured Performance

Parameter	Measured
Receive side gain	5–8 dB
Transmit side gain	2–5 dB
IF bandwidth	320 MHz
Noise figure—receive	18.5–20.5 dB
Single tone spurious—receive (−5 dBm in)	−60 dBc
OIP_3 (two tone)—receive	15 dBm
LO leakage to RF output	−74 dBm Max
RF in to RF out isolation	70 dB
Size	4.6″ × 3.9″ × 0.9″
Weight	1.1 lbs
Power consumption	22 Watts

approach uses three VCOs and a doubler to cover the same bandwidth. The demanding specifications for frequency drift, tuning linearity, and channel isolation require multiple VCO architectures which result in complex microwave assemblies. Also, controlling frequency drift due to temperature variations requires using separate proportional heaters, making the multiple VCO subsystem even more complex.

A simple MMIC-based approach which meets general purpose wideband EW source requirements uses a single VCO. This approach in combination with on-chip heaters would allow for temperature stabilization while reducing overall size and complexity [44–46].

6.6.7 Wideband Millimeter-Wave VCO

Several successful techniques have been reported in the literature demonstrating significant VCO tuning bandwidth [47,48]. Most of these techniques concentrate on only the direct VCO tuning bandwidth.

The present approach combines multiplied and heterodyne frequency conversion schemes, implemented with MMICs to achieve even wider tuning bandwidth than could be achieved with a VCO alone. A block diagram of the system is shown in Fig. 40.

The method starts by designing a Ku-band VCO which tunes from 15 to 20 GHz using an external silicon varactor. A silicon diode is used due to its low post-tuning drift characteristics. The VCO is followed by an on-chip frequency doubler to provide an output frequency of 30–40 GHz. Using a microstrip edge coupled filter, the output from the doubler is then filtered rejecting the VCO fundamental signal. The filtered 30- to 40-GHz doubled output is then downconverted using a wideband millimeter-wave MMIC mixer. The local oscillator used for the downconversion is a MMIC metal

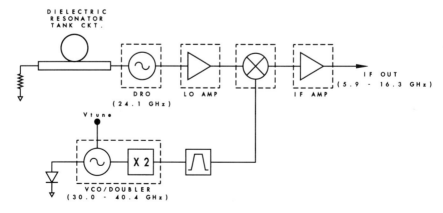

FIG. 40 Voltage-controlled source block diagram.

semiconductor field effect transistor (MESFET) oscillator operating at 24.1 GHz which uses a dielectric resonator for good frequency stability and phase noise. The DRO output drives a K-band two-stage MMIC amplifier which increases the output level to that required for the mixer to achieve low conversion loss. The IF output of 5.9–16.3 GHz from the mixer is the desired output band. By choosing the LO frequency the output band can be shifted up or down in frequency to meet a wide variety of needs. This oscillator topology results in a source capable of a bandwidth exceeding an octave from a single RF port without any need for band switching multiple VCOs.

6.6.8 VCO Circuit Design

A high-frequency VCO chip which utilizes 0.5-μm MESFET technology to implement a wideband negative resistance oscillator and an integrated buffer–doubler circuit was developed. The oscillator active device was modeled using a small signal equivalent FET circuit. A linear S-parameter analysis program was used to predict the region of negative resistance and optimize this parameter over the range of 15–25 GHz. A 150-μm gate width FET was chosen to enhance the upper frequency capability of the oscillator. A common-source topology with capacitive feedback was used to provide the necessary negative resistance. The feedback was set by using MMIC bond pads which can be connected with a wire bond to adjust the value of feedback capacitance. Since the capacitance required is typically .05 to .15 pF, this is an accurate way to control feedback. The capacitance of the bond pad is less sensitive to process variation than is of the capacitance of MIM capacitors.

Wideband tuning of the VCO is achieved with the use of an external discrete silicon hyperabrupt-doping-profile tuning varactor, which is wire bonded to the MMIC VCO/doubler chip. The gate of the oscillator device resonates with the connecting bond wire and varactor to provide the wide tuning range. Both the oscillator and the varactor bias circuits are integrated onto the MMIC chip to minimize additional external component requirements. Since the varactor is wire bonded to the MMIC the operating band can be adjusted either up or down. The distance between the varactor and the MMIC VCO chip was set to 25 mils to provide a 15.0- to 20.2-GHz fundamental tuning range. A distance of 15 mils produced fundamental output frequencies up to 24.9 GHz.

The VCO chip shown in Fig. 41 uses a FET buffer amp which provides a dual function of load isolator and frequency multiplier. The buffer amp consists of a 75-μm FET biased at I_{DSS} which provides a strong second harmonic due to the forward conduction of the gate during positive signal excursions. The small device periphery also allows the buffer amplifier to saturate at a low drive level estimated to be about +6 dBm. The buffer amp was intentionally designed with small gate width to allow the buffer amp to

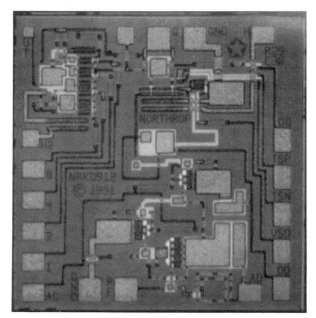

FIG. 41   MMIC VCO chip photograph.

act as a limiter to flatten output power at the saturated output level. This results in good power flatness at the second harmonic. The VCO chip was tested alone and its performance is shown in Fig. 42.

Using the second harmonic, output frequencies up to 50 GHz have been achieved at -10 dBm. The VCO chip requires a single 5-V supply and draws 48 mA of current. The MMIC VCO chip measures 46 × 46 mils (1.2 × 1.2 mm).

6.6.9 Millimeter-Wave MMIC Mixer Design

A wideband MMIC-millimeter wave double-balanced mixer was designed and fabricated to cover an RF band from 15 to 50 GHz and an IF band from 3 to 21 GHz. The actual MMIC chip, shown in Fig. 43, uses two distributed baluns for the RF and LO ports respectively, MIM capacitors for the IF diplexing circuit, and 4 interdigited diodes in a ring quad configuration.

The diodes have a junction capacitance of 0.05 pF and a series resistance of 20 ohms, providing an average conversion loss of 12 dB. The double-balanced configuration gives 20 dB LO to IF isolation and 30 dB RF-to-IF isolation as well as very good suppression of high-order spurious products. The wide bandwidth of the IF port makes it possible to downcon-

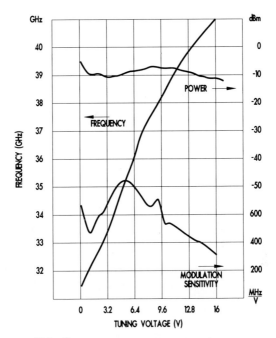

FIG. 42 MMIC VCO measured performance.

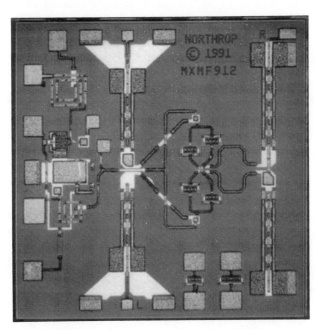

FIG. 43 MMIC millimeter-wave mixer chip.

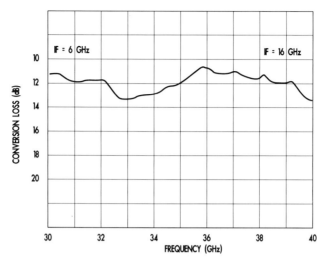

Measured conversion loss of MMIC mixer using an LO at 24 GHz with +10 dBm drive level.

FIG. 44 MMIC millimeter-wave mixer performance.

vert a millimeter frequency band into the full 6- to 18-GHz microwave band. The conversion loss of the mixer, shown in Fig. 44, was measured over the RF band of 30 to 40 GHz using an LO frequency of 24 GHz. The MMIC mixer chip size is 46 × 71 mils (1.2 × 1.8 mm).

6.6.10 DRO Circuit Design

A DRO was designed using a 150-μm MESFET which delivers a typical output power of 8 to 10 dBm at 24 GHz. A series capacitive feedback topology provides a wideband negative resistance suitable for resonating with an external 50-ohm microstrip line which is coupled to a dielectric resonator for frequency stability. The MMIC DRO chip measures 46 × 46 mils (1.2 × 1.2 mm). An 18- to 26-GHz LO amplifier was used to provide buffering and increase the LO drive signal of the DRO.

6.6.11 Wideband Agile Millimeter-Wave Source Performance

A working prototype consisting of the VCO, mixer, DRO, and buffer amplifier chips, mounted onto a metal carrier plate, was built and tested. A photo of the VCO assembly is shown in Fig. 45.

All the MMIC chips have been designed to operate at 5 V and the total current is 375 mA. The MMIC subassembly has been measured for RF frequency and power output. The results are shown in Fig. 46. The subassembly demonstrates a 2.76:1 tuning ratio, which compares favorably to previously reported work.

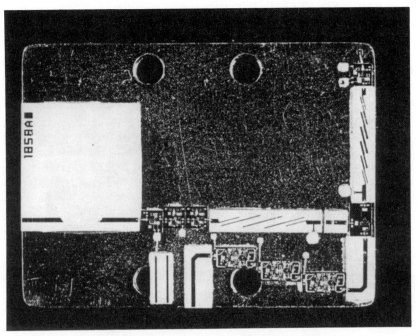

FIG. 45 Wideband VCO assembly photograph.

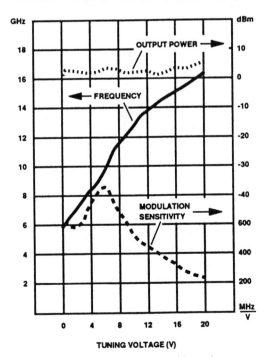

FIG. 46 Wideband VCO assembly performance.

6.7 OTHER EW RECEIVER TYPES

Previous sections of this chapter reviewed the crystal video, channelized, superheterodyne, and IFM receiver types. The application of MMIC technology to these receivers has also been emphasized. Other types of receivers currently in use in EW systems are compressive and acoustooptic [52]. The use of MMIC technology in these types of receivers is not yet well established.

The compressive receiver or "microscan" receiver is a scanning superheterodyne receiver. The RF bandwidth is swept in a period which is less than the pulse duration of the received signal. This type of receiver has high dynamic range but has bandwidth limitations which may render the receiver useless for certain pulse trains.

Acoustooptic (AO) receivers perform signal processing using Bragg cells. AO receivers are the newest receiver type and are one of the most dynamically changing types because of this. A laser beam illuminates the Bragg cell (transducer) with coherent light which deflects the light in proportion to the received signal through a lens onto a photo-diode detector. The bandwidth of a Bragg Cell receiver is limited by the acoustic attenuation and diffraction characteristics of the Bragg cell. In general, AO receivers are currently in a developmental phase and are not used widely in production EW systems to date.

6.8 SUMMARY AND FUTURE TRENDS

The concept of EW is an important part of all military strategy in today's world. Continually changing threat scenarios make the field of EW very dynamic. EW is broken into three very broad categories—ESM, ECM, and ECCM. This chapter has reviewed the various types of receiver architectures which are most commonly used in modern EW systems.

MMIC technology is being utilized in EW systems now in development and promises to have a profound impact on future EW systems [53]. Upcoming trends for EW systems include higher performance, smaller size, and lower cost systems. MMIC technology offers all of these advantages. Higher levels of integration such as combining amplifiers, oscillators, and mixers on a single chip can be readily accomplished as shown in this chapter. Implementation of receiver functions on a single device directly reduces subsystem cost and increases reliability. This in turn has a significant impact on reduction of overall system size. Higher levels of circuit integration also leads to higher performance systems by elimination of unwanted parasitics which degrade performance of discrete hybrid type circuits. The ability to produce a smaller size system allows any digital processing functions to be performed in closer proximity to the sensing device (receivers), thus reducing RF leakages and RF ringing in the aircraft or ship.

Signal intelligence (SIGINT) encompasses all communications intelligence (COMINT), electronics intelligence, and telemetry intelligence. The hardware needed to gather this intelligence information consists of antennas, receivers, and digital processors. EW systems of the future will be impacted by MMIC technology mainly in the areas of receivers and digital processing. Receivers will become smaller and lightweight through the implementation of MMICs. Digital processing of signals will also play a major role in the evolution of future EW systems. More and more analog functions, such as demodulation of information on a signal, will become digital functions as the advancement of high-speed A/D converters progresses. The identification of threats will require more advanced digital signal processing techniques as the hostile environment signal densities increase. In general, the new breed of SIGINT receivers will one day consist of a tiny receiver section along with a very sophisticated digital processor. MMIC technology will be critical to the realization of these advanced receivers.

REFERENCES

[1] J. B. Tsui, *Microwave Receivers and Related Components*, Chap. 1. Avionics Laboratory, Air Force Wright Aeronautical Laboratories, NTIS, 1983.
[2] M. I. Skolnik, *Introduction to Radar Systems*, Chap. 10. McGraw–Hill, New York, 1962.
[3] R. Goyal, *Monolithic Microwave Integrated Circuits*, Chap. 1. Artech House, Dedham, MA, 1989.
[4] D. Curtis Schleher, *Introduction to Electronic Warfare*, Chap. 1. Artech House, Dedham, MA, 1986.
[5] J. Boyd, *Electronic Countermeasures*. Penninsula, Los Altos, CA, 1965.
[6] L. Van Brunt, Applied ECM. In *EW Engineering*, Vol. 1. EW Engineering, Dun Loring, VA, 1978.
[7] D. Steer, *Airborne Microwave ECM*. Military Microwave Conference, London, 1978.
[8] P. Grant, and J. Collins, Introduction to Electronic Warfare. *IEE Proc.* **129**(F) (3), June 1982.
[9] J. B. Tsui, *Microwave Receivers and Related Components*, Chap. 3. Avionics Laboratory, Air Force Wright Aeronautical Laboratories, NTIS, 1983.
[10] W. E. Ayer, *Characteristics of Crystal Video Receivers Employing RF Pre-amplification*, Technical Report 150-3, Stanford Electronics Laboratories. Stanford University, Sept. 20, 1956.
[11] L. Klipper, Sensitivity of crystal video receivers with RF pre-amplification, *Microwave J.* **8**, 85–92, 1965.
[12] J. C. Harp, What does receiver sensitivity mean? *Microwave Systems News*, 54, July 1978.
[13] D. L. Adamy, Calculate receiver sensitivity. *Electronic Design*, 118, Dec. 6, 1973.
[14] S. E. Lipsky, Calculate the effects of noise on ECM receivers. *Microwaves* 65, Oct. 1974.
[15] J. B. Tsui, *Microwave Receivers and Related Components*, Chap. 3. Avionics Laboratory, Air Force Wright Aeronautical Laboratories, NTIS, 1983.
[16] C. E. Dexter, and R. D. Glaz, *HF Receiver Design*, Tech Notes, Vol. 5, No. 2. Watkins Jonson Co., March/April 1978.
[17] R. C. Dixon, *Spread Spectrum Systems*. Wiley, New York, 1976.
[18] H. Fudem et al., A high performance 6 to 18 GHz MMIC converter chip for EW systems. *IEEE GaAs IC Symp.* 1990.
[19] S. Maas, *Microwave Mixers*, Chap. 7. Artech House, Dedham, MA, 1993.

[20] R. A. Pucel, Design considerations for monolithic microwave circuits. *IEEE MTT-S* **MTT-29**(6), June 1981.
[21] G. Garbe et al., A 6.8 to 10.7 GHz EW module using 72 MMICs. *IEEE MTT-S Dig.* 1329–1332, 1993.
[22] C. L. Ruthroff, Some broadband transformers. *Proc. IRE* **47**, 1337, 1959.
[23] G. L. Matthai, L. Young, and E. M. T. Jones, *Microwave Filters, Impedance Matching Networks, and Coupling Structures*. Artech House, Dedham, MA, 1980.
[24] B. R. Hallford, A designer's guide to planar mixer baluns. *Microwaves* Dec. 52, 1979.
[25] H. Fudem et al., A high performance 6 to 18 GHz MMIC converter chip for EW systems. *IEEE GaAs IC Symp.* 1990.
[26] S. Moghe et al., High performance GaAs C-band and Ku band MMIC oscillators. *IEEE 1987 Int. Microwave Symp. Dig.* 911–914.
[27] B. Scott, M. Wurtele, and B. Cregger, A family of four monolithic VCO MIC's covering 2-18 GHz. *Microwave Millimeter Wave Monolithic Circuits Symp.* 58–61, 1984.
[28] A. P. S. Khanna, R. T. Oyafuso, R. Soohoo, and J. Huynk, Microwave quenchable oscillators—A new class. *IEEE MTT-S Int. Symp. Dig.* 515–518, June 1989.
[29] P. Hennessy, and J. D. Quick, The channelized receiver comes of age. *Microwave Systems News* 36, July 1979.
[30] C. B. Hoffmann, and A. R. Baron, Wideband ESM receiving systems. *Microwave J.* Part I, p. 24, Sept. 1980; Part II, p. 57, Feb. 1981.
[31] D. Allen, Channelized receiver, a viable solution for EW and ESM systems. *IEE Proc.* **129**(F) (3), June 1982.
[32] D. Lundy, SAW filters shrink channelized receivers, *Defense Electron.* Sept./Oct. 1977.
[33] J. B. Tsui, *Microwave Receivers and Related Components*, Chap. 5. Avionics Laboratory, Air Force Wright Aeronautical Laboratories, NTIS, 1983.
[34] P. Grant, J. Collins, Introduction to electronic warfare. *IEE Proc.* **129**(F) (3), June 1982.
[35] M. Smirlock, Binary beam IFM: Multiple discriminators improve frequency resolution, *Electron. Warfare* May/June, 1976.
[36] D. Barton and H. Ward, *Handbook of Radar Measurement*. Artech House, Dedham, MA, 1969 (1984).
[37] W. Sullivan, Digital IFM: Intercepting complex signals in a dense environment. *Electron. Warfare* May/June, 1976.
[38] P. East, Design techniques and performance of digital IFM. *IEE Proc.* **29**(F) (3), June 1982.
[39] M. W. Wilkens, and W. R. Kincheloe Jr., *Microwave Realization of Broadband Phase and Frequency Discriminators*, Technical Report 1962/1966-2, SU-SEL-68-057. Stanford Electronics Laboratories, November, 1968.
[40] D. Heaton, The systems engineer's primer on IFM receivers. *Microwave J.* 71, Feb. 1980.
[41] J. B. Tsui, and G. H. Schrick, Instantaneous frequency measurement (IFM) receiver with capability to separate CW and pulsed signals. U.S. Patent 4,194,206, March 18, 1980.
[42] U. H. Gysel, and J. P. Watjen, Wideband frequency discriminator with high linearity. *IEEE MTT-S. Int. Microwave Symp. Dig.* 1977.
[43] K. B. Niclas, Reflective match, lossy match, feedback, and distributed amplifiers: A comparison of multi-octave performance characteristics. *IEEE MTT-S Symp. Dig.* 215–127, 1984.
[44] G. Vandelin, Design of amplifiers and oscillators by the S-parameter method. Wiley, New York, 1982.
[45] A. Sweet, *MIC and MMIC Amplifier and Oscillator Circuit Design*, Chap. 5. Artech House, Norwood, MA, 1990.
[46] G. Dietz, R. Becker, R. Haubenstricker, S. Moghe, and G. Giacomino, A 10–14 GHz quenchable MMIC oscillator. *IEEE Microwave Millimeter-Wave Monolithic Circuits Symp.* 23–26, 1991.

[47] A. Adar, and R. Ramachandran, An HBT MMIC wideband VCO, *IEEE Microwave Millimeter-Wave Monolithic Circuits Symp.* 73–76, 1991.
[48] L. Cohen, A millimeter-wave, third harmonic, Gunn VCO with ultra-wideband tuning. *IEEE MTT-S Int. Microwave Symp. Dig.* 937–938, 1991.
[49] B. Mitchell, European markets dictate the need for better oscillators. *Microwaves RF* 57–60, Aug. 1987.
[50] J. Kitchen, Octave bandwidth varactor-tuned oscillators. *Microwave J.* 347–353, May 1987.
[51] J. Andrews, T. Holden, K. Lee, and A. Podell, 2.5–6.0 GHz broadband GaAs MMIC VCO. *IEEE MTT-S Dig.* 491–494, 1988.
[52] D. Curtis Schleher, *Introduction to Electronic Warfare*, Chap. 2. Artech House, Dedham, MA, 1986.
[53] D. Herskovitz, SIGINT/DF systems of the next century. *J. Electron. Defense* 39–68, Oct. 1994.

7
Instrumentation

Val Peterson
Hewlett-Packard
Santa Rosa, California

7.1 Introduction
7.2 Technology Choices and Merits of Using GaAs ICs
    7.2.1 GaAs vs Silicon in Instrumentation Applications
    7.2.2 Monolithic Microwave ICs (MMICs) vs Hybrid MICs
    7.2.3 GaAs MMIC Cost and Reliability
7.3 Typical Instrumentation Block Diagrams and GaAs IC Opportunities
    7.3.1 Microwave Sources
    7.3.2 RF and Microwave Spectrum Analyzers
    7.3.3 Pulse and Pattern Generators
7.4 Application Examples
7.5 Conclusions and Future Trends
    References

7.1 INTRODUCTION

Instrumentation applications of monolithic gallium arsenide integrated circuits (GaAs ICs) have become commonplace since 1985. Although instruments are produced in relatively small volumes, the performance demands of the instrumentation market have driven a great deal of GaAs IC development activity.

Instruments are invariably required to provide performance which exceeds that of the systems or components to be measured. A microwave source, for instance, might be expected to offer better bandwidth, harmonic distortion, and noise performance than that of the device under test (DUT). Similarly, a pulse generator would be expected to provide faster edge speeds, less jitter, and less overshoot and ringing than the DUT. As a result instruments have been forced to aggressively pursue performance advantages from discrete GaAs field effect transistors (FETs) and subsequently from GaAs monolithic ICs to stay ahead of customer needs.

The advantages and disadvantages of GaAs monolithic ICs will be contrasted with silicon ICs and with hybrid microwave integrated circuit (MIC) assemblies. Three types of instrumentation will be reviewed, with block diagrams presented and opportunities for GaAs IC illustrations discussed. Finally, three examples of successful GaAs IC applications in instrumentation will be presented.

7.2 TECHNOLOGY CHOICES AND MERITS OF USING GaAs ICs

In many instrumentation applications where functions have been realized with silicon monolithic ICs there is now an opportunity to design similar functions using monolithic GaAs ICs. In many microwave instruments the technology of choice has been hybrid MICs using discrete GaAs FETs. The relative merits of using GaAs monolithic microwave integrated circuits (MMICs) are presented.

7.2.1 GaAs vs Silicon in Instrumentation Applications

For many instrument applications the primary technology contenders are GaAs processes using FETs or high electron mobility transistors (HEMTs) and Si processes using bipolar junction transistors (BJTs). For the purposes of this discussion GaAs FETs and HEMTs will mostly be treated interchangeably since most of the advantages and disadvantages discussed apply to both FETs and HEMTs. Heterojunction bipolar transistors (HBTs) in GaAs-based or silicon-based material structures are not discussed here as little real experience with instrument applications is available. Early data suggests, as one might expect, that GaAs-based HBTs resemble Si BJTs much more than GaAs FETs in their attributes. This leads to the hope that GaAs HBTs will address some of the shortcomings of GaAs FETs for future high-performance instrumentation.

Table 1 lists a number of attributes and illustrates which technology offers superior performance.

GaAs FETs and HEMTs and Si BJTs continue to make rapid improvements in speed. Device cut-off frequency (F_t) has doubled at least every 5 years and is currently increasing more rapidly. GaAs devices have maintained roughly a factor of two advantage in speed over silicon devices.

In addition to the advantage in raw speed, GaAs devices are better suited to handling large amplitudes at high frequencies. The highest performance silicon bipolar processes have greatly compromised breakdown voltages as the device geometries are scaled downward; 5- or 6-V breakdowns, or even less, are common in such processes. By contrast, GaAs FET or HEMT breakdown voltages commonly exceed 10 V. Whether using these devices to produce microwave power amplifiers with 1 W output power at 20 GHz or using them to produce a pulse amplifier with 5 V output amplitude and sub-100 ps edge speeds, GaAs processes have a unique advantage.

Noise is an area both of strength and of weakness for GaAs FETs and

TABLE 1 Comparison of GaAs FETs
with Si Bipolar Transistors

	GaAs FET	Si BJT
F_t, F_{max}	*	
Breakdown voltage	*	
$1/f$ noise		*
Shot noise	*	
Voltage gain		*
Switches and attenuators implemented easily	*	
High input Zin	*	
Pulse fidelity		*

*, Superior performance.

HEMTs. These devices have notoriously high $1/f$ corner frequencies (high flicker noise). High $1/f$ noise degrades amplifier low-end noise figure (NF) and GaAs FET amplifiers commonly show degraded NF at frequencies as high as 10 MHz. Also, $1/f$ noise is converted to phase noise in oscillators and frequency dividers which gives bipolar transistors some real advantages in these applications. On the other hand, the shot noise of GaAs FETs and especially HEMTs is superior to that of Si BJTs. For many receiver applications a GaAs device is unquestionably the first choice.

GaAs FETs are extraordinarily well suited for use as control elements such as switches or attenuators. GaAs monolithic switches have been a commercial success because over wide bandwidths relatively high isolations can be achieved in the off state with relatively low insertion losses in the on state. Similarly, FETs used as variable resistors in a variety of series-shunt topologies make excellent continuously variable attenuators. Many instrument block diagrams require switches and attenuators for several functions. The superior performance of FETs for these control elements gives GaAs processes a real advantage in system-level integration.

GaAs FETs offer very high input impedances. This can be exploited, for example, in designing a probe for making high-frequency measurements within a component. A simple common-source amplifier design provides high input impedance to avoid loading the node being tested while the output is capable of driving a 50-ohm input impedance measurement instrument.

Voltage gain is a significant disadvantage for GaAs FETs when compared with bipolar transistors at frequencies where both technologies are competitive. For a GaAs FET a self-gain of as little as 10 is common and a practical gain stage may have a voltage gain of only 3 or 4. This precludes use of many design techniques commonly used in precision analog design. Also, for a design requiring a given amount of gain, it will probably be

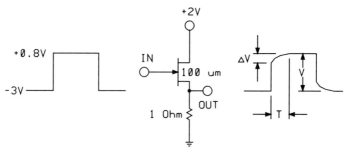

FIG. 1 Incomplete switching transient caused by gate lag resulting from surface traps. In response to a step in the gate-source voltage the drain current shows both a short and a long time constant in its response.

necessary to use more area and dissipated power for the GaAs FET realization than for the comparable silicon bipolar design.

One final area of competitive weakness for GaAs FETs is exposed in time domain circuits where pulses are amplified. When a Class A amplifier implemented in a GaAs FET process is driven with a step function the output current will commonly be observed to switch most of the way very quickly and then to slowly complete the transition with a time constant several orders of magnitude longer than the 20/80 edge speed [1]. This incomplete switching, also known as a lag effect or as a slow tail, can greatly degrade pulse fidelity. Two mechanisms bring about this effect; surface traps create an incomplete switching when the gate-source voltage is stepped, and traps at the substrate/epitaxial layer interface cause incomplete switching when the drain-source voltage is stepped. An example of the gate lag effect is shown in Fig. 1.

Another manifestation of these lag effects is a variable delay in transition which is dependent on the pattern of data; this pattern-dependent edge movement can be a major problem for many applications. Fortunately, optimization of design and of process can minimize these effects.

7.2.2 Monolithic Microwave ICs (MMICs) vs Hybrid MICs

Hybrid microcircuits using discrete transistors and thin or thick film matching circuits have been used in almost all microwave instrumentation for many years. A typical gain stage would require a transistor, blocking capacitors, supply bypass capacitors, and many wire bonds or ribbons used for interconnection and for tuning. A microphotograph of a typical hybrid MIC, in this case a 2- to 7-GHz amplifier, is shown in Fig. 2. This component contains 6 FETs, 6 thin-film circuits, 38 metal–insulator–metal (MIM) capacitors, and 233 wire bonds. It was built in production for over 10 years and required 6 h of assembly time at maturity. In addition, given the com-

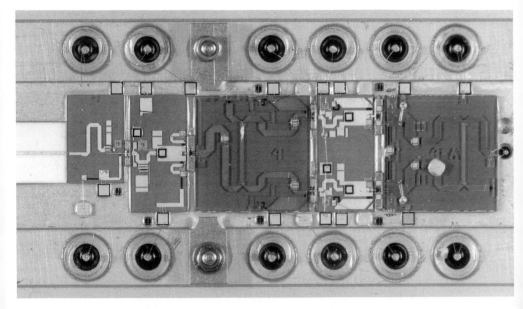

FIG. 2 A hybrid MIC amplifier covering 2- to 7-GHz bandwidth. This hybrid assembly contains 50 components which must be attached and wire bonded.

plexity and occasional need for tuning, 1 h of technician time was required per unit.

The great complexity of such a hybrid illustrates one of the advantages for monolithic devices—significantly simpler assemblies. There are several other compelling reasons to use monolithic ICs as well as some reasons that discourage their use; these are summarized in Table 2.

A monolithic IC can be designed to be quite tolerant of assembly variations. Conversely, in the hybrid MIC the interconnections become a very significant part of the matching structures at high frequencies. As assembly variations result in widely varying bond lengths, the matching structures are greatly changed and circuit performance is affected. For the monolithic IC

TABLE 2 Advantages and Disadvantages of Monolithic ICs vs Hybrid MICs

Advantages of ICs	Greatly reduced component count
	Better tolerance of assembly variations
	Fully pretestable in wafer form
	Potentially improved reliability
Disadvantages of ICs	Longer development times
	Lack of tuning flexibility

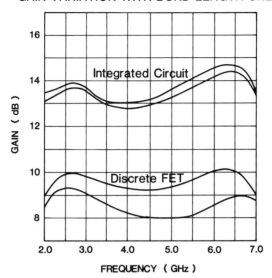

FIG. 3 Gain variation of hybrid MIC and monolithic MMIC amplifiers as bond wire lengths are varied over a ±25% range. The MMIC shows substantially more tolerance of assembly variations since critical matching structures are controlled on-chip.

one can place the most critical matching structures on chip and place the assembly-dependent interconnections in 50-ohm paths off-chip. The resulting improvement in tolerance is illustrated in Fig. 3, which compares the effects of 25% bond wire length variations for a hybrid MIC gain stage and a simple IC gain stage.

Monolithic ICs can also be fully functionally tested in wafer form. This can assure nearly 100% turn-on yield in the final assembly. By contrast, wafer level testing of a discrete FET in a 50-ohm environment is much less predictive of the final MIC performance. Finally, use of monolithic ICs holds the promise of significant reliability improvements. It is empirically found that reliability is a strong function of the number of active components in an assembly. It is also observed that the reliability of a given die is a weak function of the complexity of that component. Therefore reliability is consistently improved when several active components can be replaced with one or more highly integrated components.

A significant disadvantage of designing a new custom MMIC vs a new custom hybrid MIC is the development time. Commonly a thin-film or thick-film circuit can be turned around in a matter of days and an MIC can thick-film circuit can be turned around in a matter of days and an MIC can be assembled very quickly. On the other hand, typical MMIC turnaround times, while improving, are still several months, requiring mask layout and procurement and complex wafer fabrication.

The ability to tune a hybrid MIC is also seen as a significant advanto the structure. Unfortunately, these tuning adjustments are often left in the final product and become a real liability in the manufacturing environment.

In summary, monolithic ICs can offer real cost savings through simplicity of assembly, greater tolerance of assembly variations, and improved turn-on yield. Also, they can improve the reliability of a component. However, development times for custom MMICs compare unfavorably with hybrid assemblies and the lack of tuning flexibility increases development risks.

7.2.3 GaAs MMIC Cost and Reliability

While there are exceptions, in general GaAs MMICs are more costly than Si ICs. Some of the reasons for the cost difference include lower volumes shipped from GaAs facilities, higher starting material costs, and smaller wafer sizes. Although some high-volume commercial parts such as direct broadcast satellite (DBS) low-noise amps have been produced in GaAs for only a few dollars, most instrumentation-grade chips sell in much smaller volumes and at prices from a few tens of dollars to many hundreds of dollars.

The more relevant comparison is at the next (assembly) level, between the cost using GaAs MMICs and the cost to accomplish comparable performance using other technology. As previously discussed, the alternative technology is often a complex hybrid MIC requiring time-consuming assembly, tuning, and test. In one example where a microwave source was converted from hybrid MIC technology to one based primarily on GaAs ICs the redesign achieved a cost reduction of 10% for the complete instrument [2].

Reliability has been excellent in most instruments which are heavy users of GaAs MMICs. At Hewlett-Packard, instrument failure rates attributed to high-frequency components dropped by almost an order of magnitude in the past decade. Ten years ago products such as microwave sources used many discrete FETs and annual component failure rates as high as 2% were observed. Comparable instruments using GaAs MMICs now commonly show component annual failure rates substantially less than 0.25%. Much of the improvement is due to progress in GaAs FET reliability which applies equally to discrete FETs or complex ICs. In addition, however, the reduced parts count in IC-dominated components directly improves end-product reliability.

7.3 TYPICAL INSTRUMENTATION BLOCK DIAGRAMS AND GaAs IC OPPORTUNITIES

Three classes of instruments are considered in this section. Microwave sources are covered in the greatest depth since they have traditionally been the driving force for technology development. Sources also have the largest GaAs device content of any instrument category. Spectrum analyzers are discussed as a representative receiver application. Finally, high-speed pulse and/or data generators are briefly reviewed as representative time-domain instruments.

7.3.1 Microwave Sources

Basic Microwave Source

The essential elements of a microwave source include a tunable oscillator, a modulator, an output amplifier, and a directional detector. Such a source is illustrated in Fig. 4.

Oscillator To be useful as a general-purpose instrument, the oscillator should offer at least an octave of tuning range. The desire for good phase noise from the source suggests a high-Q resonator for the oscillator. The combined needs of wide tuning range and good phase noise usually result in the choice of yttrium–iron–garnet (YIG) as the tunable resonator element. The resonant frequency of YIG can be tuned over several octaves by varying the applied magnetic field.

Most YIG oscillators have used a discrete silicon bipolar transistor or a discrete GaAs FET or HEMT as the negative resistance device [3]. The superior $1/f$ characteristics of silicon bipolar transistors provide significant phase noise advantages over GaAs FETs, but the lower F_t of Si limit the use

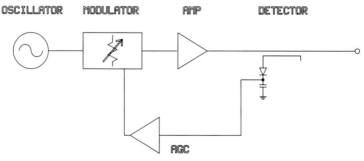

FIG. 4 A basic microwave source. A broadband oscillator, typically a YIG component, is modulated, amplified, and leveled at its output.

of these devices. One or two additional gain stages are added to isolate the oscillator device from load impedance changes; otherwise a change in load impedance would result in a small shift of the oscillator frequency. Additionally, these buffer amplifier stages amplify the output signal and provide higher output power levels. The combination of oscillator device and isolation amplifiers onto a single monolithic die presents an excellent opportunity for Si or GaAs MMICs within their respective frequency ranges. At least one such oscillator design has been successfully completed.

Modulator (Variable Attenuator) The modulator varies output amplitude. This device can be used for (1) AM modulation, (2) pulse modulation, and/or (3) output signal level adjustment. This function has traditionally been realized using p–i–n diodes whose radio frequency (RF) impedance is controlled by applied DC current. These devices offer wide dynamic range and relatively fast adjustment and are available at low cost. Disadvantages of p–i–n modulators include the need for precise positioning and bonding techniques, widely varying port impedances as amplitude is varied (in some topologies), and a strongly nonlinear attenuation vs control current characteristic.

As an alternative to using the variable resistance of a diode, one can use the variable drain-source resistance of an FET to accomplish variable attenuation. Usually some combination of series and shunt FETs are used in such a design. Monolithic GaAs attenuators offer exceptionally broad bandwidth and can be designed to offer linear attenuation vs control voltage characteristics [4]. Generally bandwidth is traded off against on/off ratio since larger device geometries will give lower insertion loss and greater attenuation at the expense of poorer high-frequency performance. A single monolithic attenuator IC might provide up to 50 dB or more of isolation; significantly larger values are hard to achieve due to cross-talk from input to output bond wires.

The primary disadvantage of monolithic GaAs IC attenuators is added distortion at large signal levels. Although the drain-source resistance is fairly linear for small signals at $V_{DS} = 0$, the nonlinearity increases as amplitudes become large and the FET begins to saturate during part of the cycle. This distortion is improved when the FET pinchoff voltage is greatly increased; however, increasing V_p may be incompatible with integration of other active circuits onto a single die unless a process offering multiple implants is used (which permits integration of different optimized FETs). Distortion is typically worst for intermediate levels of attenuation; the fully-on or fully-off states will handle higher power levels for a given degree of distortion. Monolithic switches have been used in several applications for pulse modulation where continuously variable attenuation is not required [5]. In spite of this limitation, GaAs MMICs have been widely applied in both continuously variable and step attenuation applications in instruments.

Output Amplifier The output amplifier is required to provide substantial broadband output power without adding significant harmonic distortion. Maximum output power requirements might fall between 10 mW and 1 W and often must extend over several octaves of bandwidth. Discrete GaAs FETs replaced discrete Si bipolar transistors in the early 1980s, and hybrid amplifiers using FETs are now being widely displaced by monolithic GaAs amplifiers.

Amplification covering the one or two octaves of bandwidth available from a single YIG oscillator is easily accomplished either monolithically or in hybrid MIC form. For this function the choice will often be made based on total cost. The monolithic IC option usually wins since the tolerance to assembly variations is much greater and the number of components to be assembled is much smaller. Also, the wide availability of monolithic microwave amplifiers means that an off-the-shelf option is available for almost any application, accelerating the product development cycle time.

Directional Detector Directional detection is critical for two purposes: (1) providing a specified output power level which is held constant for all frequencies and (2) improving the source match of the instrument. A simple diode detector without a directional coupler can do an acceptable job of providing an indication of output amplitude, in which case the signal is then processed in an automatic gain control (AGC) circuit and fed back to the modulator. However, without a directional coupler, the detector diode output will be strongly affected by the load impedance presented to the source. By operating the diode detector off a directional coupler one can monitor and control the power incident upon the load with minimal errors caused by nonideal load impedances. If the directional coupler is ideal, the AGC loop is unaffected by load impedance and reflections off the load. In this ideal case the instrument's output level is completely independent of the load impedance and the instrument has a perfect source match.

This directional detector function has traditionally been realized by using a broadband directional coupler and attaching a diode detector. A recent development allows substitution of a GaAs monolithic IC to provide directionality and detection in a single chip. The bridge circuit, shown in Fig. 5, lends itself extremely well to monolithic integration [6]. This approach offers extremely broad bandwidths with excellent flatness in a very compact size. A disadvantage of this monolithic approach is its slightly higher insertion loss. Also, the directivity is very good but not equal to the best performance achieved with the highest performance (and most expensive) directional couplers. On balance, these monolithic detectors are useful in a wide variety of instrumentation applications.

Advantages and disadvantages of using GaAs ICs in each function of the basic microwave source are summarized in Table 3.

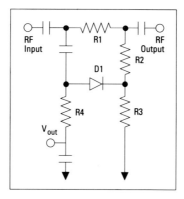

FIG. 5 A monolithic bridge detector. This compact MMIC can replace a bulky and costly directional coupler and detector assembly.

Synthesized Sources

The basic source previously described operates with open-loop frequency control. While this is acceptable for some measurements, the precise control of source frequency and improvement in phase noise afforded by frequency synthesis is becoming increasingly essential. Synthesis means that the output signal is some multiple of a stable, well-known reference oscillator. There are three primary types of synthesis which may be found in high-frequency instrumentation: (1) indirect synthesis, (2) direct analog synthesis, and (3) direct digital synthesis [7].

TABLE 3 Advantages and Disadvantages of Using GaAs ICs in the Basic Microwave Source, Compared with Hybrid MIC Technology

Block diagram element	Advantages	Disadvantages
Oscillator active device	Higher operating frequencies possible Opportunity to integrate buffer amp	Inferior $1/f$ noise of GaAs degrades phase noise of oscillator
Modulator/level control	Very fast pulse modulation Potential for linear control	Isolation limited on-chip Distortion added for high power levels
Output amplifier	Lower cost and broader bandwidth	
Detector	Very broad bandwidth Excellent flatness	Slightly higher loss Slightly poorer directivity

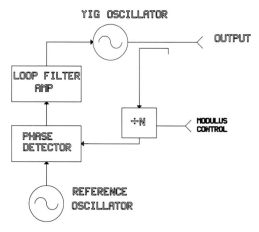

FIG. 6 A simple indirect synthesis loop. This is the approach most commonly used for precise control of frequency in microwave instrumentation.

Indirect Synthesis Indirect synthesis uses a phase-locked loop to stabilize the tunable oscillator, in most cases a YIG oscillator. The output from the YIG is divided or downconverted to a lower frequency where it can be compared with a stable reference oscillator. This comparison occurs in a phase detector which produces a voltage output related to the phase difference of the downconverted output signal and a reference oscillator (Fig. 6). This voltage in turn is filtered and amplified to drive a tuning adjustment on the YIG oscillator (typically a small additional coil above the YIG sphere). When this phase-locked loop is operational, the two input signals to the phase detector have a constant phase relationship and therefore are at the same frequency. The example shown fixes the output at a single frequency determined by the reference frequency. In practice, variable modulus frequency dividers and additional loops will be added to provide fine tuning resolution for the synthesizer.

Two possible methods for downconversion in this block diagram are sampling and frequency division. A high-speed sampler combines frequency downconversion and phase detection in a single component. Sampling can be accomplished with a set of diodes which are turned on and off by an impulse. A sampler might resemble a mixer where the local oscillator (LO) port is driven by a narrow impulse at the sampling frequency. When the YIG oscillator's output frequency is almost equal to an exact multiple of the sampling frequency, the low-frequency output signal from the sampler can be used to phase lock the oscillator to the reference oscillator driving the sampler.

Sampling is commonly employed since it can be used over very broad bandwidths covering frequencies as high as 50 GHz. The maximum frequency range of the sampler will be determined by the quality of the sampling diodes, the sharpness of the impulse which drives them, and the parasitics of the assembly. Excellent bandwidth has been demonstrated in an integrated sampler using an IC process optimized for diode performance; the termination resistors and hold capacitors are integrated with the sampling diodes onto one die [8].

High-speed frequency division offers an alternative to sampling. Frequency dividers are required to operate over the complete tuning range of the YIG oscillator. With the recent availability of commercial GaAs IC frequency dividers well beyond 10 GHz, this technique becomes a viable option for many instruments.

The highest-speed frequency dividers generally operate at a fixed divide-by-two or divide-by-four. In this case, the divider acts as a prescaler. The simple synthesizer shown in Fig. 6 is modified by adding a high-speed prescaler in front of the lower speed variable modulus dividers. Prescaling allows high-frequency operation but sacrifices resolution since the basic synthesizer resolution is reduced by the prescaler division ratio.

Dynamic GaAs IC frequency dividers offer the highest frequencies of operation. They resemble a ring oscillator with two series pass transistors controlled by clock and clockbar; these dividers have bandwidths limited typically to one or two octaves without bias adjustments [9]. Static frequency dividers, realized using conventional toggle flip-flops, can offer decades of bandwidth, but their upper frequency limit is typically limited to about 40% of the F_t of the devices in a given IC process. For application in a synthesizer, the YIG bandwidths are small enough and high enough in frequency that dynamic dividers are the most likely candidates.

A concern about use of GaAs monolithic frequency dividers is the degradation of the phase noise performance of the synthesizer. The design of a synthesizer to optimize phase noise performance is extraordinarily complicated. Performance at various offsets from the carrier can be dominated by the reference oscillators, the frequency translation devices, or the YIG oscillator itself (outside the bandwidth of the phase-locked loop). The relatively high $1/f$ noise of GaAs FETs causes GaAs IC frequency dividers to have residual phase noise which is substantially higher than that of silicon bipolar dividers [10]. It appears that scaling up device geometries improves noise performance; this gives dynamic dividers a potential phase noise advantage since the few critical devices can more easily be enlarged. Such a divider has been used in place of a sampler to achieve superior broadband phase noise performance in a broadband source instrument.

Direct Analog Synthesis Direct analog synthesis uses many frequency dividers, frequency multipliers, and mixers to synthesize a range of frequencies from a single, fixed stable reference oscillator. This type of synthesizer has an advantage over indirect synthesizers in switching speed since both

YIG settling time and phase-locked loop (PLL) settling time are avoided. This technique uses a very large number of components which can result in a bulky instrument when hybrid MIC technology is used. Since all of the components required for direct analog synthesis have been widely demonstrated in monolithic GaAs ICs, there should be opportunities to integrate for size and cost reasons.

Direct Digital Synthesis Direct digital synthesis (DDS) is a technique where GaAs ICs have recently made major contributions. Using DDS a waveform is synthesized directly from digital data which drives a very fast digital-to-analog converter (DAC). Although Nyquist sampling theory suggests that an output frequency equal to 50% of the system clock rate can be achieved, practical numbers appear to be closer to 45% of the clock frequency. With the availability of GaAs DACs and GaAs Accumulator ICs which clock at 1.0 GHz, a digitally generated output frequency of 450 MHz can be achieved [11].

Key high-speed elements of a direct digital synthesizer include the phase accumulator, a read-only memory (ROM) used as a sinusoid lookup table (or a logic implementation of an algorithim), and a digital-to-analog converter as shown in Fig. 7. Instantaneous phase is represented digitally in the phase accumulator. These data are fed to the ROM (or appropriate logic) which yields a digital representation of the instantaneous amplitude. These data in turn drive the DAC which produces the desired analog sinusoidal output. The number of bits in the phase accumulator and sinusoid lookup affects the phase and frequency resolution; the number of bits in the DAC affects the amplitude resolution and the spectral purity [12].

DDS systems offer many advantages as illustrated in Table 4.

The ability to abruptly change frequency while maintaining phase continuity is one example of the power of DDS. This technique is perceived as the fastest frequency switching synthesis technique. Phase noise performance is very good since the clock frequency can be derived from a very stable fixed oscillator. Finally, with some additional digital logic it is possible to achieve complex modulation.

Spectral impurity is the biggest limitation of current high-speed DDS systems. The output of the DAC is low-pass filtered to remove the clock

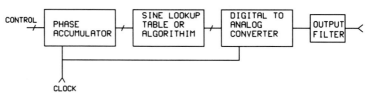

FIG. 7 A direct digital synthesizer. The waveform is synthesized directly from digital data which drives a digital-to-analog converter.

TABLE 4 Advantages and Disadvantages of Direct Digital Synthesis

Advantages of DDS	Extremely fast switching speed
	Low phase noise
	Excellent frequency resolution
	Frequency change with phase continuity
	Complex phase modulation implemented digitally
Disadvantages of DDS	Bandwidth limited by logic and DAC clock rates
	Spurious signal content

frequency component, but other nonidealities induce in-band discrete spurious signals. Both phase and amplitude quantization errors can generate these discrete spurs. At progressively lower frequencies it becomes possible to approximate the ideal sinusoid much more accurately, reducing spurious signals.

The bandwidth limit imposed by clock speed and Nyquist sampling theory makes this technique useful mainly for RF frequencies at this time. A powerful combination uses indirect synthesis and replaces part of the PLL circuitry with a DDS block for narrowband frequency coverage. Although the speed performance of existing DDS systems based on GaAs ICs is impressive, Si bipolar systems have been reported with comparable or superior performance [13]. Design and production of very fast DACs and complete DDS systems appears to be a prime battleground between GaAs and high-speed silicon.

Frequency Coverage Extensions

Multiple-Band Sources A typical YIG oscillator might cover one or perhaps two octaves of bandwidth. By contrast, useful microwave source instruments are commonly required to cover several decades of bandwidth (0.01–26.5 GHz being typical). Extending the upper end of the frequency range is commonly achieved either by (1) adding additional octave (or less) bandwidth YIG oscillators, or (2) multiplying the output frequency of the YIG oscillator.

Adding YIG oscillators, as illustrated in Fig. 8, is the most straightforward approach to extending microwave frequency coverage. A low insertion loss, low distortion switch is used to select the appropriate source. An advantage of this approach is that each band of frequencies can be low-pass filtered to improve the harmonic distortion content. Disadvantages are that the total cost of microwave components is high and the phase-lock circuitry is required to work over the full source bandwidth. Conversely, multiplica-

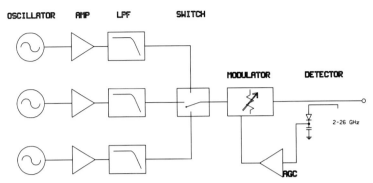

FIG. 8 Wide bandwidths can be covered by combining several narrower bandwidth oscillators and amplifiers.

tion of the output of a single YIG simplifies the phase-locked loop component design.

Broadband frequency multipliers can be realized in several ways. A full-wave rectifier can be built with Schottky diodes and used for frequency doubling. Similarly, one can substitute FETs for the diodes and reduce the conversion loss of the doubler [14]. Construction of a full-wave rectifier requires the design of a broadband balun. Hybrid construction using thin-film circuits affords many options in balun design. While the planar constraint of an IC process limits flexibility, the repeatability of the monolithic baluns can offer improvements in uniformity and manufacturability. To optimize the efficiency and spectral purity of such a multiplier it is desireable to use filter sections which may not be easily integrated. The ideal doubler component would probably use a combination of monolithic IC and hybrid assembly techniques.

Another alternative which has been successfully used is shown in Fig. 9. In this case one constructs a comb generator using a step recovery diode.

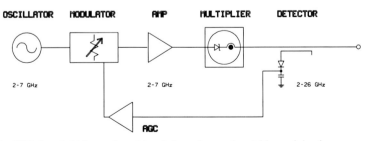

FIG. 9 Wide bandwidths can be achieved through use of variable-modulus frequency multiplication of a single YIG oscillator.

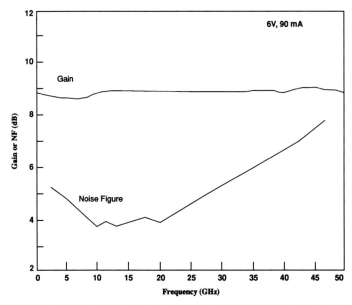

FIG. 10 Gain and noise figure plotted vs frequency for an ultrabroad bandwidth traveling wave amplifier (TWA).

Since this approach produces multiple output frequencies it is necessary to follow it with a tunable bandpass filter which is typically another YIG device. Placing the YIG filter in very close proximity to the step recovery diode also reflects all the energy in undesired frequency components and improves the conversion loss of the multiplier. The need to combine semiconductor and resonator devices in near proximity makes this approach a poor application for monolithic GaAs ICs.

Advantages of using frequency multipliers are reduced component count, simplified phase-lock circuitry design, and improved tuning linearity. Disadvantages include the presence of subharmonically-related signals on the output. Also, for the comb generator approach, there is some effort required to assure that the YIG filter tracks the YIG oscillator precisely.

Whether using multiple YIGs or a multiplication technique it is desireable to be able to use a single broadband amplifier at the output of the source. The GaAs traveling wave amplifier (TWA) MMIC lends itself extremely well to this application by providing ultrabroad-bandwidth gain with very good gain flatness, medium output power levels, and reasonable harmonic distortion. A plot of gain and noise figure vs frequency for a 0.5- to 50-GHz TWA [15] is shown in Fig. 10; the layout of this MMIC is shown in Fig. 11.

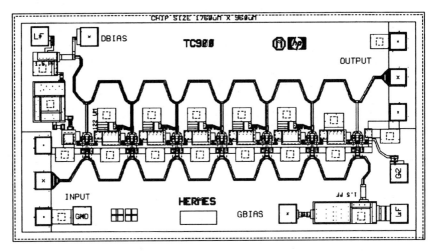

FIG. 11 A layout drawing of the ultrabroad bandwidth traveling wave amplifier. The chip size is less than 1.8 mm².

Low Frequency Extensions Typically a broadband microwave source will extend frequency coverage well below that of the YIG oscillator. For example, it is common to extend a 2- to 26.5-GHz source to a 10-MHz low-end. This can be accomplished either by use of heterodyne downconversion or by the use of frequency dividers.

The heterodyne approach (Fig. 12) covers a very broad range of frequencies with a small number of components. By sweeping the YIG oscillator (which drives the LO port of the mixer) and driving the RF port from a

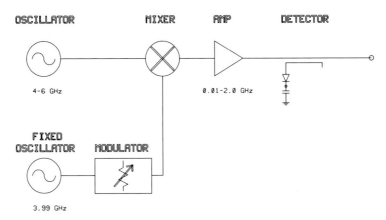

FIG. 12 A heterodyne approach generates broadband RF from an octave (or less) bandwidth microwave oscillator.

fixed-frequency oscillator, one can derive a difference frequency which covers many decades of bandwidth from a YIG with less than an octave of bandwidth. Also, by placing the amplitude modulator between the fixed-frequency oscillator and the mixer, one can use a simple, narrowband modulator design to cover the full output bandwidth. A disadvantage of this technique is the presence of undesired mixing products which can appear as spurious signals; careful choice of the fixed and swept frequency ranges can minimize this problem.

This heterodyne technique provides a wealth of opportunities for monolithic GaAs IC applications such as the downconverting mixer and the IF amplifier. Traditionally, this type of mixer has been built in a hybrid MIC assembly using passive baluns and Schottky diodes as the mixer devices. Now monolithic mixer ICs which compete favorably with hybrid MIC mixers are available. Key parameters for comparison include conversion efficiency, added noise, and spurious suppression.

Conversion efficiency for any passive mixer is directly related to the on-resistance of the mixer devices whether diodes or FETs. An active mixer cell or a monolithic passive mixer using active baluns can add gain such that the conversion gain is observed. The noise figure of a broadband monolithic mixer using active baluns will typically be substantially worse than that of a purely passive structure. The total added noise in the system, however, will not necessarily be worse since there is less gain required after the active balun mixer to achieve a given output power. Suppression of spurious mixing products is largely a function of how ideally the baluns function. Over a wide bandwidth it is believed that active baluns can exhibit better phase and amplitude tracking than can a hybrid passive structure which may be subject to significant assembly variations. In all cases, the system designer must take care to optimize the choice of the YIG's frequency sweep bandwidth and the frequency of the fixed LO to minimize the number of in-band spurs.

An alternative to the heterodyne approach uses a variable-modulus frequency divider to extend the low frequency coverage (Fig. 13). This

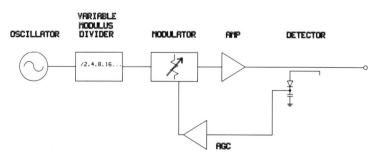

FIG. 13 A variable-modulus frequency divider can generate a wide range of RF frequencies from an octave (or less) bandwidth microwave oscillator.

approach offers significant phase noise advantages since each divide-by-two ideally improves the phase noise by 6 dB. If the primary oscillator covered frequencies from 2 GHz upward, division would only be necessary from 2.0 to 4.0 GHz. This is well within the range of silicon bipolar frequency dividers which offer better phase noise performance than their GaAs counterparts. On the other hand, one could choose to ease the bandwidth requirements of the fundamental YIG by moving the YIG's lower limit up to 3.0 or 4.0 GHz. This now requires division to 6.0 or 8.0 GHz where GaAs frequency dividers are stronger competitors. Ideally one could choose to implement a 2- to 20-GHz source with a single octave bandwidth YIG covering 10–20 GHz and a set of frequency dividers capable of operating to 20 GHz.

Ultrabroadband Oscillators YIG filters which tune over substantially more than a decade of bandwidth are readily available. The lower frequency limit of most YIG resonators is commonly between 1 and 2 GHz while the upper limit of the tuning range may well exceed 30 GHz. Achieving comparable oscillator bandwidths is very difficult; it is very hard to get a single active device to provide the appropriate negative impedance to the resonator over very broad bandwidths. An oscillator covering a decade of bandwidth has been demonstrated, but this bandwidth pushes the limits of manufacturability and cost.

This problem creates potential opportunities for monolithic integration. One might switch in and out various active devices optimized for operation over specific frequency bands. Alternatively, one might switch in different microwave matching structures to extend the useable bandwidth. The design of a robust 2- to 26.5-GHz oscillator in a single magnetic structure would greatly simplify source block diagrams of the future, since all other key elements (amplifiers, modulators, detectors) exist in MMIC form for this bandwidth.

7.3.2 RF and Microwave Spectrum Analyzers

A spectrum analyzer displays signal amplitude vs frequency. One approach to the design of a spectrum analyzer would be to parallel many bandpass filters, each driving a unique level detector. This approach gives real-time information on spectral content, but the realization is hopelessly expensive. At relatively low frequencies one can use an analogue-to-digital converter (ADC) to capture the signal in the time domain and then use Fourier Transform techniques to display the spectral content. The most common approach at RF and microwave frequencies is to mix the signal with a swept LO and detect the mixing products at a fixed IF frequency.

Typically a YIG oscillator is used to drive the mixer's LO port. For good frequency accuracy and good phase noise performance, this oscillator is stabilized with one or more phase-locked loops (section 7.3.1). Since the LO is swept, the synthesis commonly happens at the beginning of the sweep and the YIG is tuned open-loop thereafter.

RF Spectrum Analyzers The typical block diagram for an RF spectrum analyzer is shown in Fig. 14. The input signal is mixed with a swept LO which covers a frequency band well above the range of input frequencies. An IF frequency is picked which is also above the range of input frequencies. Applying a low-pass filter to the RF input signal avoids potential downconversion of a much higher frequency signal. Adding a bandpass filter in the intermediate frequency (IF) path similarly avoids image products in the second frequency converter. This approach provides an image-free signal but is limited to frequencies below the coverage of the YIG oscillator.

Microwave Spectrum Analyzer A microwave spectrum analyzer typically uses harmonic mixing to extend the useful bandwidth of the instrument. The input signal is downconverted in contrast to being upconverted in the RF analyzer. A typical microwave spectrum analyzer block diagram is shown in Fig. 15. Since the harmonic mixer will produce numerous mixing products, there is ambiguity about which IF output is the desired signal and which are undesired images. For example, if an 18-GHz signal drives the mixer RF input and the LO input is swept from 3 to 6 GHz, the IF port will produce signals at the desired IF frequency of 300 MHz for five different LO frequencies: 3.54 and 3.66 GHz (mixing with the fifth harmonic of the LO), 4.425 and 4.575 GHz (mixing with the fourth harmonic of the LO), and 5.9 GHz (mixing with the third harmonic of the LO). This confusion can be resolved by placing a tracking filter at the input and sweeping it with the LO. In this way it is possible to suppress the RF input for all but the desired combination of RF and LO frequencies. Use of the tracking filter is com-

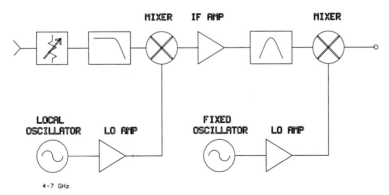

FIG. 14 A typical RF-frequency spectrum analyzer. The first mixer uses upconversion to avoid images without expensive tuned input filters.

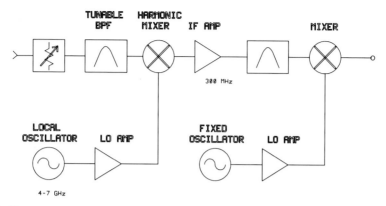

FIG. 15 A typical microwave spectrum analyzer. The first frequency conversion occurs in a downconverting harmonic mixer. This allows an octave-bandwidth LO oscillator to cover more than a decade bandwidth of input frequencies.

monly called preselection and while it dramatically improves the user friendliness of the instrument it also adds cost and complexity. The tracking filter usually uses a YIG device and may be a separate component or may be integrated into the mixer assembly. A preselector also allows measurement of low-level signals in the presence of much larger amplitude signals which might otherwise overload the instrument.

Input Mixer Design of the input mixer for a spectrum analyzer involves complex trade-offs between noise figure, intermodulation performance, conversion loss, bandwidth, and cost. Traditionally this type of mixer has been realized as a hybrid microcircuit using Schottky diodes. To minimize unwanted feedthrough of LO and RF signals to the IF a singly or doubly balanced topology is desireable. To accomplish this topology over reasonably broad bandwidths requires very careful design of baluns.

Two types of monolithic broadband mixers have been realized in GaAs. Passive mixers have been designed using either planar balun structures or Lange couplers and using diodes or FETs as the switching devices [16,17]. Active balun mixers which replace the passive baluns with inverting and noninverting amplifiers have also been designed. A comparison of key characteristics of these devices is shown in Table 5. Note that any mixer type can be optimized to peak one or two figures of merit; the table represents a best effort to compare similar octave LO bandwidth designs.

The active balun mixer adds gain at the RF input and potentially on the IF output as well. As a result, it is easy to achieve conversion gain from such a device. For a conventional passive mixer, conversion loss is found to be directly related to the on-resistance of the diodes or FETs used as switching

TABLE 5 Comparison of Conventional Diode Mixers and Monolithic Mixers

	Hybrid Schottky	Passive IC	Active IC
Conversion efficiency	0	−	+
Noise figure	+	0	−
Bandwidth	0	0	0
Intermodulation distortion	0	+	+
Size	−	+	+
Power dissipation	+	+	−

Note. +, Best; 0, intermediate; −, poorest.

devices. Since diodes can be optimized for lowest on-resistance in a diode-only process, the hybrid diode mixer should offer better conversion loss than one designed in an IC process optimized for best FET performance.

For an ideal passive mixer there would be no addition of noise in the mixer and consequently the noise figure of the mixer would be equal to its conversion loss. In practice, the noise figure is usually only slightly worse, perhaps typically 1 dB. For the active balun mixer, noise figure is set primarily by the amplifier stages and does not have the same relationship to conversion loss as for the passive mixers. In most existing active balun mixer designs the noise performance is measurably worse than in passive designs.

The bandwidths of hybrid assemblies and of monolithic passive mixers are both limited by the balun structures used. Typically, input mixers for spectrum analyzers have been designed for about one octave of LO bandwidth; this bandwidth requirement is reasonably challenging. The availability of double-sided substrates, recessed ground planes, etc., in hybrid assemblies offers several more degrees of freedom for the balun designer. On the other hand, assembly variations and mismatches in a physically large structure may make consistent, well-behaved performance harder to achieve than in the monolithic passive mixer. In an active balun mixer, the baluns can operate over decades of bandwidth but have upper frequency limits set by the speed of the IC devices.

Comparisons of intermodulation distortion (IMD) between mixer types is extremely difficult. The primary sources of IMD are the nonlinearity of the switching devices and the finite time required for switching to occur. In general, the LO drive level is increased to minimize the time the diodes or FETs are in transition. This suggests that use of either an active balun or an on-chip LO amplifier driving a passive balun could improve intermodulation distortion. It has been shown that using FETs as the switching devices in lieu of diodes can yield better IMD [18]. The channel resistance of the FET is used to realize the time-varying resistance, and the improved IMD is attributed to the superior linearity of the FET channel resistance. Combining such a FET mixer with on-chip LO amplification should yield superior IMD performance and realize an advantage for monolithic integration of the mixer.

At this time the vast majority of instrumentation mixers are hybrid MICs. Only now are the performance, size, and possibly cost advantages of MMIC mixers beginning to win over new applications.

Preamplifiers The use of a preamplifier can greatly improve the sensitivity of the spectrum analyzer, allowing detection of substantially lower level signals [19]. Ideally the bandwidth of a preamplifier would identically duplicate the useful bandwidth of the spectrum analyzer. Since this typically covers many decades, the ultrabroad bandwidth of GaAs IC traveling wave amplifiers is a perfect device for this function.

In addition to extremely wide bandwidth, noise figure and gain flatness are key specifications for a preamplifier. When the gain of the preamplifier is roughly equal to or greater than the noise figure of the spectrum analyzer the noise figure of the preamplifier will become the effective system noise figure. Since a typical instrument might offer a 29-dB noise figure, a preamplifier consisting of several stages of traveling wave amplifiers can easily improve the system noise figure by more than 20 dB. Any deviation from perfect gain flatness can lead to amplitude measurement inaccuracy since one is generally programming a fixed gain offset into the spectrum analyzer. Again, the exceptional gain flatness available from a monolithic GaAs traveling wave amplifier is a perfect fit for this application. Such preamplifiers, using GaAs MMIC TWAs, are commercially available now.

LO Amplifiers The primary need for an LO amplifier is to deliver a relatively constant amount of power to the LO port of the mixer. Often the range of power levels desired is smaller than the range of levels coming out of the YIG oscillator. As a result, one requires some properties of a limiting amplifier for this function. To accomplish the limiting function one can either tailor the compression characteristics of the amplifier or build in an (AGC) loop. Also, there is commonly a need for the YIG output to appear at several places in the instrument (fed to a sampler for the frequency synthesis, for example). A monolithic IC affords a very easy way to divert signal to several outputs with signal levels adjusted appropriately for each port. Such an MMIC is currently in use in Hewlett-Packard's high-performance spectrum analyzer product line.

7.3.3 Pulse and Pattern Generators

A basic pattern generator is presented as an example of time domain instrumentation. The goals for such an instrument are to put out digital data

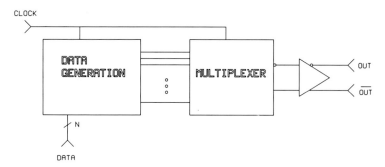

FIG. 16 A typical pattern generator block diagram. Data are generated in words at moderate clock rates and then multiplexed into a single bit stream at clock rates as high as 10 GHz. Output data amps requiring fast edges with large amplitudes are uniquely well suited to GaAs MMIC capabilities.

with very fast edge speeds at very high clock rates, to provide good pulse fidelity, to minimize jitter, and to maximize output amplitudes.

A major driving force for such instrumentation is the design and production of systems using increasingly higher frequency clock rates for digital data transmission on optical fiber. Research and development laboratories have moved beyond 0.625 and 2.5 Gb/s systems to work on 10 and 20 Gb/s systems. Data rates have been moving upward much faster than speeds of commercially available semiconductor processes. Consequently these systems have moved from use of relatively straightforward silicon parts to state-of-the-art GaAs designs in just a few years.

A typical pattern generator block diagram is shown in Fig. 16. Specific bit sequences (which may be user generated or may be part of a pseudorandom sequence) are derived at lower frequencies and then multiplexed up to form a single, high-speed stream of data. The output amplifier improves the edge speeds as well as providing larger output amplitudes. In addition, the output amplifier is often required to offer variable attenuation and variable DC voltage offsets. Customers may use such an instrument to drive emitter coupled logic (ECL) circuits, a variety of GaAs logic types with logic swings ranging from a few tenths of a volt to several volts, or other devices such as an optical modulator or a laser itself. Thus, the output levels must be extremely flexible.

When very fast edge speeds and relatively large output amplitudes are required, GaAs ICs make a unique contribution owing to significant F_t and breakdown voltage advantages over silicon. GaAs MMICs become the technology of choice when the anomalous lag effects discussed earlier are dealt with through a combination of process and circuit design improvements. Several commercial pulse generators shipping today use GaAs MMICs for their output amplification.

7.4 APPLICATION EXAMPLES

Three examples of instrument applications are briefly presented here. Two are representative of microwave instruments where monolithic ICs are replacing much more complex hybrid MICs. The third illustrates a high-speed, time-domain instrument where a GaAs IC has improved performance well beyond what had previously been achieved with Si ICs.

The "TAMP" Project—A 2- to 7-GHz Amplifier This work, by Andrew Teetzel, involved the retrofit of a complex hybrid MIC 2- to 7-GHz amplifier with a much simpler assembly using a pair of GaAs MMICs. [20] The amplifier is used in microwave sources to amplify the output of a YIG oscillator and drive a broadband frequency multiplier. A four-stage IC design using primarily lumped reactive matching was completed in 1988. This design used a first-generation IC process which used ion-implanted material, 1-μm gate lengths, and offered 14 GHz F_t. The amplifier, shown in Fig. 17, has 1 × 2 mm chip dimensions.

The IC was designed to offer high gain (20 dB) and relatively high broadband output power (½ W). Emphasis was placed on achieving large output amplitudes without excessive harmonic distortion. In early 1989 the project to replace the hybrid MIC/discrete FET design was started. In September of 1990 shipments began using the new design. The primary problems which delayed shipments were systems issues where subtle, often unspecified, performance of the original hybrid MIC differed slightly from that of the monolithic retrofit.

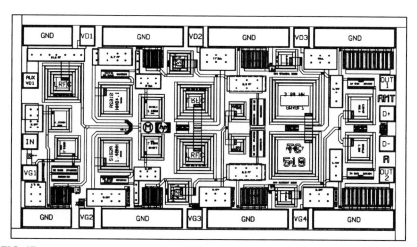

FIG. 17 An MMIC 2- to 7-GHz power amplifier. This IC delivers greater than ½ W output power with 20 dB gain from 2 to 7 GHz. Chip size is 2 mm².

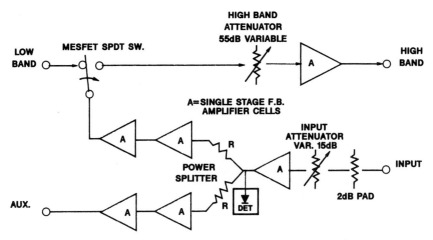

FIG. 18 The block diagram for a 2- to 7-GHz distribution amplifier ASIC. This single IC combines numerous amplifiers, an SPDT switch, a variable attenuator, power splitting, and level detection on a single die.

Cost savings for this retrofit were so great that the break-even time for this retrofit project was only 4 months.

A GaAs Distribution Amplifier Applications-Specific Integrated Circuit (ASIC) This design, by Scott Trosper, demonstrates the potential for high-level integration of microwave systems. The instrument application, a new microwave tracking generator, required a component which would amplify the output of a 2- to 7-GHz YIG oscillator and split it into three paths. No single specification presented major technical challenges, but the integration of so much microwave functionality onto a single chip was unprecedented at Hewlett-Packard.

The design was done by first designing and characterizing several functional blocks for this bandwidth. These blocks included a single-stage feedback amplifier, a variable attenuator, a diode detector, and an single-pole double-throw (SPDT) switch. A block diagram for the complete MMIC is shown in Fig. 18. The design worked on the first iteration, meeting all specifications with margin, except for an isolation problem which was resolved without redesign.

The functions of this chip were performed in a predecessor instrument by hybrid MICs. The cost of the custom IC approach was 47% of the hybrid MIC cost. Also of great importance was that the area required by the packaged IC was only 33% that of the hybrid MIC space requirement. Finally, this design demonstrated that many functions can be integrated in relatively near proximity on a single die using standardized cells. The chip layout is illustrated in Fig. 19.

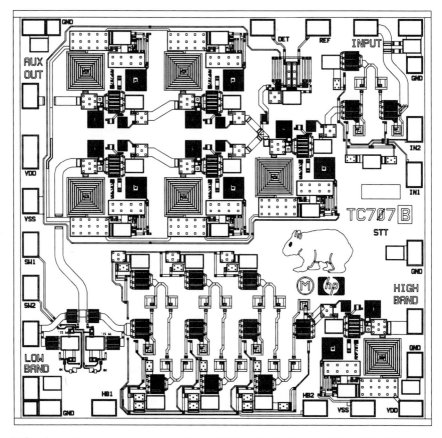

FIG. 19 A layout drawing for the 2- to 7-GHz distribution amplifier ASIC. Chip size is 6 mm². This is believed to be the highest level of microwave integration achieved in instrumentation.

A 500-MHz Pulse Generator Output Amplifier This final design, by Hans-Juergen Wagner and Allan Armstrong, used a first-generation GaAs IC process to duplicate a function previously designed in monolithic silicon bipolar processes [21,1]. GaAs was chosen to provide faster edges with large output amplitudes. Unlike the previous two examples, this design was motivated primarily by performance needs rather than cost needs.

The GaAs chip is driven by a 1.5-V p–p input signal with 350-ps edges. Three differential gain stages are used to deliver an output of 5 V p–p with 150 ps typical edge speeds. The output stage is weighted to allow variable output amplitudes over a 50:1 range. A block diagram for this chip is shown in Fig. 20.

The primary design challenge (and primary source of design delays) was coping with lag effects in GaAs FET current switching. A first-pass design

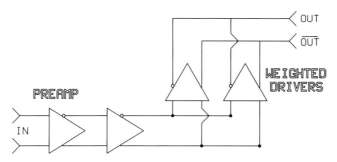

FIG. 20 The block diagram for an MMIC pulse amplifier. Chip size is 3.4 mm². This IC produces 5-V p–p pulses with 150-ps 10–90 edge speeds.

using a linear amplifier had to be scrapped because of incomplete switching attributable to the lag effects. The final design used fully switched differential pairs with cascoded source followers and cascoded current sources to minimize the lag effects. Although these problems delayed the design and shipment of this design, it has been very successful in all other respects with good performance and excellent reliability in the field.

7.5 CONCLUSIONS AND FUTURE TRENDS

Instrumentation applications inspired much of the earliest work in monolithic GaAs ICs. Since 1985, insertions of GaAs ICs in instruments have become widespread. The performance requirements of instrumentation have proven to be a powerful driver of GaAs IC technology. Instrument shipment volumes, however, are modest and make development of application-specific ICs hard to justify in many cases. Favorable return on investment (ROI) may be particularly hard to demonstrate where integration is being pursued for cost reduction in low-volume instruments.

As the RF/microwave market shifts from a military focus to a commercial one there will be both more opportunities and more competition for GaAs ICs. The larger shipment volumes of commercial instrumentation will offer new opportunities. At the same time, the lower frequencies and lower cost expectations will force GaAs ICs to compete more aggressively with silicon. The changes resulting from these market shifts will make a very dynamic marketplace for the GaAs IC industry.

REFERENCES

[1] A. Armstrong and H. J. Wagner, A 5 Vp-p 100 ps GaAs pulse amplifier IC with improved pulse fidelity. *IEEE J. Solid State Circuits* **SC-27**, 1476, 1992.
[2] J. Imperato, Low-cost MMICs replace hybrids. *Microwaves & RF* 115–117, Mar. 1986.

[3] G. Basawapatna and R. Stancliff, A unified approach to the design of wideband microwave solid state amplifiers. *IEEE Trans. Microwave Theory Tech.* **MTT-27,** 379, 1979.
[4] H. Kondoh, DC—50 GHz MMIC variable attenuator with a 30 dB dynamic range. *IEEE MTT-S Symp. Dig.* 499, 1988.
[5] R. Goodall, Fast pulses modulate L-band synthesizer. *Microwaves & RF* 161–164, Apr. 1987.
[6] Hewlett-Packard Co., *HP 33336C Coaxial GaAs Directional Detector Datasheet.* Dec. 1989.
[7] W. Egan, *Frequency Synthesis by Phase Lock.* Wiley Interscience, 1981.
[8] S. Gibson, Gallium arsenide lowers cost and improves performance of microwave counters. *Hewlett–Packard J.* 4–9, Feb. 1986.
[9] S. Long and S. Butner, *Gallium Arsenide Digital Integrated Circuit Design.* McGraw–Hill, 231–238, 1990.
[10] M. Bomford, Selection of frequency dividers for microwave PLL applications. *Microwave J.* 159–167, Nov. 1990.
[11] J. Browne, Hybrid circuit sets DDS clock beyond 1 GHz. *Microwaves RF* 128–130, Feb. 1990.
[12] R. Zarvel, DDS provides an alternative to PLL synthesizer design. *Microwaves RF* 145–148, Sept. 1990.
[13] I. Fobbester, One-chip direct digital synthesizer clocks past 1.6 GHz. *Microwaves RF* 192–202, June 1990.
[14] R. Stancliff, Balanced dual-gate GaAs FET frequency doublers. *IEEE MTT-S Symp. Dig.* 143, 1981.
[15] J. Perdomo et al., A monolithic 0.5 to 50 GHz MODFET distributed amplifier with 6 dB gain. *IEEE GaAs IC Symp. Tech. Dig.* 91, 1989.
[16] D. Kruger, Monolithic dual-quadrature mixer using GaAs FETs *Microwave J.* 201–206, Sept. 1990.
[17] Hittite Microwave Corporation, *Preliminary Data Sheet, HMC115.* Dec. 1991.
[18] S. Maas, A GaAs MESFET mixer with very low intermodulation distortion. *IEEE Trans. Microwave Theor. Techniques* **MTT-35,** 425, 1987.
[19] R. Rausch, Boost the sensitivity of spectrum analyzers using preamplifiers. *Microwaves RF* 166–175, Apr. 1990.
[20] A. Teetzel, The TAMP GaAs IC microwave amplifier project. *IEEE GaAs IC Symp. Short Course Notes,* 1991.
[21] H. J. Wagner et al., A 500 MHz pulse generator output section. *Hewlett-Packard J.* 79–84, Aug. 1990.

8
Personal Communications Service

J. Mondal, S. Ahmed, and Sanjay B. Moghe
Advanced Microwave Technology,
Electronics System and Integration Division,
Northrop Grumman Corporation,
Rolling Meadows, Illinois

8.1 Introduction
    8.1.1 Historical Perspective
    8.1.2 MMIC Components for PCS
8.2 Requirements for a PCS Voice/Data System
8.3 MMIC Components for PCS
    8.3.1 MMIC Receiver
    8.3.2 MMIC Transmitters
    8.3.3 Highly Integrated Transceivers
8.4 Future Trends in PCS
References

8.1 INTRODUCTION

Personal communication service (PCS) refers to advanced wireless communication services such as cellular voice, wireless data, and messaging, which are offered separately today. All of these services will be offered on the personal communications network (PCN), which will be an advanced network that retains the best features of the current cellular network. The goal of PCS is to offer communication services anywhere, anytime, and anyplace, even if the person is mobile. That means, for example, that people will have a phone number which will follow them whether at work, at home, on vacation, or even on an airplane. This requires an intelligent wireless network which can locate a person at any given time.

    Significant research has been done in the past few years to define the nature of PCS wireless services. At present, a spectrum is being allocated and auctioned by the Federal Communications Commission (FCC) for PCS. Services are expected to be available in approximately 2 years. There have been many experimental licenses awarded and a considerable body of experimental work has been done to assess the kind of services that can be offered with a new wireless network. The standards for PCS are evolving. Therefore, it is hard to predict what shape PCS will take in the next few years.

In this chapter we describe the historical perspective of PCS, the technology requirements for PCS, and the role that MMIC technology can play in PCS. We specifically concentrate on the 900- and 1800-MHz PCS band. Applications like wireless telephone systems, data systems for computer networks, one-way and two-way messaging, electronic mail, multimedia, and video services are discussed. In Section 8.2 we discuss the system requirements for voice and data systems. We discuss in detail one system architecture based on cellular digital packet data (CDPD) technology. In Section 8.3 we discuss examples of monolithic microwave integrated circuits (MMICs) for PCS. These examples include receivers and transmitters as well as more highly integrated MMICs. The last section of this chapter discusses future trends in PCS. Here we forecast how the PCS will evolve and what technologies will play a vital role in PCS evolution in the future.

8.1.1 Historical Perspective

Ever since cellular services started in the early 1980s there has been a strong interest in wireless communication services. The cellular industry has enjoyed enormous growth in the past 10 years; it has averaged 30% annually and had more than 19 million subscribers in late 1994. This significant growth has stimulated the development of other wireless voice and data services for personal communications. The interest in PCS has been worldwide. Europe, Japan, and the United States have moved in somewhat different directions and only in the past 2 years has the world community agreed on a common set of frequencies and standards for PCS. While the standards for PCS are still evolving, the worldwide frequency spectrum allocation is similar. Even though there are slight differences, most countries are considering wideband PCS services in the 1.8- to 2.2-GHz frequency range.

Ever since the FCC held its first hearing in 1991, there has been strong interest in PCS. Over the past few years more than 75 companies have applied for and received an experimental FCC license to develop versions of PCS. Services have included one-way and two-way messaging and telephone services combined into one device, data services, video and multimedia, and various combined services. The frequency spectrum is generally at a premium, especially at lower frequencies. Thus, the FCC has allowed experiments to be performed over a wide range of frequencies from 800 MHz to 50 GHz, with the majority of the effort below 10 GHz.

The PCS experiments performed to date have verified customer interest in PCS. Some of these services offered cellular voice, data, and paging-like services in one device at a cost lower than that for current cellular voice only. They also offered, in some cases, phone number portability, where the service follows the person. This means that wherever the person goes their personal phone number goes with the person and the network keeps track of where the person is.

These wireless services are similar in some ways to cellular and paging services, but in other ways provide more advanced services. Even though they have been extremely successful, cellular services today have a number of problems including poor handoff when going from one cell to another cell, dropping the call at times, limited capacity, and high price. These problems can be overcome with the newer PCS technology. Its smaller cells require lower power transceivers. Also the newer digital spread spectrum technology makes the handoff from cell to cell much smoother and greatly reduces the chances of call dropping. The cost for PCS is expected to be much lower because of the lower power requirements, the availability of monolithic technologies, and the cost benefits of the advanced intelligent network. Thus, PCS is expected to be affordable to the general population and thus be more widely used.

The technical challenges facing the current cellular industry can be overcome with PCS. In 1994, the FCC gave preference to four companies that have done pioneering work in PCS, one of which is developing a wireless two-way messaging service. The FCC has allocated frequencies for narrowband PCS at 900 MHz for two-way messaging service and wideband PCS at frequencies around 1800–2200 MHz for voice and data services. The FCC is also allocating an unlicensed PCS band for use by cordless telephones and wireless computer communication equipment. These new spread spectrum cordless phones will have a longer range than do current cordless phones. For these unlicensed services, specific FCC approval is not required as long as certain rules regarding spectrum etiquette are followed.

A number of companies have experimented with frequencies at 5, 12, 18 GHz, and other higher frequency bands for PCS. However, the FCC has concluded that the technology at those frequencies is not mature enough to immediately apply to PCS. Therefore, in October of 1993, the FCC implemented the policy for spectrum allocation and auction schemes for PCS in the 1.8- to 2.2-GHz range where the technology is more mature. The FCC has auctioned a number of frequency bands in each of the areas and the winners of those auctions are given a license to build the PCS network for 10 years. These auctions took place in May/June 1994 for the narrowband PCS (900 MHz) and fall of 1994 for wideband PCS (1.8- to 2.2-GHz band).

There are a number of committees involved in setting standards for PCS. Key contributors include the Telecommunications Industries Association (TIA), the Cellular Telephone Industries Association (CTIA), and the

Institute of Electrical and Electronics Engineers (IEEE). As these standards become defined, many equipment manufacturers will be developing equipment for PCS.

8.1.2 MMIC Components for PCS

PCS is expected to be popular worldwide and there are estimates of a few hundred million users within the next few years. These quantities are so large that many people feel the equipment for PCS, including the telephone handsets, wireless modems, and paging devices will become commodity items. There will be fierce competition to offer these devices with high performance and low price. In particular, the market for low-cost MMIC components required for the radio frequency (RF) radio subsystems will be very large.

An important driver in PCS RF component selection is the need to support spread spectrum technology. This increases the complexity of the RF radio. A number of technologies will compete for the radio portion of the PCS transceiver. There are several technologies, including discrete devices, monolithic GaAs devices, and monolithic silicon devices, that are currently being used to build wireless transceivers. The discrete transistor approach has a somewhat lower discrete component cost, but has a larger size and requires a higher assembly and tuning cost for a simple PCS transceiver. MMIC technology can result in the integration of the radio on one to three MMIC chips, resulting in small size, good performance, low current consumption, and high reliability.

Silicon MMIC technology offers small size and potentially low cost, however, highly integrated radios have not yet been realized for specific PCS products. Silicon MMIC technology is well established and offers good performance and low cost at a few hundred MHz to 1 GHz. However, at higher frequencies (>1 GHz) silicon device performance degrades.

GaAs MMIC technology has high performance at the higher frequencies. In addition, it is sufficiently cost competitive to be the technology of choice for the 1.8-, 2.4-, and 5.7-GHz bands. Many companies have offered products such as cordless phones and wireless local area networks (LANs) in the 0.9-, 2.4-, and 5.7-GHz industrial, scientific, and medical (ISM) bands and some of these products already utilize gallium arsenide (GaAs) MMICs. Currently, FCC permission is not required to operate in these ISM bands as long as Part 15 of the FCC regulations is followed.

Another important application of MMIC technology can be found in wireless data communications for portable notebook and hand-held computers. The Personal Computer Memory Card International Association (PCMCIA) card has become a standard in the notebook PC industry for connecting external devices, such as modems and memory cards, to the computer. This offers an opportunity to establish wireless communication for computers through a wireless modem which can be plugged into a

PCMCIA slot. A number of companies have already developed a PCMCIA card wireless modem that attaches to a cellular phone. The PCMCIA card size is so small (credit card sized) that it is hard to integrate complete radio modem and paging functions onto a single card. This challenge can be met by using GaAs MMIC technology.

In the future, we expect that by using MMIC technology all three functions (the cellular phone, the modem, and the pager) can be integrated into a single PCMCIA card. To realize this, the radio function must be integrated onto one or two chips. This can be done with the level of integration currently realized in both GaAs and silicon MMICs. The successful technology will be the one which meets the required performance goals at the lowest cost.

Currently, GaAs MMIC technology has the edge at higher frequencies. Silicon technology is restricted to a multichip solution because it is more difficult to integrate all of the radio components, i.e., low-noise amplifier, switch, mixer, oscillator, power amplifier, modulator, etc., on a single silicon chip. This level of integration has been achieved with GaAs MMICs and, later in this chapter, we will discuss examples of MMICs for PCS application. In the next section we discuss the system architectures for voice and data systems with focus on cellular digital packet data systems (CDPD).

8.2 REQUIREMENTS FOR A PCS VOICE/DATA SYSTEM

PCN, developed in Europe, has prompted interest in PCS in the United States. The FCC has embraced a broad definition of PCS which has included basic wireless communication devices such as cordless phones, WPBX, pagers, and cellular phones [1].

Cellular technology currently provides the mobility required for PCS, but with a reduced voice quality and high interconnect charges that limit its ability to meet the full definition of PCS [2].

The narrowband frequency modulation (FM) modulation used in analog cellular does not allow the bandwidth required for high-quality speech and data transmission nor does it allow the increased channel capacity required to bring the cost of use down. The schemes, such as Gaussian minimum shift keying (GMSK), maximize the information density in a given channel bandwidth. In addition to achieve spectrally efficient modulation, channel utilization is maximized by time sharing of channels in a complex manner transparent to the user.

Two systems have evolved using cellular packet switched communication systems technology. These are Mobitex, and CDPD. Originally, the Mobitex system was placed in commercial operation in Sweden in 1986. Today, RAM Mobile Data (RAM) is licensed to provide Mobitex service coverage in the United States.

In 1992, a consortium of U.S. cellular carriers created a uniform standard for sending data over existing cellular telephone channels. They selected CDPD technology [3–5].

This section provides an overview of CDPD and the application of MMICs in CDPD mobile end systems (M-ES).

CDPD is a packet switched network. This means that data are transmitted in discrete chunks or packets rather than in a continuous stream. The modem attached to the sending device needs to break the data into packets before sending the data to the network. CDPDs connectionless packet switched service differs from other data communication services by allowing messages to be sent without actually having established and maintained a direct connection. CDPD is based upon Internet protocol (IP) and connectionless network protocol (CLNP) can be used to access services through the CDPD network. Therefore, both sending and receiving devices have IP addresses, not telephone numbers. There are no connect time charges associated with CDPD. Instead of paying for the connect time used, one pays for only the amount of data transmitted, regardless of the time it takes to transmit that data.

CDPD was designed to exploit the capabilities of the advanced mobile phone service (AMPS) cellular infrastructure that is in place throughout the United States and Canada. The CDPD services available today handle short, "bursty" data at rates up to 19.2 kbps. It was recently proposed in the CDPD Forum that code division multiple access (CDMA) may be incorporated with CDPD for short messaging which will result in a data throughput of up to 64 kbps in the existing cellular system [6].

CDPD is open and usable by anyone, but it is particularly suitable to mobile professionals who want e-mail and wireless Internet access. It offers a high-speed, high-capacity, low-cost system with the greatest possible coverage. CDPD communication provides seamless intercellular roaming with user-transparent connection continuity. CDPD also offers robust transmission security to its subscribers. Its AMPS cellular ancestry gives CDPD a combined voice and data feature. Voice communications can be provided by adding analog interfaces and a handset.

The CDPD communication system is essentially a wireless wide-area-network. The communication links consist of two segments, the RF segment known as the airlink and the wired segment known as the network.

Subscribers access the CDPD system by initiating an airlink with an M-ES to a local mobile data base station (MDBS). The MDBS is hardwired to a local mobile data intermediate system (MD-IS) which routes packets among the M-ESs in its subdomain of one or more cells or to other MD-ISs as required.

In the CDPD voice/data system discussed here, a credit-card-size PCMCIA Type-II extended card is used, which has a transceiver and all modem functions. This PCMCIA Type-II extended card and a rechargeable battery, all packaged into one assembly, is shown in a conceptual CDPD PCMCIA modem in Fig. 1.

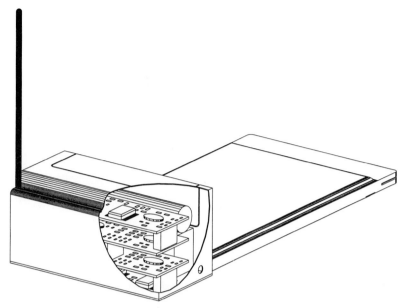

FIG. 1 A conceptual wireless modem on extended PCMCIA card.

GaAs MMICs are ideally suited for mobile wireless modem applications due to the following performance advantages:

—High intercept point
—Low noise figure
—High power-added efficiency
—Low voltage operation
—Low current consumption.

Small PCMCIA plug-in cards used in portable computers, notebooks, and PDAs (Personal Digital Assistance) as memory extension are also suitable for wireless modem applications [7].

Figure 2 shows the block diagram for a mobile CDPD radio modem system. The key system parameters are listed in Table 1. The system is divided into subsystems based on functionality. A transmitter block diagram is shown in Fig. 3.

Power efficiency is the most important issue in a CDPD transmitter. The modulation technique GMSK is a simple approach to improving power efficiency. Filtering the baseband modulation signal can greatly reduce the spectrum growth of the transmit signal [8]. GMSK uses a Gaussian low-pass filter (LPF) to smooth this frequency changes. The 3-dB bandwidth of this filter is generally the bit rate. This tends to round off the nonreturn-to-zero (NRZ) or Manchester coded data and results in some degree of intersymbol interference (ISI). The Gaussian filter chosen in CDPD usually de-

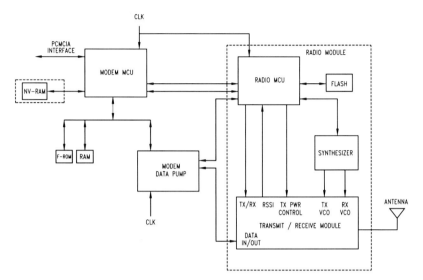

FIG. 2 CDPD PCMCIA radio modem system block diagram.

noted by 0.5 GMSK has 3 dB bandwidth equal to half of the bit rate (BT = 0.5). The output of this filter has a narrow spectrum and uses a simple modulation scheme (as shown in Fig. 4). This data modulation scheme also has good bit error rate (BER) performance.

The 0.5-GMSK modulation has the following advantages: First, it has a constant envelope nature. This allows the 0.5-GMSK modulated signal to operate with class-C power amplifiers (PAM) without introducing spectrum regeneration; therefore, lower power consumption and high power efficiency can be achieved [9]. Second, it has a narrow power spectrum, narrow

TABLE 1 CDPD PCMCIA Modem System Requirements

Frequency (MHz)	Receiver, 869 to 894; Transmitter, 824 to 849
Access method	TDMA, Half/full duplex
Modulation	0.5 GMSK(BT = 0.5), Modulation index = 0.5
Protocol	TCP/IP/TP4
Data throughput	19.2 kbps
Channel bandwidth	30 kHz
Receive sensitivity	−113 dBm at 5% block error rate
Error control	Reed–Solomon
Connectivity	To host via PCMCIA-II extended card
Transmit power (ERP)	1 W
Channel switching time	40 ms
TX level control	0 to −22dB (6dB steps)
RSSI	−113 to −53 dBm

8. Personal Communications Service   263

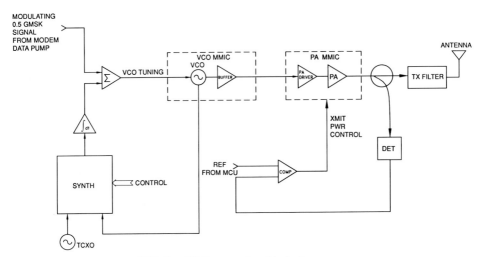

FIG. 3   CDPD transmitter block diagram.

main lobe, and low spectral tails, keeping the adjacent channel interference at low levels and achieving higher spectral efficiency. Third, 0.5 GMSK is really a frequency modulation. It can be noncoherently demodulated by simple limiter/discriminator circuitry. The simple structure leads to low-cost CDPD receivers.

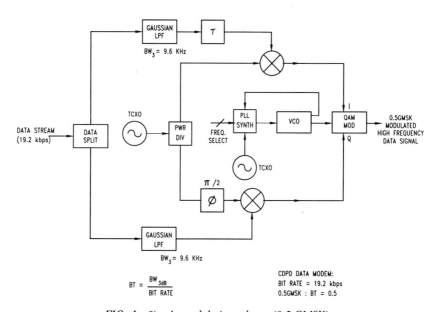

FIG. 4   Simple modulation scheme (0.5 GMSK).

After passing through the Gaussian LPF, the data stream is applied to the input of the voltage-controlled oscillator (VCO) which is part of the integrated MMIC. Typically, a modulation index of 0.5 is implemented for CDPD systems. The VCO is part of a phase-locked loop (PLL) and is controlled by a monolithic synthesizer. The output of the VCO is a frequency-modulated signal with Gaussian response. The data in this case will be transmitted at a rate of 19.2 kbps. This high data rate is facilitated by 30-KHz channel spacing of the existing cellular network and spectral conservation due to 0.5 GMSK. Voice has priority over data transmission for cellular systems, forcing the CDPD system to seek idle cellular channels.

Demodulation of the GMSK signal requires as much attention to the preservation of an unmodulated waveform as does modulation of the signal. The choice of a Gaussian-shaped premodulation filter is justified for three reasons: (1) narrow bandwidth and sharp cutoff, (2) lower overshoot impulse response, and (3) preservation of the output pulse area.

The first condition gives GMSK modulation its spectral efficiency. It also improves its noise immunity during demodulation. The second condition offers GMSK low-phase distortion. This is a major concern when the receiver is demodulating the signals to baseband and care must be taken in design of the IF filtering to preserve this characteristic. The third condition ensures the coherence of the signal which is required for some applications. While this is quite stringent and not realizable with a physical Gaussian filter, the phase response can be kept linear for coherent demodulation.

In most systems the constraints on the above goals include:

—data rate
—transmitter filter bandwidth
—channel spacing
—allowable adjacent channel interference
—peak carrier deviation
—transmitter and receiver carrier accuracy
—modulator and demodulator linearity
—receiver intermediate frequency (IF) filter frequency and phase characteristics.

The data rate, transmit bandwidth time (BT) product, peak carrier deviation, and carrier frequency accuracy between receiver and transmitter contribute to the necessary width of the IF filter. The receiver IF filter should have sufficient width to accommodate the maximum variation in the above parameters so that the received signal will not run into the skirts of the filter. The skirts of the IF filter can introduce an excessive amount of group delay in the higher frequency components of the received data.

The pass band of the IF filter should have very little or no group delay. The more group delay introduced, the more degraded the BER performance of the receiver will become. Rules of thumb for group delay dictate that less

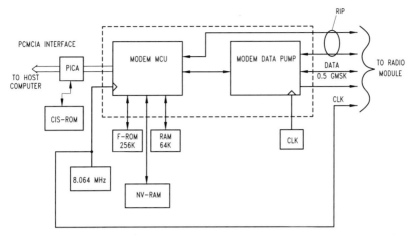

FIG. 5 CDPD data modem PCMCIA type-II card.

than 10% of the bit time is tolerable. This level of performance is also dependent on other factors that influence the BER of the system. These are BT product, signal strength, fading, etc. Phase equalization can also reduce group delay, but if there is no control over the receive-IF filter design, this step can be avoided.

At the heart of this CDPD PCMCIA data communication system is a modem chip set which must be power efficient and fully compatible with the dimensional requirements of the PCMCIA Type II card. The complete chip set contains a modem data pump, a microcontroller, and firmware (which resides in the flash read-only memory (ROM) and is shown in Fig. 5. The data pump delivers the data bits to the CDPD transceiver unit. It can also pump the bits into telephone lines when a wire line is available. The firmware performs error correction and data compression and provides modem attention-type command interface to a personal computer (PC) terminal. Since CDPD was designed to work with standard transmission control protocol/Internet protocol (TCP/IP) products, CDPD must accommodate M-ES mobility, yet keep this movement transparent to the TCP/IP products. CDPD can essentially untether Internet use since the TCP/IP protocol is understood by all Internet computers. Typically, the data rate can be adjusted for 19.2kbps/14.4kbps/9.6kbps.

The transmitter power output is approximately 1 W for CDPD M-ES. This may require isolating the transmitted signal from the local oscillator because synthesizers can lock to the modulated transmit signal resulting in excessive spurious emission. The VCO must have low phase noise to keep the reciprocal mixing to an acceptable level. A typical transmit MMIC chip consists of integrated VCO and power amplifier. The VCO output is sam-

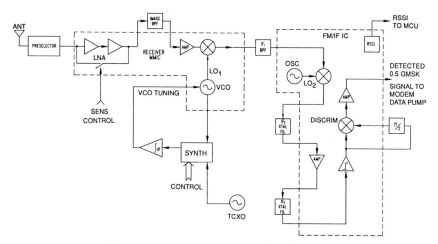

FIG. 6 CDPD modem receiver block diagram.

pled at the local oscillator (LO) monitor port for PLL phase comparison. An additional MMIC PAM is used when higher transmit power is needed. An integrated attenuator on this MMIC PAM provides the transmit power control required for CDPD systems. Power management control is implemented via three distinct operating modes (transmit, receive, and standby), thus making this chip set an ideal choice for CDPD M-ES applications where extended battery life is desirable. The Reed–Solomon coded 0.5-GMSK modulated input data stream directly modulates the transmit MMIC VCO to create high-frequency data signals. The VCO output signal of an approximately +8 dBm is applied to the MMIC PAM. This device has an output power level (P_{1dB}) of 30 dBm. The PAM boosts the high-frequency data signal by approximately 24 dB. The internal voltage-controlled attenuator reduces the power output signal to the required level. The amplifier output is then connected to a diplexer. The diplexer has a bandpass filter (BPF) to the transmit side, limiting the out-of-band energy broadcast by the transmitter. The diplexer feeds the signal to the antenna. Antenna diversity can be implemented when a multipath environment is too severe.

The block diagram of the CDPD PCMCIA modem receiver is shown in Fig. 6. The receive channel consists of a 900-MHz MMIC receiver and FM/IF chip.

The signal from the antenna enters the receiver through the diplexer. The diplexer also has a BPF to the receive side. The purpose of this filter is to select a specific frequency band of interest and reject potentially interfering signals. The selected signal is then connected to a low-noise amplifier (LNA) on the receiver MMIC chip. The LNA boosts the signal and passes it to an external image rejection filter prior to downconversion. The downconverter consists of a VCO and mixer. A 45-MHz IF is used. The mixer needs low power to drive the LO port and provides some conversion gain to the signal path. The receiver VCO is part of a PLL, controlled by an external synthesizer. A command from the radio controller chip will select the LO frequency for the desired signal. A monolithic FM/IF circuit incorporates a second mixer, second LO, limiting amplifier, quadrature detector, received signal strength indicator (RSSI), and audio amplifier. After demodulation, the detected GMSK signal is passed to the modem data pump which delivers data to the host computer via a PCMCIA interface card adapter (PICA).

8.3 MMIC COMPONENTS FOR PCS

In this section we discuss the present ability of MMIC technology to implement the RF transmit/receive (T/R) block in wireless personal communication systems. A general architecture for the T/R block that is implemented in a hand-held wireless unit using time division multiple access/time division duplex (TDMA/TDD) type communications is shown in Fig. 7. The architecture shown is for a quadrature phase shift keying (QPSK) system; other RF T/R front-ends will use the same building blocks, with specifica-

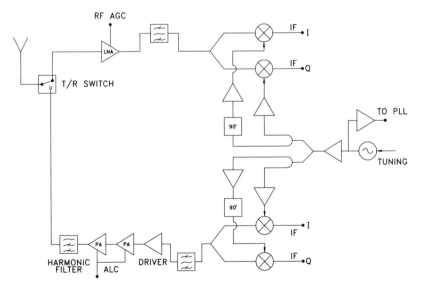

FIG. 7 A general architecture for the RF T/R block that is implemented in hand-held unit in TDMA/TDD systems. If both I and Q are not necessary in either receive or transmit side, one of the I and Q channels can be shut off.

tions determined by the particular system. In the receive path the antenna is followed by a single-pole double throw (SPDT) switch for signal routing and a high gain LNA. The gain of this amplifier is determined by the total system noise figure and a 1-dB compression point of the IF output. The LNA is followed by a quadrature downconverter that may give an IF signal for a second conversion or go directly to baseband signals for digital application-specific integrated circuits (ASICs). In the transmit path, the IF mixer will upconvert the IF signal (30–100 MHz) to the transmit frequency. The transmit filter will allow to pass the transmit frequency which is amplified first through a driver amplifier and finally through a power amplifier to establish the power level. A transmit filter after the power amplifier reduces the harmonic levels produced by the power amplifiers.

MMIC development for PCN/PCS applications is progressing rapidly in the following areas:

1. Low-voltage operation (<5 V)
2. High-efficiency power amplification
3. Low-current receiver section
4. Increased levels of integration
5. Low-cost plastic packaging.

In the following subsections, a specific architecture for a hand-held wireless unit will be discussed and specifications for each MMIC block in that architecture will be presented. The current state of each MMIC block is also given.

The general architecture for the T/R module for a hand-held wireless unit is shown in Fig. 7. The main bands of interest are the following:

1. 0.8–1.0 GHz. Applications: AMPs, ADC, GSM-900, DSRR, CT-2, ISM.
2. 1.46–1.52 GHz. Applications: JDC.
3. 1.70–1.91 GHz. Applications: DECT, DCS-1800, wireless LAN/modem, PHP.
4. 2.4–2.5 GHz. Applications: ISM, LAN.
5. 5.7–5.9 GHz. Applications: ISM, LAN.

For an analog system (like AMPs) and simpler BPSK digital systems, both I and Q ports (Fig. 7) are not needed. This architecture is general enough to be adapted to more complex QPSK schemes or to simpler frequency shift keying (FSK) and BPSK modulation schemes.

8.3.1 MMIC Receiver

Low-Noise Amplifier (LNA)

Typical performance specifications for a low-noise amplifier, in hand-held applications, are given in Table 2. The specifications are derived from

TABLE 2 Typical Electrical Performance Specification for Low Noise Amplifiers in Hand-Held Set of Wireless Applications

Frequency	900 MHz	1.8 GHz	2.4 GHz	5.8 GHz
I_{ds} (mA)	6–10	6–12	8–16	10–20
V_{DS} (V)	3–5	3–5	3–5	3–5
Gain (dB)	10–16	8–14	8–14	8–12
OP1 (dBm)	0 to 2	0 to 3	0 to 3	−3 to 2
OP3 (dBm)	10–14	10–15	10–15	8–12

the operating system needs in the bands of interest. MMIC LNAs are available that can meet present low-power (both low voltage and low current) requirements. Meanwhile, higher gain and lower noise figure (NF) components are being designed for PCS applications, using both metal semiconductor field effect transistor (MESFET) and pseudomorphic high electron mobility transistor (PHEMT) devices. The gate lengths for these advanced devices vary between 0.15 and 0.5 μm. A design trade-off must be made between low dc power, gain, NF, and 1 dB compression point for the LNA. An S-parameter analysis of MESFETs shows that, for the same dc power consumption, the lower the device gate periphery, the smaller the gain circle [10]. The gain circle will also grow smaller in some region of the bias if the dc power consumption is reduced for a given gate periphery. The reason for this is the variation in the device input/output Q with a change in bias. The higher the input Q, the smaller the input gain circle will be [10]. The smaller the gain circle, the greater the difficulty in designing matching circuits will be. Figure 8 shows the formulae for calculating the input and output Qs of a FET device.

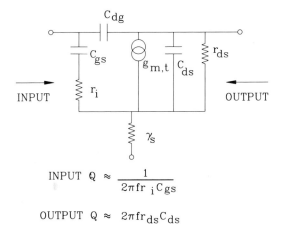

$$\text{INPUT Q} \approx \frac{1}{2\pi f r_i C_{gs}}$$

$$\text{OUTPUT Q} \approx 2\pi f r_{ds} C_{ds}$$

FIG. 8 Simplified equivalent circuit of a MESFET. The higher the Qs, the smaller the gain circles will be and the more difficult it is to match the transistors.

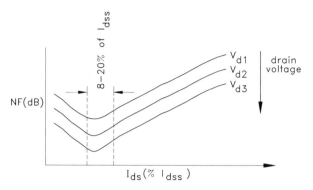

FIG. 9 Minimum noise figure as a function of drain-source current and voltage. The lower the drain voltage, the lower the noise figure will be. The lowest noise figure occurs at a drain voltage slightly higher than the knee voltage. The valley becomes narrower, as Vds approaches the knee voltage.

Noise figure is a function of drain current and drain voltage. For MESFET devices, minimum noise figure occurs at approximately 8–25% of Idss for a Vds = 2–4 V. The lower the drain voltage, the lower the minimum noise figure (Fig. 9) (but with a narrower valley). This illustrates that once the device is biased for low power and low noise figure, the 1-dB compression point is to be compromised as well. This places severe restrictions on the isolation specification of the T/R switch in a full duplex system. In general, the designer must consider the given specifications for gain, NF, and 1-dB compression and determine the required gate periphery with the lowest dc power consumption and highest circuit yield. The circuit yield will increase as the gain circles and noise circles become bigger because the circuit is less sensitive to device parameter variation (mainly in G_m and C_{gs}).

A circuit schematic for a broadband low-noise amplifier is shown in Fig. 10. The amplifier is designed to operate over the band 0.8–1.0 GHz with a noise figure of 3.0 dB and a gain of 19–20 dB from a supply voltage of 3–5 V. The amplifier draws a nominal drain current of 10 mA and has a 1-dB compression point of −2 dBm. The LNA has two stages of amplification and uses resistive loads at the drain. The active devices are enhancement type MESFETs and operate close to the knee voltage to achieve lower dc power consumption and low noise figure.

Figure 11 shows the frequency response of the LNA. Table 3 summarizes the key parameters of the LNA. A feature of this LNA is the use of single polarity supply by utilizing resistive self-biasing. This slightly degrades the efficiency of the device. The MMICs using GaAs MESFET need a negative (−Ve) supply to the gate. The negative supply can be created by a

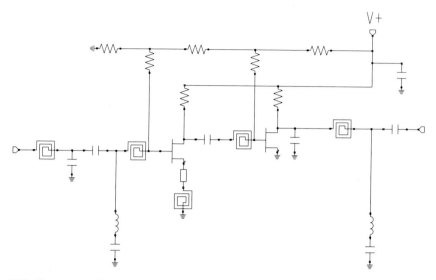

FIG. 10 Circuit schematic of a low-power PCS LNA designed to operate over 0.8–1.0 GHz. The LNA consumes 40–50 mW of dc power and has a gain of 19 dB, NF of 3 dB, and 1-dB compression point of −2 dBm at the output.

simple switched capacitor inverter so long as the −Ve supply does not draw much current (≤2 mA). Low-noise amplifiers for a hand-held phone have been reported in the literature [11–13]. The objective in all of these amplifiers is to achieve low noise and high intercept at the lowest possible dc power.

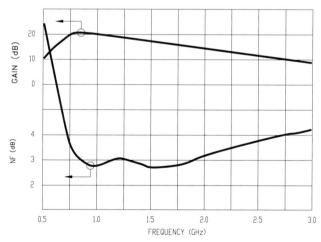

FIG. 11 Response of the LNA over frequency. The responses show gain and noise figure. The LNA is meant for 0.8–1.0 GHz band. V = +5V, I = 10 mA.

TABLE 3 Summary of Key Parameters of the LNA Shown in Fig. 10

Frequency	0.8–1.2 GHz
Gain	18–20 dB
Noise figure	3–3.5 dB
DC voltage	3–5 V
DC current	8–12 mA
OP1	−2 dBm
OP3	10 dBm

Downconverter

The next MMIC in the receiver path after the LNA is a downconverter. Figure 7 (referenced earlier) shows a downconverter with the LO applied at 90° out-of-phase to create I and Q channels. In many applications, such as double-down conversion schemes or simple BPSK and FSK schemes, I–Q channels are not necessary, at least at high IF frequency. In those applications one of the channels is disabled. The downconverter can use one of two topologies, the active Gilbert cell or the passive FET/diode quad, to realize a balanced mixer.

The IF frequency is in the 30- to 500-MHz range. Figure 12 shows the schematic for a Gilbert cell mixer [14]. The operation of this type of mixer is thoroughly described in reference [15]. Under small signal conditions, the output current ($i_{out}(t)$) at the IF port is given by

$$i_{out}(t) \sim K \cdot V_{rf}(t) V_{Lo}(t),$$

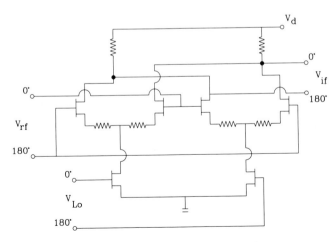

FIG. 12 Schematic of an active Gilbert cell mixer [5].

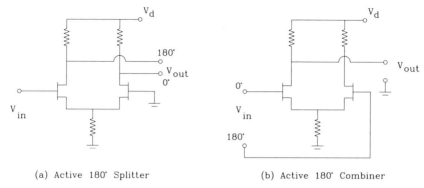

(a) Active 180° Splitter   (b) Active 180° Combiner

FIG. 13 Active 180° splitter and combiner using differential amplifiers.

where, K is a scaling factor, V_{rf} is the voltage across the RF port, and V_{Lo} is the voltage across the LO port. The LO and RF voltages are applied through a 180° splitter. The 180° splitter at this low microwave frequency (≤ 2.5 GHz) can take the form of an active balun or differential amplifier. The active combiner at the IF ports can also be a differential amplifier with single ended output as shown in Fig. 13. The main advantage of an active mixer is its lower LO power requirement and its ease of multiplication at low power levels (both RF and LO). On the other hand, the passive mixer, using diode or FET quads, consumes less dc power, but needs more LO power for mixing. Figure 14 shows a passive mixer using FET quads. The main advantages of using FET quads over the diode quads are:

1. A MESFET channel without drain bias is a very linear resistor and thus leads to lower intermodulation distortion.
2. FET ring quads provide inherent isolation between LO, RF, and IF.
3. FET ring quads needs less LO power than diode mixers.

The diode quad double-balanced mixers are discussed in reference [16]. The topology and operation principle of diode quad mixers are the same as those of FET quad mixers. There are other active mixers that use dual-gate MESFET devices, in which LO and RF are applied to two gates (LO to control gate, RF to the other gate) for mixing. In this case, LO to RF isolation is easily achieved due to dual-gate MESFET properties. For personal communication applications, the main goal is to achieve the mixing with high third-order distortion with as low a dc power as possible. Table 3 shows a summary of key parameters for a downconverter that will accommodate the system requirements. For example, a downconverter with a NF of 15–17 dB, conversion gain of 6–8 dB, and P_{IP3} at input of -8 to -6 dBm is sufficient for DECT system requirements. It should be noted that the dc current consumption (Table 4) is for only one mixer.

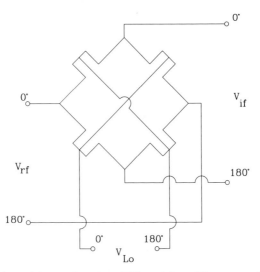

FIG. 14 FET quads used for passive mixer. Differential amplifiers are used as 180° splitter and combiner.

If both I and Q channels are needed, the dc power consumption will be twice as high, provided that the P_{IP3} requirement remains unchanged. Figure 15 shows the conversion gain of a downconverter [17] using FET quads and differential amplifiers as the active 180° splitter and combiner. Table 5 details the key measured parameters of the downconverter. The parameters are optimized for a wireless LAN application. Figure 16 shows the photograph of a MMIC LNA and downconverter. The total chip size is 1.25 × 2.50 mm.

TABLE 4 Key Parameters of the MMIC Downconverter That Are Needed in Different PCS Systems

RF frequency range	0.8 to 1.0 GHz	1.8 to 2.0 GHz	2.3 to 2.5 GHz
IF range	30 to 200 MHz	30 to 300 MHz	30 to 300 MHz
Lo power	0 to 5 dBm	0 to 5 dBm	0 to 5 dBm
Conversion gain	12 to 14 dB	10 to 12 dB	10 to 12 dB
Noise figure	12 to 15 dB	12 to 15 dB	12 to 15 dB
OP_{1dB}	−10 to −5 dBm	−10 to −5 dBm	−10 to −5 dBm
V_{DS}	3 to 5 V	3 to 5 V	3 to 5 V
I_{ds}	10–15 mA	10–20 mA	10–20 mA

The actual specifications will be determined by the system needs.

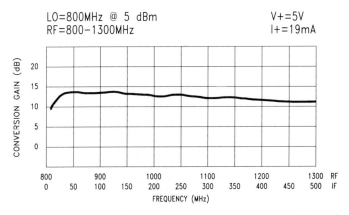

FIG. 15 Measured conversion gain of an MMIC downconverter using FET quads as mixer and differential amplifiers as 180° splitter and combiner.

8.3.2 MMIC Transmitters

Upconverter (UC)

The first block in the transmitter path from the IF side is the upconverter. An IF frequency of 30–500 MHz is upconverted to the transmit frequency. The same mixer topology discussed for the downconverter can be used for upconversion. The Gilbert cell active mixer or diode/FET quad passive mixer is used for the UC. In some applications (such as DECT), the VCO itself can be directly modulated with the baseband data. In the TDMA/TDD system (or any other system where transmit and receive do not occur at the same time), it is possible to use a directly modulated VCO instead of a UC. Under such circumstances, the VCO has to switch between the receive and the transmit frequency within a specified period of time. For the DECT system, a change of 130 MHz in VCO frequency (worst case) has

TABLE 5 Measured Key Parameters of the MMIC Downconversion

RF range	0.8–1.2 GHz
IF range	0–500 MHz
Lo power	3 dBm
Conversion gain	12–14 dB
Noise figure	15 dB
DC voltage	5 V
DC current	20 mA

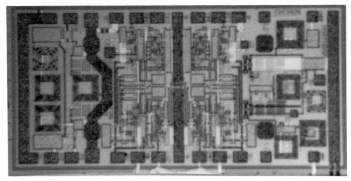

FIG. 16 MMIC photograph of LNA and downconverter together. It measures 1.25 × 2.5 mm.

to happen in 400 ms. In frequency-hopping spread-spectrum techniques the VCO is also directly modulated by a pseudorandom code sequence. The VCOs in frequency-hopping systems have a fast settling time (≤ 0.3 ms) and research is still ongoing to make a fast direct digital synthesizer (DDS) chip at frequencies >900 MHz that are cost effective [18].

Here, we will discuss a UC where the VCO frequency remains the same for transmit and receive modes. Figure 17 shows the conversion gain of a UC MMIC that uses FET quads for mixer. The UC has IF, RF, and LO amplifiers. The UC shows a gain of >10 dB at RF frequencies >800 MHz. The 1-dB compression point at the output is approximately 0 dBm. The key parameters of the UC are given in Table 6. The parameters are optimized for wireless LAN type applications. The critical parameters at other frequencies are similar to those given in Table 4, with the RF and IF interchanged.

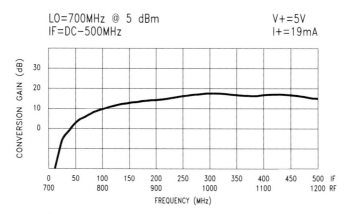

FIG. 17 Measured conversion gain of an MMIC upconverter using FET quads as mixer and differential amplifiers as 180° splitter and combiner.

TABLE 6 Measured Key Parameters of the MMIC Upconversion

RF range	0.8–1.2 GHz
IF range	100–500 MHz
Conversion gain	10–12 dB
Lo Power	3 dBm (min)
Noise figure	15 dB (typical)
DC voltage	3–5 V
DC current	20 mA

Power Amplifier

The PAM at the transmitter output gives the necessary gain and output power. The main function of the power amplifier is to produce the required amount of linear power (high 1-dB compression point) with the highest possible power-added efficiency in order to increase the battery life of the system. The power amplifier in the RF block is the largest user of battery power and is generally turned on only when it is transmitting. MMIC components used in the PA are not high Q elements, especially the on-chip coils. This is a disadvantage, since inductors (or coils) are the single greatest power dissipator in the amplifier circuits. One way to improve the coil Qs is to broaden the line width. This solution, however, requires expensive GaAs real estate. Alternatively, in order to improve the efficiency of a MMIC power amplifier, it is desirable to place the output circuit on a high Q substrate that can accommodate high Q matching elements (for example, transmission lines with greater metal thickness). These circuits, even though strictly not MMIC, are a combination of MMICs and hybrids and give the best of both worlds by improving the efficiency of the amplifiers and reducing the size of the chip.

Figure 18 shows the photograph of an MMIC power amplifier designed for 0.8- to 1.0-GHz applications. The amplifier uses two stages; the first stage has 450 μm of gate periphery and the second stage has 1200 μm. Approximate power output is 200–250 mw/mm at 1-dB gain compression. The chip uses on-chip coils for matching circuits, but bypasses the on-chip coils for dc biasing. In this way, the unnecessary dc power consumption is avoided in low-Q MMIC coils. The measured power and efficiency data are plotted in Fig. 19. At higher gain compression (≥ 2 dB), the efficiency goes over 50%. The power output at a 1-dB gain compression is approximately 250 mW. This shows that the GaAs MESFET MMIC can surpass the performance of silicon technology at this lower microwave frequency.

Currently, research is focused on making cost-effective, high-power MMICs using GaAs technology. It appears that at low-voltage operation (<5V) GaAs MESFETs will give higher power at a higher efficiency than

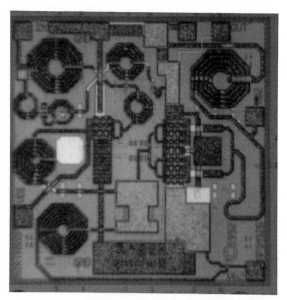

FIG. 18 MMIC layout of a 0.8- to 1.1-GHz power amplifier for PCS applications.

will silicon bipolar at devices. One disadvantage with GaAs MESFETs is the requirement of a negative supply for gate biasing. Since the gate does not draw much current (≤ 1 mA), it is possible to use a switched capacitor to provide this negative supply. Other types of on-chip negative voltage generators are also possible to mitigate this disadvantage. GaAs–heterostructure bipolar transistor (HBT) devices which are suitable for low-voltage, high-

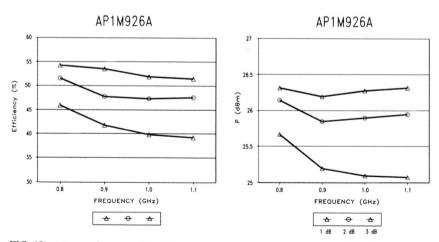

FIG. 19 Measured power-added efficiency and power output for the amplifier shown in Fig. 18. The amplifier shows more than 50% efficiency at higher compression points (≥ 2 dB).

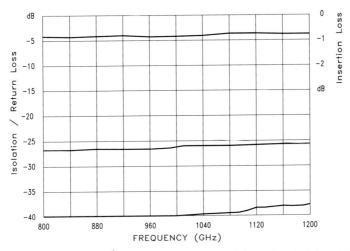

FIG. 20 Measured performance of an MMIC T/R switch for PCS applications. The switch has an SPDT topology.

efficiency applications and require only a single positive supply, will provide an alternative as the technology matures.

Transmit/Receive Switch

The T/R switch routes the signal between receive and transmit paths. The required characteristics of the T/R switch are low insertion loss, high isolation, and low voltage standing wave ratio (VSWR). The T/R switch has an SPDT topology. Figure 20 shows the measured performance of a T/R switch over a 0.8- to 1.2-GHz band. The switch itself has an input 1-dB compression of 24 dBm. The measured response shows an insertion loss of better than 1 dB and input/output return loss of better than −20 dB.

The isolation between ports is better than 25 dB. The higher the isolation, the better it is for the LNA in the receive path. Even though the LNA is switched off during transmission for TDMA systems, there may be enough leakage from the transmit to the receive path to damage the LNA at high power transmission. Since GaAs MESFET switches consume negligible DC power, the isolation can be achieved by adding more switching sections (series and shunt FETS) in the two paths; the increase in insertion loss can be compensated by an increase in gain. The loss in NF can be tolerated in many systems requiring high input IP3.

Voltage-Controlled Oscillator

The MMIC VCO (in some systems) has to tune to the frequency of interest with low phase noise and high settling speed. Figure 21 shows the measured performance of a VCO tuned with an off-chip varactor. The VCO draws a dc current of 18–22 mA from a supply voltage of 3–5 V. The

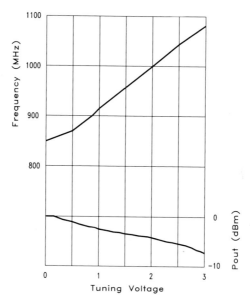

FIG. 21 Measured VCO tuning over a frequency range 850 to 1090 MHz. The varactor voltage is varied from 0 to 3 V. The power output is approximately 0 dBm.

power output of the VCO is approximately 0 dBm. The VCO can be tuned in various bands of interest up to 2.4 GHz by using different varactors. The frequency is tuned with a maximum voltage of 3 V on the varactor. The VCO has a phase noise of -110 dBc at 1 MHz offset from the carrier frequency and settling time of less than 4 μs. For frequency-hopping spread-spectrum applications, (according to the FCC Regulation Part 15.24F), the VCO setting time should be less than 200 ns. Much interest has been focused on developing fast VCOs using digital techniques such as the DDS.

8.3.3 Highly Integrated Transceivers

Single-Chip MMIC Transceiver

In the long term, the cost-effective solution lies in the total integration of the RF front-end on a single chip. This approach, however, has formidable technical challenges in the areas of isolation and leakage. The direct VCO to antenna leakage (carrier leakage) can pose a big problem. Figure 22 shows a MMIC front-end chip [10] for wireless personal communication. The functional block diagram of the MMIC is shown in Fig. 23. Except for the I–Q ports, the MMIC chip contains all other RF functions described above in the receiver and transmitter sections.

Work is progressing toward mounting this complex chip inside a low-cost plastic package without any performance degradation. One specific

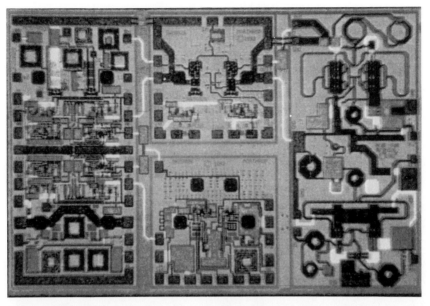

FIG. 22 A complete GaAs MMIC RF front-end for PCS applications with all the necessary building blocks. The chip measures 2.250 × 3.375 mm².

area with regard to plastic packaging which is of prime concern is heat dissipation.

The performance of a complete transceiver MMIC in a plastic/ceramic package have recently been published. One such transceiver chip at 2.4 GHz downconverts to an IF of 915 MHz [19]. It is shown in Fig. 24. The MMIC is shown in a 28-lead surface mount technology (SMT)-compatible cofired

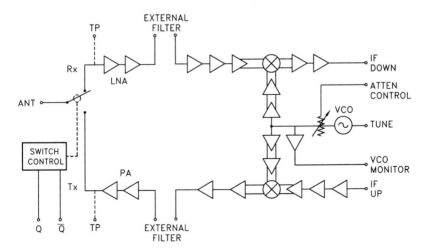

FIG. 23 The RF functional block for the integrated MMIC RF front-end shown in Fig. 22.

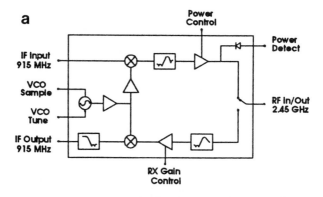

FIG. 24 2.4-GHz transceiver chip for PCS application (a). Functional block diagram (b) of an approach of packaging the MMIC in SMT-compatible 28-lead cofired ceramic package. The MMIC itself measures 5.9 × 3.8 mm. (Reprinted with permission from T. Apei *et al.* [19], and IEEE, © 1994.)

ceramic package. The MMIC itself can deliver 20–24 dBm of RF power at 2.4 GHz with 350–550 mA of current at 5 V. In the receiver chain the MMIC draws 90 mA at 5 V, including all the supporting and VCO circuits.

Figure 25 shows the block diagram of a two-chip solution for PCS applications [20]. Each chip shown within the dashed line boxes of Fig. 25 measures 1.4 × 2.8 mm. By partitioning the total chip in an appropriate

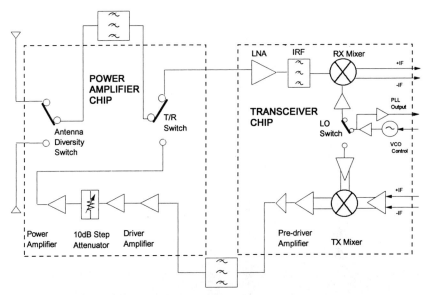

FIG. 25 A two-chip set block diagram for PCS transceiver applications. (Reprinted with permission from B. Khabbaz et al. [20], and IEEE, © 1994.)

way, it is possible to use different processes for the two chips. For example, a low-current process can be used for the transceiver chip and a high-current process for the power amplifier chip. The actual chips are shown in Fig. 26. All of the components are densely packed on the MMIC surface. Both chips have been mounted in quadrature small outline package (QSOP)-24 plastic packages. The power amplifier chip in QSOP-24 pin plastic package is shown in Fig. 27. The power amplifier exhibits an output of 21 dBm (P_{1dB}) with 25% power-added efficiency. This includes the loss due to the T/R switch on the power amplifier chip. The overall receiver single sideband (SSB) noise figure and gain are 4.7 and 11 dB, respectively.

8.4 FUTURE TRENDS IN PCS

There has been an explosive growth of PCS-related wireless products over the past 6–7 years. This growth is likely to continue in the next 10–15 years until untethered communication becomes a worldwide reality. The PCS infrastructure is being developed to accommodate more user capacity by switching over from simple FDMA analog communication to digital communication.

At present, the simple TDMA along with standard digital modulation (QPSK, BPSK) techniques can accommodate a maximum of three users per

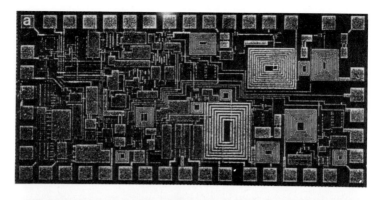

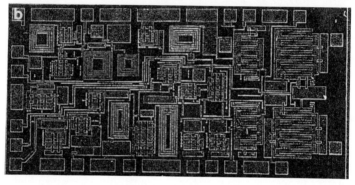

FIG. 26 Photographs of the two chip sets for PCS applications. (a) MMIC transceiver. (b) MMIC power amplifier. (Reprinted with permission from B. Khabbaz et al. [20], and IEEE, © 1994.)

channel, which is approximately 30 kHz wide in a 6.66-ms time slot. The future trend seems to be leading more toward CDMA, where all of the channels in all of the cells can be reused simultaneously without going over a certain interference level (typically, the interference level is kept not more than 18 dB above the required signal level). The use of TDMA or CDMA with different digital modulations increases the system design complexity over that of the simple analog (FDMA) schemes. New techniques, like CDMA, call for an increased level of signal processing steps as compared to the analog system [21]. Hence, the new generation of PCS products will be built around digital signal processing (DSP).

As the spectrum allocation for low Earth-orbit global communication (including projects like Iridium or Globalstar) becomes final, there will be rapid growth and demand all over the world for light, small-sized, complex, hand-held units. PCS units that cover the bands 1.8–2.0 GHz (specifically allocated for PCS), 2.4 ± 0.05 GHz, 0.8–1.0 GHz, and 5.8 ± 0.05 GHz, will require highly integrated front-ends that are power efficient so as to

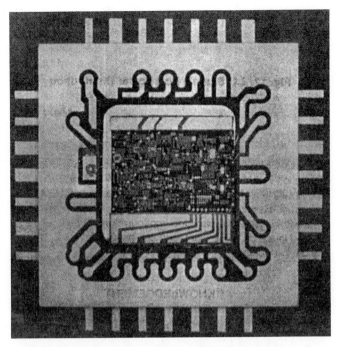

FIG. 27   Power amplifier chip in QSOP-24 pin plastic package.

increase continuous usage. However, the conflict between the two requirements of (1) the larger number of building blocks to process complex signal and (2) the lower power consumption must be considered. The system architecture for RF front-ends and the accompanying MMIC design must incorporate the device technology that will utilize low voltage, low current, and possibly a single polarity supply. A prime candidate for this new generation of MMICs is HBT technology. A nearer-term candidate is a MESFET PAM with an on-chip negative voltage generator. A higher level of integration with proper partitioning of MMICs will result in achieving complex functions in small-size, hand-held units. The MMICs must be designed with minimum dependence on discrete external components which compromise size and power dissipation.

It should be understood that the evolving system standards for PCS will drive RF front-end specifications, such as frequency range, channel bandwidth, output power level, receiver sensitivity, minimum signal to noise ratio (including adjacent channel), and cochannel interference. For example, lower power requirements in digital cordless applications (see Table 7) will require integration of the power amplifier and the receiver on the same chip with T/R routing switch.

To make the units low cost and light weight, packaging technology will

TABLE 7 System Standards for the United States, Europe, and Japan [21]

System	Application	Channel	Access/duplex	TxFr	RxFr	Tx power
IS 54/55	Digital cellular (USA)	30 kHz	(a) FDMA/FDD (b) TDMA/TDD	824–849 MHz	869–894 MHz	0.6 W
IS 95	Digital cellular (USA)	N/A	CDMA	824–849 MHz	869–894 MHz	0.6 W
PDC	Digital cellular (Japan)	25 kHz	TDMA/TDD	810–826 MHz	940–956 MHz	0.3–0.8 W
GSM	Digital cellular (Europe)	200 kHz	TDMA/TDD	890–915 MHz	935–960 MHz	0.8–2 W
DECT	Digital cordless (Europe)	1.7 MHz	TDMA/TDD	1.88–1.9 GHz	1.88–1.9 GHz	0.25 W
PHP	Digital cordless (Japan)	300 kHz	TDMA/TDD	1.88–1.9 GHz	1.88–1.9 GHz	0.01 W

play an important role. Plastic packages must be developed to cover frequencies up to 5.8 GHz. The future trend will be to use a metalized plastic package to improve heat dissipation and provide better RF grounding. RF front-ends will be in either a leadless, ball grid array (BGA) plastic package or simply on substrates. These new emerging technologies must be used to lower the cost, save space, and reduce the assembly time. The substrate with BGA will be a multilayer MIC technology (like low-temperature cofired ceramic (LTCC) that can accommodate all off-chip components including bypass capacitors, RF chokes, and resistors. One simple LTCC substrate will accommodate all the interconnections. Figure 28 shows the layout of a LTCC substrate for a complex PCS transceiver at 2.4 GHz. In the future, the RF connections may be brought out at the bottom of the BGA and will use chip on board (COB) technology for environmental protection. Figure 29 shows a road map for the future trend in the development of highly integrated MMIC front-ends for PCS.

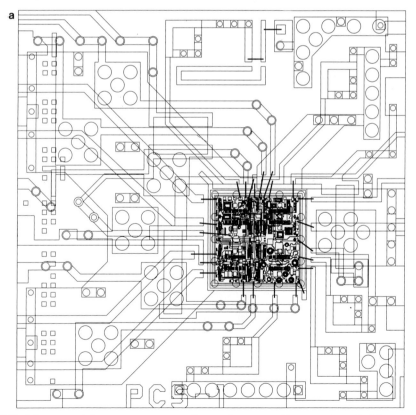

FIG. 28 (a) Multilayer MIC (LTCC) substrate for a 2.4-GHz transceiver. (b) The transceivers block diagram with on chip MMIC and off-chip LTCC components.

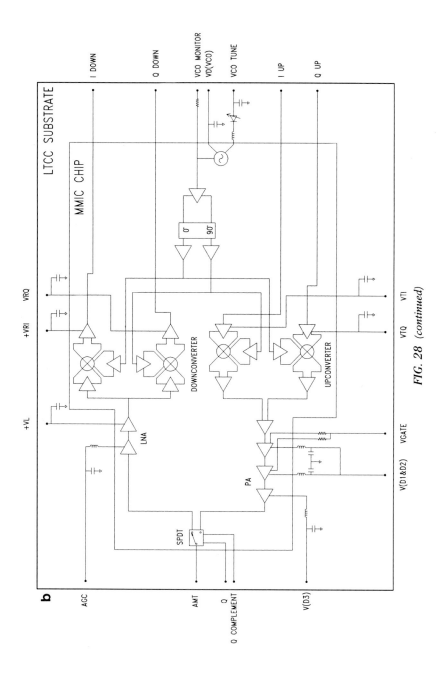

FIG. 28 (continued)

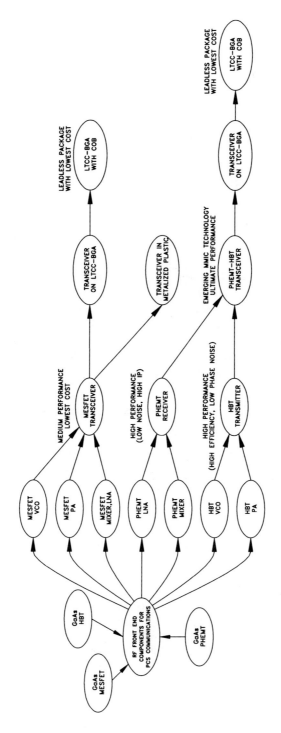

FIG. 29 Roadmap for integrating various functions of PCS transceivers in a package.

There are presently three competing MMIC technologies—MESFET, HBT, and PHEMT—that are used to realize the components in the RF front-ends for PCS applications. The MESFET technology is the most mature and provides the lowest cost approach, though it has lower performance when compared with HBT in terms of PA efficiency and VCO phase noise. Also, the HBT needs only one power supply, which is an added convenience. MESFETs also has higher noise figure in LNAs and mixer when compared with PHEMT technology. An emerging MMIC technology is attempting to achieve an integration of HBT and PHEMT by selective MBE growth [22]. This will give an ultimate performance in terms of lowest noise figure (LNA, mixer, and VCO) and highest power-added efficiency. It is expected to be the technology of future MMIC chips (PHEMT and HBT) that will meet all of the stringent specification requirements of a PCS RF front-end.

The integrated MMIC transceiver will be mounted on a multilayer substrate (LTCC) that will support other off-chip supporting circuitry and also provide an interface through leadless BGA to the rest of the signal processing circuitry. Work is in progress to use COB technology for encapsulating the transceiver to protect it from harsh environments. It is expected that the COB technology will have an effect similar to that that the plastic has on the RF performance of the MMICs. It is important to remember that the goal of the MMIC is to achieve the ultimate RF performance at the lowest cost in future PCS applications.

REFERENCES

[1] Michael Paetsch, *The Evolution of Mobile Communications in the U.S. and Europe*, Chap. 9, pp. 225–248. Artech House.

[2] Mark Williams and Tom Schlitz, *A 1.96 GHz Chipset for Personal Communications*, pp. 468–478. Proceedings of the Second Annual Wireless Symposium Sponsored by Microwave and RF Magazines, February 1994.

[3] Pacific Communication Sciences, Inc., 10075 Barnes Canyon Road, San Diego, CA 92121. *Technical Notes on CDPD, The Future of Cellular Data Communications.*

[4] Cincinnati Microwave, Inc., *Cellular Digital Packet Data*, (A Primer on How CDPD Can Change the Way Your Organization Does Business), May 26, 1994.

[5] CDPD Industry Input Coordinator, P.O. Box 97060, Kirkland, WA 98083-9760. *Cellular Digital Packet Data System Specification*, carrier release V.1.0, April 30, 1993.

[6] *TRENDS, PC Magazine*, March 14, 1995.

[7] J. Mondal, S. Geske, T. Kotsch, A. Laundrie, G. Dietz, E. Bogus, J. Blubaugh, S. Moghe, *Highly Integrated GaAs MMIC RF Front End for PCMCIA PCS Applications*, pp. 169–174. Conference Proceedings, RF Expo East, October 19–21, 1993.

[8] K. Murota and K. Hirade, GMSK modulation for digital mobile radio telephony. *IEEE Transactions on Communications*, Vol. COM-29, No. 7, pp. 1044–1050, July 1981.

[9] F. Kostedt and J. C. Kemerling, Practical GMSK data transmission. *Wireless Design and Development*, Vol. 3, January 1995.

[10] Kenneth R. Cioffi, Ultra-low DC power consumptions in monolithic L-band components.

IEEE Transactions on Microwave Theory and Techniques, Vol. 40, No. 12, pp. 2467–2472, December 1992.
[11] Michael Frank et al. *Wireless Applications for GaAs Technology,* pp. 143–153, Proceedings of the 1st Annual Wireless Symposium, January 12–15, 1993.
[12] Charles Huang, *GaAs MMIC for PCS Applications,* pp. 368–371. Proceedings of the 1st Annual Wireless Symposium, January 12–15, 1993.
[13] David Williams, *An Integrated Microwave Radio Transceiver for WLAN Applications,* pp. 305–313. Proceedings of the 1st Annual Wireless Symposium, January 12–15, 1993.
[14] Jose I. Alonso and Juan C. Sanchez, A simple design technique for the Gilbert-cell in MMIC technology; Application to a DSB modulator. *1993 GaAs IC Symposium Digest,* pp. 223–226, San Jose, October 10–13, 1993.
[15] Paul R. Gray and Robert A. Meyer, *Analysis and Design of Analog Integrated Circuits,* Chap. 10. Wiley, 1977.
[16] Stephen A. Maas, *Microwave Mixers,* Chaps. 4–8. Artech House, 1986.
[17] Jyoti P. Mondal et al., *Highly Integrated GaAs MMIC RF Front-End for PCMCIA PCS Applications,* pp. 169–174. RF Expo East, Tampa, October 19, 1993.
[18] Robert C. Dixon, *Spread Spectrum Systems,* Chap. 22. Wiley, New York, 1984.
[19] Tom Apel et al., *A GaAs MMIC Transceiver for 2.45 GHz Wireless Commercial Products,* pp. 15–18. IEEE 1994 Microwave and Millimeter-Wave Monolithic Circuits Symposium, May 22–25.
[20] B. Khabbaz et al., *A High Performance 2.4 GHz Transceiver Chip Set for High Volume Commercial Applications,* pp. 11–14. IEEE 1994 Microwave and Millimeter-Wave Monolithic Circuits Symposium, May 22–25.
[21] P. Bonner, *Architecting RF Subsystems Using Integrated R.F. Circuits for Wireless Communications,* pp. 70–79. Proceedings of the Second Annual Wireless Symposium, Santa Clara, February 15–18, 1994.
[22] D. W. Streit et al., Monolithic HEMT-HBT integration by selective MBE. *IEEE Transcripts on Electronic Devices,* Vol. 42, No. 4, pp. 618–623, April 1995.

9
Satellite Communications

Ramesh K. Gupta
COMSAT Laboratories, Clarksburg, Maryland

9.1 Introduction
9.2 Evolution in Communications Satellites
9.3 MMIC Technology in Satellite Transponders
9.4 Space-Qualified MMIC Components
    9.4.1 C-Band MMIC Amplifier
    9.4.2 Ku-Band MMIC Amplifier
    9.4.3 C- and Ku-Band Attenuators for Transponder Gain Control
9.5 Space-Qualification of MMICs
9.6 MMIC Subsystem Design Examples
    9.6.1 Low-Noise Receivers
    9.6.2 Microwave Switch Matrices (MSM)
    9.6.3 Phased-Array Transmit Antennas
    9.6.4 Solid-State Power Amplifiers
9.7 Future Direction
    References

9.1 INTRODUCTION

Gallium Arsenide (GaAs) monolithic microwave integrated circuits (MMICs) are increasingly being used in the design of microwave subsystems for on-board communications satellite and ground terminal applications. Key technology factors contributing to this trend are the relative maturity of MMIC design techniques, improvements in fabrication techniques resulting in higher yields, uniform performance characteristics for MMIC components, space qualification for selected MMICs, and demonstrated radiation resistance of GaAs field effect transistor (FET) devices. Furthermore, reduction in size and mass of satellite transponder subsystems by using reproducible GaAs MMICs is crucial in decreasing overall subsystem integration costs and spacecraft launch costs [1–3]. Enhanced component reliability because of minimum bond wire connections and the resulting increase in communications payload life is another key driver for GaAs MMIC insertion in satellite transponders. Although the impact of rapidly emerging MMIC technology on communication satellite payloads was assessed in the early 1980s [4], adaptation of MMIC technology in satellite transponders has been relatively slow because of long technological development cycle time.

Accurate device and circuit models [5], design techniques for fabrication process tolerant designs [6], improvement in process yields [7] by using statistical process control (SPC), and space-qualification of MMICs [8] using acceptable standards have been some of the key areas in which significant developments have taken place during the past several years. These developments have contributed to performance demonstrations of several transponder subsystems for on-board satellite applications, such as receivers [9], microwave switch matrix arrays [10–12], active single- and multibeam phased-array antennas [13,14] for beam reconfigurability and scanning, and high-efficiency solid-state power amplifiers (SSPAs) [15]. MMIC components have already been flown in space in the in-orbit test transponder (IOTT) for the ITALSAT spacecraft launched in 1991 [16], and several types of other MMICs have been space qualified for various commercial satellite programs [8].

Changes in the regulatory environment for communications satellites and competition with optical fibers has made economic factors, such as satellite development, launch costs, and procurement cycle, major drivers for definition and design of future spacecraft [17]. Emphasis has shifted toward technologies that enhance the competitive position by minimizing cost per circuit. In this chapter, these economic and technology trends are explored and the evolution of communications satellite systems is presented. Key factors for use of MMIC technology in future spacecraft transponders are discussed. Finally, examples of some of the transponder subsystems designed by using MMIC technology are presented.

9.2 EVOLUTION IN COMMUNICATIONS SATELLITES

Communications satellites have witnessed an explosive growth in traffic over the past 25 years [18] because of rapid increase in global demand for voice, video, and data traffic. For example, the International Telecommunications Satellite Organization (INTELSAT) I (Early Bird) satellite, launched in 1965, carried only one wideband transponder operating at C-band frequencies (6-GHz uplink and 4-GHz downlink) and could support 240 voice circuits. A large 30-m-diameter antenna was required. In contrast, the communications payload of INTELSAT VI series of satellites, launched from 1989 onward, provides 50 distinct transponders operating over C- and Ku-bands [19], providing the traffic carrying capacity of approximately 33,000 telephone circuits. By using digital compression and multiplexing techniques, INTELSAT VI is capable of supporting 120,000 two-way telephone channels and three television channels [20]. The effective isotropic radiated power (EIRP) of these satellites is sufficient to allow use of much smaller Earth terminals at Ku-band (E1: 3.5 m) and C-bands (F1: 5 m) together with other larger Earth stations. Several technological innovations have contributed to this increase in number of active transponders required to satisfy growing traffic demand. A range of voice, data, and facsimile services is available today through modern communications satellites to fixed and mobile users. The satellite services may be classified into three broad categories [21].

Fixed Satellite Services (FSS)

The signals are relayed between the satellite and relatively large fixed Earth stations. Terrestrial land lines are used to connect long-distance telephone voice, television, and data communications to these Earth stations. Figure 1 shows the fixed satellite service configuration. The FSS were the first services to be developed for global communications. The International Telecommunications Satellite Organization (INTELSAT), with headquarters in Washington, DC, was formed to provide these services. INTELSAT has more than 130 signatories today, the largest being the US signatory—COMSAT Corporation. With satellites operating over the Atlantic, Indian, and Pacific ocean regions, INTELSAT provides a truly global communication service to the world. Several regional satellite systems have been de-

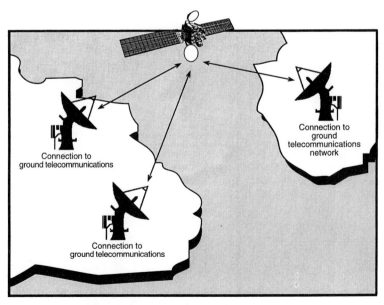

FIG. 1 Fixed satellite communications services.

veloped around the world for communications among nations within a continent or nations sharing common interests. Such systems include Eutelsat, Arabsat, and Asiasat. For large countries like the United States, Canada, India, China, Australia, and Indonesia (with approximately 5000 widely scattered islands) fixed satellite services are being used to establish communications links for domestic use. The potential for global coverage, and the growth in communications traffic as world economies are becoming global, has attracted many more private and regional satellite operators with several new systems being planned for operation in years 2000 and beyond.

Direct Broadcast Satellite (DBS) Services

Direct broadcast satellites use relatively high-power satellites to distribute television programs directly to subscriber homes (Fig. 2) or to community antennas from which the signal is distributed into individual homes by cable. The use of medium-power satellites has stimulated growth of available TV programs in countries like India and China through Asiasat. Launch of DBS services with DIRCTV will provide multiple pay channels via an 18" antenna. The growth of this market will be driven by the availability of low-cost receive equipment consisting of an antenna and a low-noise block downconverter (LNB). GaAs MMICs are likely to play a major role in the low-cost realization of high-performance LNB units.

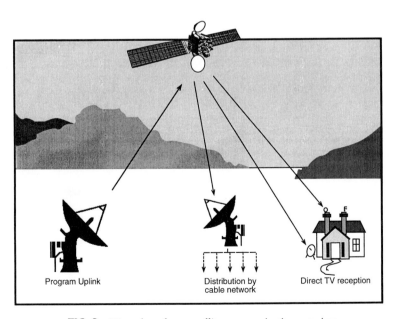

FIG. 2 Direct broadcast satellite communications services.

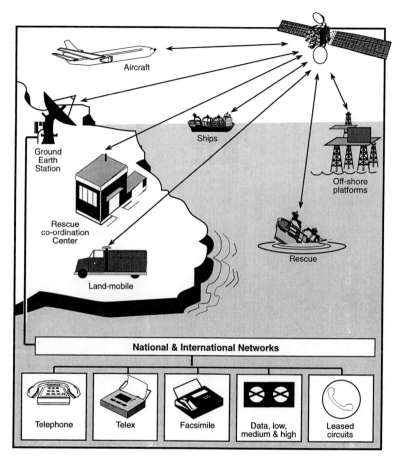

FIG. 3 Mobile satellite communications services.

Mobile Satellite Services (MSS)

For mobile satellite services, the communications takes place between a large fixed Earth station and a number of smaller Earth terminals fitted on vehicles, ships, boats, and aircraft (Fig. 3). The International Maritime Satellite Organization (Inmarsat), headquartered in London, was created in 1976 to provide space segment for improving maritime communications, improve distress communications, and enhance the safety of life at sea. Subsequently, the mandate was extended to include aeronautical services and land-mobile services to trucks and trains. Launch of the Inmarsat 2 series of satellites, started in 1989 and improved communications capacity by providing a minimum of 150 simultaneous maritime telephone calls. This number can be further increased by using digital technology. With the launch of Inmarsat 3 spacecraft, which cover the Earth with seven spot beams, the communications capacity will increase 10-fold.

Frequency Allocations

The frequencies allocated for commercial FSS in popular S-, C-, and Ku-bands [21] are listed below. Additional frequencies are also available in the 30/20-GHz band.

Frequency band (GHz)	Band designation	Service
2.500–2.655	S	Fixed regional systems, space-to-earth
2.655–2.690		Fixed regional systems, earth-to-space
3.625–4.200	C	Fixed, space-to-earth
5.925–6.245		Fixed, earth-to-space
11.700–12.500	Ku	Fixed, space-to-earth
14.000–14.500		Fixed, earth-to-space

At the 1992 World Administrative Radio Conference (WARC) [22,23], the L-band frequencies were made available (Fig. 4) for mobile communications. The L-band allocations have been extended because of the rapid increase in mobile communications traffic. Recently, a number of new satellite systems such as Odyssey, Ellipso, Iridium, Globalstar, and Inmarsat-sponsored Project-21 [22] have been proposed to provide personal communications services to small hand-held terminals. These proposed systems use low Earth-orbits (LEO) between 600 and 800 km (Iridium, Ellipso, Globalstar) or an intermediate (10,000 km) altitude circular orbit (ICO). A summary of these proposed systems is available in Ref. [22].

INTELSAT System Example

The evolutionary trends in INTELSAT communications satellites are shown in Fig. 5. This illustration shows an evolution from single-beam global coverage to multibeam coverages with frequency reuse. The number of transponders increased from 2 on INTELSAT 1 to 50 on INTELSAT VI and with a corresponding satellite EIRP increase from 11.5 dBW to 30 dBW in the 4-GHz band. During the same time, the size of the Earth stations has decreased from 30 m (Standard A) to 1.2 m (Standard G) for VSAT data services (Fig. 6). The average cost of designing, fabricating, and testing 1 kg of satellite payload (dry mass) is shown in Fig. 7 in 1992 dollars for INTELSAT satellites [17,24]. Although the cost per kilogram of dry mass decreased by an order of magnitude during the first decade, it has remained nearly constant ($60K to $100K/kg) over the past decade. On the other hand, telephone channel capacity for 1 kg of payload has increased dramatically over the same period. This trend in communications satellites has been largely due to rapid technology advances. These efficiency improvements have been obtained together with an increase in useful satellite

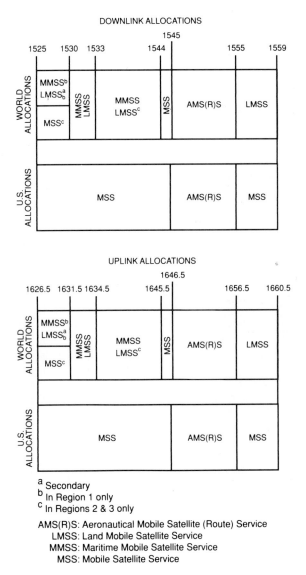

FIG. 4 WARC-92 L-band mobile satellite service allocations [22].

design life for 1.5 years for Early Bird to 10 years for INTELSAT VI and INTELSAT VII designs. However, as seen in Fig. 7, the cost of satellite/kg has leveled off. Therefore, for the satellites to remain competitive, it is even more important today to include technologies and techniques which will significantly improve the overall capacity as a function of payload mass and

The improved design of INTELSAT satellites has yielded increased capacity and reduced costs for service.

INTELSAT DESIGNATION	I	II	III	IV	IV-A	V	V-A	VI
Year of First Launch	1965	1967	1968	1971	1975	1980	1985	1989
Prime Contractor	Hughes	Hughes	TRW	Hughes	Hughes	Ford Aerospace	Ford Aerospace	Hughes
Width Dimensions, m (Undeployed)	0.7	1.4	1.4	2.4	2.4	2.0	2.0	3.6
Height Dimensions, m (Undeployed)	0.6	0.7	1.0	5.3	6.8	6.4	6.4	6.4
Launch Vehicles	Thor Delta	Thor Delta	Thor Delta	Atlas Centaur	Atlas Centaur	Atlas Centaur or Ariane 1, 2	Atlas Centaur or Ariane 1, 2	Ariane 4 or NASA STS (Shuttle)
Design Lifetime, Years	1.5	3	5	7	7	7	7	14
Bandwidth, MHz	50	130	300	500	800	2,144	2,250	3,300
Capacity								
Voice Circuits	240	240	1,500	4,000	6,000	12,000	15,000	120,000
Television Channels	—	—	—	2	2	2	2	3

FIG. 5 Trends in INTELSAT communications satellite configurations (Courtesy INTELSAT).

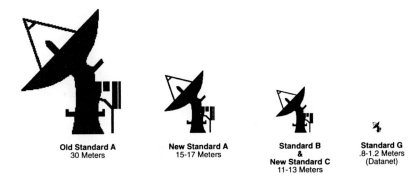

FIG. 6 Trends in INTELSAT Earth stations.

also enhance the system hardware reliability for 18- to 20-year life expectancy [25]. Miniaturization of transponder subsystems and components such as low-noise receivers, gain blocks, microwave switch matrices (MSMs), and SSPAs using MMIC technology would help in achieving these competitive goals. In addition to reducing communication payload weight and volume, MMIC implementations offer higher reliability due to batch processing and

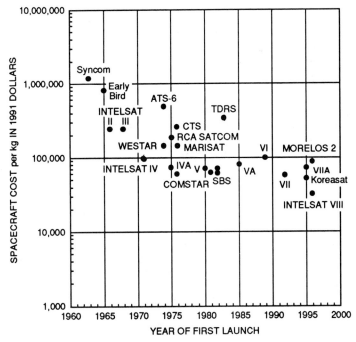

FIG. 7 Communications satellite cost per kg [24].

reduction in part counts because of higher levels of integration. Additional cost savings with MMIC technology would result from reduction in assembly and test times.

Improvements in quality of service and satellite operational flexibility can be achieved by using on-board regeneration and signal processing [1], which offer additional link budget advantages and improvements in bit error rate performance because of separation of additive up- and downlink noise. Use of reconfigurable, narrow, high-gain pencil beams with phased-array antennas offers additional flexibility of dynamic transfer of satellite resources (bandwidth and EIRP). Since these active phased-array antennas require several identical elements with high reliability, MMIC technology makes them feasible.

9.3 MMIC TECHNOLOGY IN SATELLITE TRANSPONDERS

A simplified block diagram of the INTELSAT VI communications satellite payload [26], is shown in Fig. 8. This satellite accommodates a total of 50 transponders with 10 beams at C- and Ku-bands, which provide two global, two hemispheric, and four zone coverages at C-band, and two steerable spot beam coverages at Ku-band. The multibeam antennas require the use of on-board microwave switch matrix arrays to allow interconnectivity among beams. The key candidates for insertion of MMIC technology in these satellites are the receivers, gain control elements in each channel, microwave switch matrix arrays, and high-power amplifiers (HPAs). For future spacecraft, advanced payload architecture with reconfiguration flexibility, increased EIRP for communication with low-cost Earth stations, increased satellite life, smaller size and mass to make communication satellites more cost efficient, and reduced subsystem assembly costs are some of the key factors favoring MMIC insertion in communications satellite transponders. For satellite system and circuit designers the following generic system insertion issues [27] also become important:

- Identification of systems that are most attractive for MMIC insertion,
- Partitioning of system functions and level of chip integration,
- Specification development for MMIC chips and any modifications required in system architecture,
- Reduction in MMIC development and acquisition cycle lengths and cost,
- Packaging, interface, and control requirements,
- Achievable performance and cost of development,
- Reduction in development and production costs.

There are additional issues that are specific to communications satellite payload design. These include power consumption and dissipation, linearity

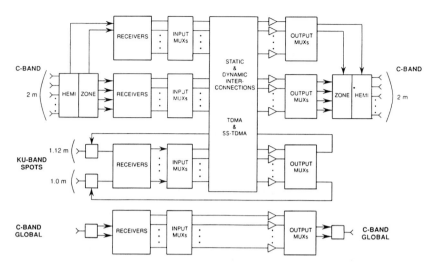

FIG. 8 Simplified block diagram of the INTELSAT VI communications payload [26].

for multicarrier operation, radiation tolerance and space qualification. Satellite systems must be immune to natural or man-made radiation in space for both commercial and military applications. Several experiments have been conducted on the effect of radiation on GaAs MMICs. The results indicate that these devices are inherently radiation-hard for space applications [28,29].

9.4 SPACE-QUALIFIED MMIC COMPONENTS

The first Ku-band MMIC amplifiers were integrated at COMSAT Laboratories in 1989 into an IOTT for the ITALSAT F-1 spacecraft. These MMICs have been in orbit since January 1991, thus demonstrating successful operation in space. Since then, several more MMICs have been space-qualified at COMSAT Laboratories for flight on more than 10 satellite payloads. The following examples illustrate some of the MMIC designs which have been space-qualified for flight [8].

9.4.1 C-Band MMIC Amplifier

Figure 9 is a photograph of a space-qualified C-band amplifier, which uses a two-stage feedback design to achieve broad bandwidth, superior gain flatness, and good linearity. For feedback amplifier implementations, devices with larger transconductance are desirable [30]. Therefore, metal semiconductor field effect transistors (MESFETs) with a gate length of 0.5

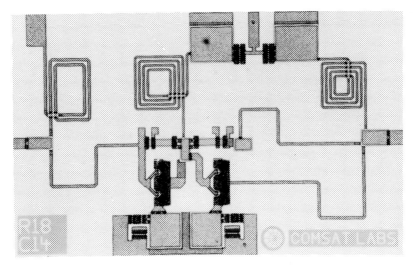

FIG. 9 Space-qualified C-band MMIC amplifier (Courtesy COMSAT Laboratories).

μm were selected. The matching networks used lumped element circuits to reduce the size and thus provide higher yields. Extensive sensitivity analysis showed that the design is tolerant to process-related parameter variations. The amplifier has a self-biasing arrangement, thus requiring a single DC operating voltage (3 to 5 V). Figure 10 shows the typical performance of a C-band two-chip cascade. The minimum gain and maximum gain flatness in the 3.7- to 4.2-GHz band are 30 and 0.2 dB peak to peak, respectively. The input and output return losses are better than 15 dB.

9.4.2 Ku-Band MMIC Amplifier

Figure 11 is a photograph of a Ku-band amplifier. This MMIC uses 0.5 × 300-μm (gate length × device width) FETs. Each FET uses a self-biased

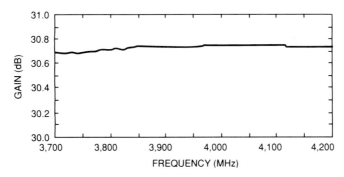

FIG. 10 Measured frequency response of the C-band MMIC amplifier [8].

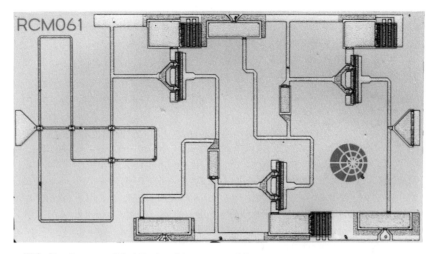

FIG. 11 Space-qualified Ku-band MMIC amplifier (Courtesy COMSAT Laboratories).

grounded gate layout and requires a single DC operating voltage (3 to 5 V). The bias current can be adjusted to control gain, gain flatness, and DC power dissipation. The performance of two-chip cascade of Ku-band amplifiers is shown in Fig. 12. The minimum gain and gain flatness in the 12.25- to 12.75-GHz band are 30 and ±0.2 dB, respectively. The output power at 1-dB gain compression is 14 dBm.

9.4.3 C- and Ku-band Attenuators for Transponder Gain Control

Gain adjustments in satellite transponders are provided through ground commandable attenuators, which can be used to compensate for channel-to-

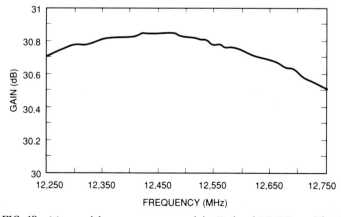

FIG. 12 Measured frequency response of the Ku-band MMIC amplifier [8].

channel gain variations and temperature and aging effects. In INTELSAT VI transponders, p-i-n diode attenuators with an attenuation range from 2 to 14 dB have been used. Additional channel ferrite step attenuators provided a single-step adjustment range of greater than 10 dB. Dual-gate FETs have also been used in hybrid circuits for gain control in satellite transponders. A study done at COMSAT Laboratories [31] indicated several benefits of using a digitally controlled MMIC step attenuator. In addition to substantial savings in mass of the components, the digital approach simplified the control network and reduced its complexity. A design based on passive FET devices, which operated as variable resistors in a non-power-consuming mode, was developed [32]. The MMIC chip consisted of 8-, 4-, 2-, 1-, and 0.5-dB attenuation steps in cascade resulting in 15.5 dB dynamic range and 0.5 dB step resolution [33]. In this implementation, the RF signal was routed between a reference path and an attenuation path using passive FETs as variable resistors and switches.

For the 2-, 1-, and 0.5-dB attenuation steps, reference and attenuation paths consisted of T-networks, and attenuation ratios were varied by changing the device widths and shunt resistor values for each step. The 8- and 4-dB attenuation steps were realized by two broadband single-pole double-throw switches with transmission lines in reference and a resistive T-network in the attenuation path. The 5-bit attenuator chip (Fig. 13) consists of 34 FETs, 36 resistors, and 26 inductive elements on a 1.3 × 2.6 mm chip. Broadband performance up to 14 GHz has been demonstrated for all five bits of the attenuator, making the MMIC useful for L-, C-, and Ku-band applications. For example, Fig. 14 shows the performance of a Ku-band attenuator that provides 24-dB attenuation range with a resolution of 1 dB. Figure 14 shows measured data for 50 chips at 12.2 GHz, selected from

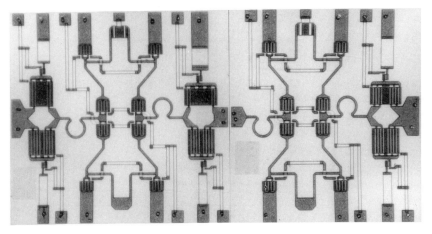

FIG. 13 Microphotograph of the 5-bit MMIC digital attenuator (Courtesy COMSAT Laboratories).

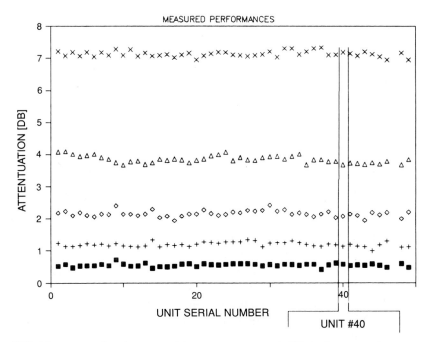

FIG. 14 Measured performance of 50 attenuators at 12.2 GHz in five attenuation states.

different wafers, in five attenuation states. These results show good chip-to-chip performance uniformity and close agreement between design and measured values. These gain-step attenuators have been space-qualified and may be directly cascaded with C-band feedback amplifiers [30] or Ku-band amplifiers to provide transponder gain control at C and Ku frequency bands.

9.5 SPACE-QUALIFICATION OF MMICs

Standardized procedures for space-qualifying the MMICs and determining MMIC reliability have been recently developed [8]. The MMIC qualification procedures consist of three major steps: process qualification tests, wafer acceptance tests, and lot acceptance tests. As the names suggest, the purpose of these tests is to space-qualify the manufacturing process, and also qualify the individual wafers as well as wafer lots, out of which the space-qualified MMICs are drawn for flight. These tests also provide information about an important reliability parameter, namely the mean time to failure (MTTF) for the MMICs.

The process qualification tests involve accelerated life tests on two sets of 10 MMICs at elevated junction temperatures of 215 and 210°C where failures are monitored. For active amplifiers, a failure is defined by a change in DC current of more than 20%. Using these tests, for example, the MTTF at 70°C junction temperature was estimated to be approximately 5×10^9 h. The wafer acceptance tests involve accelerated stress testing at 200°C, under normal operating DC voltage, for 168 h. At the end of the test, the MMICs are recharacterized to monitor changes in drain current and gain of less than 20% and 1 dB, respectively. For the C-Band and Ku-Band MMIC amplifiers, the changes were less than 12% and 0.5 dB, respectively. For digital attenuators using passive FETs, the criteria for these tests is the change in insertion loss and allowable change in leakage currents before and after the tests. The lot acceptance procedures are similar to the wafer acceptance procedures, except that the channel temperature is maintained at 125°C and the test duration is 1000 h. Detailed test results are available in Ref. [8].

9.6 MMIC SUBSYSTEM DESIGN EXAMPLES

Transponder subsystem level performance, power consumption and dissipation, packaging, interface, and control issues have been addressed at COMSAT Laboratories by realization and performance demonstrations of several transponder subsystems. Some of these designs are presented in the following subsections.

9.6.1 Low-Noise Receivers

Hybrid MIC technology has been used in practically all the commercial satellite programs at C- and Ku-bands. Discrete FET devices are selected and bonded into MIC substrates requiring a significant amount of manual labor. GaAs MMIC technology may be used to make these receivers significantly smaller and lighter in weight. For example, a C-band hybrid receiver, which typically weighs 1.7 kg and has approximate dimensions of $16 \times 10 \times 8$ cm, may be reduced to less than half those dimensions. A C-band two-stage MMIC low-noise amplifier, with chip dimensions of 2×3.5 mm was demonstrated at COMSAT (1987) by using MESFET devices [34]. This LNA provided 21 dB gain and a 1.7-dB noise figure in the 5.9- to 6.4-GHz uplink band. A noise figure less than 0.8 dB can be achieved by using pseudomorphic high electron mobility transistor (P-HEMT) technology. Similarly, at Ku-band, a gain of 32 dB and a noise figure of 2.6 dB have been demonstrated [9] across the 14- to 14.5-GHz frequency band. Using high electron mobility transistors (HEMTs), a noise figure better than 1.1 dB may be achieved. Development of an MMIC low-noise down converter has recently been demonstrated at COMSAT Laboratories [9]. It is estimated that use of MMIC technology in satellite receivers can produce 50% mass savings.

9.6.2 Microwave Switch Matrices (MSM)

The MSM technology leading up to the INTELSAT VI spacecraft has been developed in a cross-bar configuration by using hybrid MIC technology. A broadband 8 × 8 MSM for operation over the 3.5- to 6.5-GHz frequency range was developed at COMSAT Laboratories, using the power divider/combiner approach [35]. P–i–n diodes were used as switching devices resulting in an insertion loss of approximately 21 dB and path-to-path insertion loss variations of less than 1.5 dB. For INTELSAT VI, the 10 × 6 MSM (10-for-6 redundancy) was implemented in a coupler cross-bar configuration with p–i–n diodes as switching elements [26] providing 50 dB on-to-off isolation. The MSM insertion loss of 34 dB was partially compensated by a two-stage FET driver amplifier with 22 dB gain, over the operational bandwidth of the INTELSAT VI MSM, which extends over less than 200 MHz (channels 1–2 and 3–4). As the number and/or size of these MSMs increase, it is necessary to minimize the insertion loss, enhance reliability, and reduce the mass and cost of fabrication.

A miniaturized broadband (3.5- to 6.5-GHz) 4 × 4 MSM was developed at COMSAT Laboratories [10], using MMIC dual-gate FET switches as switching elements. This hardware presented significant design challenges in terms of MMIC realization with high on-to-off ratio, MSM package design, and control interface to the MMICs. A two-stage dual-gate FET design was used in order to achieve the desired 50-dB on-to-off ratio. Reactive matching with resistive loading was used to obtain a broadband match with very small changes in return loss in the two states of the switch element. The MMIC chip (Fig. 15) was fabricated on a 12.5-mil (0.3 mm) thick GaAs substrate with overall chip dimensions of 1.5 × 2.5 mm. The switch has a gain of approximately 10 dB and on-to-off isolation greater than 50 dB over the 3.5- to 6.5-GHz frequency range. A lightweight package with overall dimensions of 5 × 5 × 2.5 cm was developed in which 16 MMIC switches were integrated with 8 MIC power dividers/combiners, 16 RF feedthroughs, driver/control circuits, and a number of bias and control wires. For the control circuit implementation, commercially available silicon complimentary metal oxide semiconductor (CMOS) integrated circuit chips, which are available in radiation-hardened versions, have been used. A photograph of the fully assembled 4 × 4 MSM is shown in Fig. 16. This hardware demonstrated MMIC packaging approaches which maintained the high isolation achieved from the MMIC switch module. Measured transmission loss performance for all 16 paths shows on-state insertion loss of less than 6.25 dB and on-to-off isolation greater than 50 dB (Fig. 17) over the 3-GHz design frequency band. The path-to-path insertion loss variation was measured to be less than ±0.5 dB over 3.5- to 6.5-GHz frequency. The equivalent weight of a fully assembled 4 × 4 MSM is approximately 130 g,

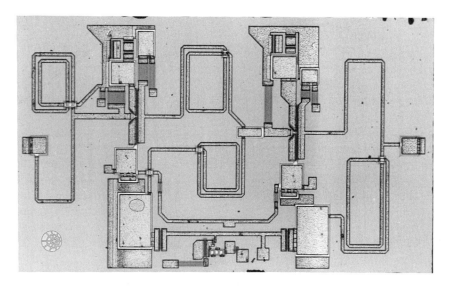

FIG. 15 Two stage MMIC dual-gate FET switch element (dimensions: 2.5 × 1.5 mm) (Courtesy COMSAT Laboratories).

which is 15% of that using INTELSAT VI hybrid MSM technology, while the design bandwidth is 15 times greater. The power consumption of the MSM is approximately 1.2 W, which is 50% of the equivalent INTELSAT VI hybrid MSM.

9.6.3 Phased-Array Transmit Antennas

Phased-array antennas using GaAs MMICs are gaining added significance for future on-board satellite applications because of their ability to form single or multiple beams and provide power sharing among the beams [13,14]. These active antennas offer improved operational flexibility by providing independent beam reconfigurability and steerability and more efficient use of satellite power resources. In addition, with the formation of narrow beams, higher EIRP can be achieved, making communication with small Earth stations possible. A single-beam low-power Ku-band transmit phased-array antenna using GaAs MMIC technology has already been successfully demonstrated at COMSAT [13]. A photograph of the fully assembled 64-element active array and an active element is shown in Fig. 18. Each active element includes a 5-bit digital phase shifter and a 5-bit digital attenuator, amplifiers to compensate for insertion loss (Fig. 19), a microstrip-to-waveguide transition, an orthomode transducer (OMT), and a radiating

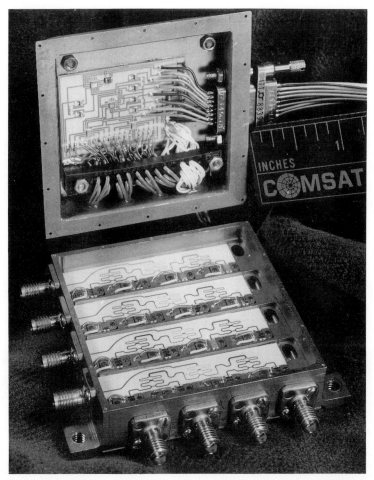

FIG. 16 Fully integrated miniaturized 4 × 4 microwave switch matrix (MSM) (Courtesy COMSAT Laboratories).

horn. The circuitry to control the interface to the spacecraft electronics is integrated in each active element. The 5-bit phase shifters provide beam steering with 360° range and 11.25° resolution, and the 5-bit digital attenuator [33] provides beam shaping with 15.5 dB dynamic range and 0.5 dB resolution. The high beam-to-beam isolation (27 dB) needed for the active phased array places stringent requirements on the uniformity of the attenuator and phase-shifter performance. For example, a sensitivity analysis using the Monte Carlo technique indicated that maximum amplitude and phase

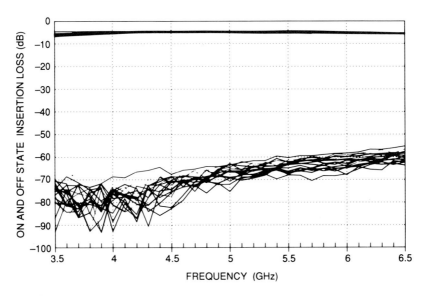

FIG. 17 Measured on- to off-state transmission loss for all 4 × 4 MSM paths [10].

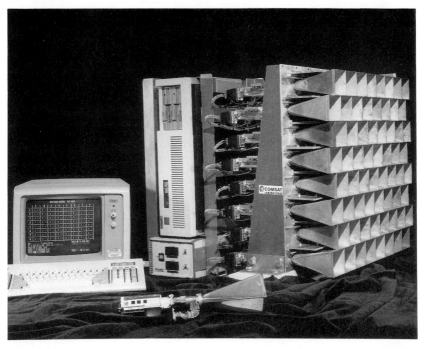

FIG. 18 Photograph of the 64-element active phased-array antenna (Courtesy COMSAT Laboratories).

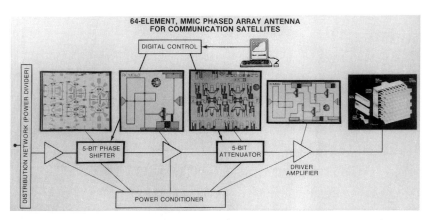

FIG. 19 Active element for the 64-element phased array (Courtesy COMSAT Laboratories).

perturbations of up to ±1.5 dB and ±10° were allowed in the active elements of the phased array. Across the 11.7- to 12.7-GHz band, this performance was achieved [13], demonstrating good chip uniformity and high run-to-run reproducibility for GaAs MMICs. The radiation patterns of the antenna were measured for a number of shaped beams. Excellent agreement was obtained between the measured and modeled azimuth cut (Fig. 20).

A multibeam Ku-band high-power phased-array antenna, capable of forming four simultaneous beams has also been demonstrated at COMSAT Laboratories [36–38]. The implementation concept for this multibeam an-

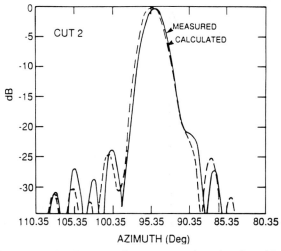

FIG. 20 Measured and computed antenna cuts through a shaped beam [13].

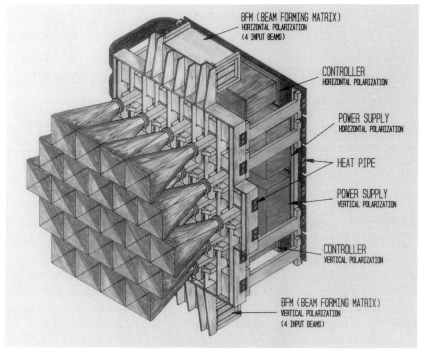

FIG. 21 Multibeam, high-power, phased-array system concept [36].

tenna is shown in Fig. 21. The high power array consists of 24 active radiating elements with 2-W SSPAs integrated behind each element. The SSPAs are cooled by a liquid cooling loop that simulates the use of heat pipes for space applications. The power sharing and independent beam steerability is accomplished by a beam forming matrix (BFM) which contains the power divider and combiner networks and GaAs MMIC phase shifters. The system also contains a controller and power supplies to provide bias and control functions to the phase shifters and amplifiers.

The BFM is a relatively complex assembly, consisting of four 1- to 24-way power dividers and 24 four-way power combiners. It is designed using three identical modular shelves each containing four 1- to 8-way dividers and eight 4- to 1-way combiners. All the control interface electronics, RF dividers/combiners, and MMIC phase shifters are assembled in a Kovar BFM shelf package with dimensions of $19 \times 10.4 \times 1.3$ cm (Fig. 22). The interface electronics consists of level-shifter boards and driver boards, which provide a simple interface to the BFM controller. Blind-mate RF connectors are used in each BFM shelf to allow easy assembly and disassembly with antenna structure and input dividers. The insertion loss of for all 32 BFM paths, corresponding to the reference and five phase-shifter states, is

9. Satellite Communications   315

FIG. 22 Photograph of the fully assembled BFM (Courtesy COMSAT Laboratories).

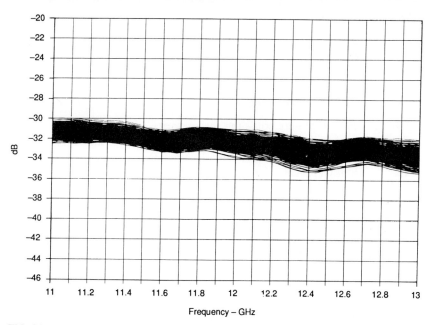

FIG. 23 Measured insertion loss of a BFM shelf in reference and five phase-shifter states [36].

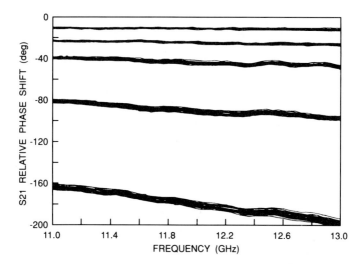

FIG. 24 Relative phase shift in five states for 32 paths of the BFM shelf [36].

within ±1.2 dB (Fig. 23). The corresponding phase shift performance for all 32 BFM paths for five phase-shifter states is shown in Fig. 24. These results demonstrate the relative uniform and reproducible performance of MMIC phase-shifters. The measured amplitude and phase variations for 24 ports corresponding to each of the input beams are within ±1 dB and ±10° in the reference phase shifter state. Integration of the BFM with the 24-element high-power array has resulted in the successful demonstration of four independent, steerable beams.

9.6.4 Solid-State Power Amplifiers

The transmit amplifiers consume the majority of the DC power in the communications satellite payloads. Therefore, their power added efficiency is of prime importance. MIC SSPAs are already being introduced into space at C-band with power levels up to 10 W. For multibeam active phased-array antennas, highly efficient broadband and reproducible SSPAs with transmit power levels of only 1 to 2 W are sufficient because of high antenna gain and the addition of power from all active antenna elements in space. Multicarrier operation, however, requires high linearity, while simultaneously achieving highest possible DC-to-RF efficiency, high reliability and small size, mass, and minimum cost. The SSPA technology was originally based on discrete component realization using packaged FETs and matching networks fabricated on a soft board or ceramic materials. In recent years, the focus has been on development of enhanced hybrid and monolithic implementations [15]. An enhanced hybrid approach in which a discrete FET device is selected, while all the passive components are integrated on an alumina substrate, offers the advantages of reduced size and component count for assembly. Device operation in Class-B or deep AB mode with second harmonic termination resulted in single-stage efficiencies of 65% at C-band. For a seven-stage SSPA at Ku-band, a small signal gain of 40 dB, an output power of 1.5 W, and efficiency of 23% have been demonstrated in a production mode. These efficiency numbers include power distribution and regulation in each SSPA module [15].

An MMIC Ku-band 1-W SSPA has been developed at COMSAT Laboratories for broadband (10.7- to 12.75-GHz) applications [39]. A photograph of the cascaded MMIC chips is shown in Fig. 25. This three-stage SSPA, with a single-stage driver followed by a two-stage output stage, demonstrates in 22.5-dB small-signal gain at room temperature, with approximately ±0.5-dB gain variations. At 50°C, the small signal gain decreases by approximately 2 dB. The module achieves greater than 31 dBm saturated output power and about 30 dBm at 2-dB gain compression. Power added efficiency between 25 and 35% was demonstrated over the full 2-GHz design bandwidth at a two-carrier third-order intermodulation (C/I3) level of 16 dB (Fig. 26). Noise-loaded efficiency at 0.6 W output power was measured to be greater than 20%. The excellent uniformity of the RF performance is summarized in the composite swept gain plot of more than 75 units shown in Fig. 27. A total of 80 modules were produced, including 64

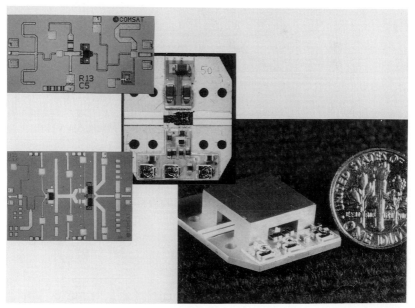

FIG. 25 1-W Ku-Band SSPA photograph showing cascaded chip configuration (Courtesy COMSAT Laboratories).

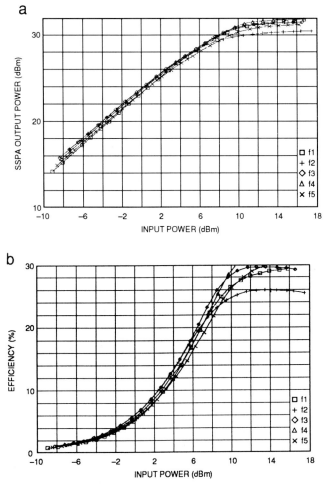

FIG. 26 Output power (a) and efficiency (b) for the 1-W MMIC SSPA [39].

for the actual antenna array, plus a number of spare units. Approximately 75 of the modules have gain within a 2-dB window across the full frequency band.

9.7 FUTURE DIRECTION

GaAs MMIC technology is very attractive for insertion in satellite transponders. Commercial fixed satellite systems have been technology driven during the early years of their evolution. Because of the changing compet-

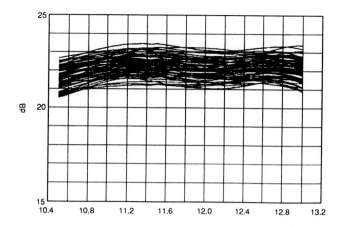

FIG. 27 Small signal gain of more than 75 MMIC SSPA amplifiers [39].

itive environment, other factors such as communication capacity, communications performance, quality of service, and cost per circuit become even more important. Transponder technologies which significantly improve overall capacity as a function of payload mass, and also enhance the system hardware reliability to increase satellite life, are particularly desirable. Significant improvements in cost efficiency can be achieved by providing onboard regeneration and active phased-array antennas for reconfiguration flexibility and sharing of satellite power resources. GaAs MMIC technology is crucial to implementation of these complex systems with high reliability. Each subsystem design with GaAs MMICs presents a number of design challenges in terms of partitioning of system functions, packaging, and control interface design. Further innovations are expected in these areas for implementation of relatively complex functions. Hardware developments have already resulted in significant progress toward solutions to packaging problems. Maturity of MMIC designs and fabrication techniques and space qualification of selected MMICs have resulted in MMICs already being flown in space. A rapid growth in applications of MMICs in future satellite transponders is expected based on present trends.

A major driver for GaAs MMIC technology for on-board satellite applications would be the large satellite constellations for mobile communications. As these systems are further developed, there will be a need to realize identical low-cost spacecrafts making the use of GaAs MMICs even more attractive. Even a larger growth application would be the handsets for mobile phones [40], in which GaAs MMICs will provide both RF front-end and power transmit functions to achieve superior performance and better system efficiencies to conserve the battery power of the handset. Additional ground applications in which GaAs MMICs are expected to play a major role are the VSAT satellite terminals and LNB downconverters for direct TV receptions.

ACKNOWLEDGMENTS

The author gratefully acknowledges the contributions of many of his colleagues at COMSAT Laboratories to the work described herein. Most of the work reported here was funded by COMSAT Corporation.

REFERENCES

[1] C. Mahle, G. Hyde, and T. Inukai, Satellite scenarios and technology for 1990's. *IEEE J. Selected Areas Commun.* SAC-5(4), 556–570, May 1987.
[2] S. Campanella and C. Mahle, *Satellite and Cables in the Future Marketplace and the Role of MMICs*, pp. 19–26. IEEE 1988 Microwave and Millimeter-Wave Circuits Symposium, New York, June 1988.
[3] C. Mahle and H. Huang, MMICs in Communications. *IEEE Commun. Magazine* 23(9), Sept. 1985.
[4] B. Geller and F. Assal, *Impact of Monolithic Microwave Integrated Circuit Development on Communications Satellites*, pp. 1E.1.1–1E.1.5. IEEE International Conference on Communications, Philadelphia, PA, June 1982.
[5] R. Pucel, Design considerations for monolithic microwave circuits. *IEEE Trans. Microwave Theory Tech.* MTT-29, 513–534, June 1981.
[6] R. Gupta, T. Smith, and J. Reynolds, GaAs MMICs: CAD, modeling and process monitoring. *Microwave J.* 30(8), 95–114, Aug. 1987.
[7] R. Gupta et al., *Manufacturing Technology and Yield Studies for MMIC 5-Bit Digital Attenuators and Phase Shifters*, pp. 305–308, IEEE GaAs IC Symposium, New Orleans, LA, Oct. 1990.
[8] J. Potukuchi, R. Mott, R. Gupta, and F. Phelleps, *Space-Qualification of Monolithic Microwave Integrated Circuits*, pp. 1013–1024. 15th AIAA International Communications Satellite Systems Conference, San Diego, CA, Feb.–Mar. 1994.
[9] H. Hung et al., *Monolithic Integrated Receiver Technology for Future Satellite Applications*. 14th International Communications Satellite Systems Conference, Washington, DC, March 1992.
[10] R. Gupta, F. Assal, and T. Hampsch, *A Microwave Switch Matrix Using MMICs for Satellite Applications*, pp. 885–888. 1990 IEEE MTT-s International Microwave Symposium, Dallas, TX, May, 1990.
[11] K. Araki, M. Tanaka, and H. Kato, Onboard large-scale monolithic IC switch matrix for multibeam communications satellites. *Acta Astronautica* 19(1), 41–45, 1989.
[12] D. Ch'en, W. Petersen, and W. Kiba, *Advanced Large Scale GaAs Monolithic IF Switch Matrix Subsystem*. 14th AIAA International Communications Satellite Systems Conference, Washington, DC, March 1992.
[13] J. Potukuchi et al. *MMIC Insertion in a Ku-Band Active Phased-Array for Communications Satellites*, pp. 881–884. 1990 IEEE MTT-s International Microwave Symposium, Dallas, TX, May 1990.
[14] R. Gupta et al., Beam-forming matrix design using MMICs for a multibeam phased-array antenna. *IEEE GaAs IC Symp.* Oct. 1991.
[15] B. Geller et al., *Production of Miniaturized Multi-Stage SSPAs for a Ku-Band Array Antenna*. 14th AIAA International Communications Satellite Systems Conference, Washington, DC, March 1992.

[16] R. Mott et al. *ITALSAT In-Orbit Test Transponder (IOTT)—Design and Performance.* 14th AIAA International Communications Satellite Systems Conference, Washington, DC, March 1992.
[17] C. Mahle and G. Hyde, *Efficiency of Communications Satellites.* 14th AIAA International Communications Satellite Systems Conference, Washington, DC, March 1992.
[18] S. Bennett and D. Braverman, INTELSAT-VI—A continuing evolution. *Proc. IEEE* 72(11), 1457–1468, Nov. 1984.
[19] N. Wong, *INTELSAT-VI—A 50 Channel Communication Satellite with SS-TDMA,* 20.1.1–20.1.9. Satellite Communications Conference, Ottawa, Canada, June 1983.
[20] A. Nakamura, R. Gupta, F. Assal, J. Narayanan, and L. Argyle, *Integration and Testing of the Satellite Switched TDMA (SS-TDMA) Subsystem for the INTELSAT-VI Spacecraft,* pp. 49–57. 8th International Conference on Digital Satellite Communications, April, 1989.
[21] B. Gallagher (Ed.) *Never Beyond Reach—The World of Mobile Satellite Communications.* International Maritime Satellite Organization, 1989.
[22] J. Evans, *Satellite and Personal Communications,* pp. 1013–1024. 15th AIAA International Communications Satellite Systems Conference, San Diego, CA, Feb.–Mar. 1994.
[23] E. Reinhart and R. Taylor, Mobile communications and space communications. *IEEE Spectrum* 27–29, Feb. 1992.
[24] C. Mahle, High end microwave technologies in global communications. *Microwave J.* pp. 22–30, Dec. 1993.
[25] F. Assal and C. Mahle, *Hardware Development for Future Commercial Communications Satellites,* pp. 332–343. AIAA 12th International Communications Satellite Conference, Arlington, VA, March 1988.
[26] R. Gupta, J. Narayanan, A. Nakamura, F. Assal, and B. Gibson, INTELSAT VI on-board SS-TDMA subsystem design and performance. *COMSAT Technical Rev.* 21(1), 191–225, Spring 1991.
[27] R. Gupta, *MMIC Insertion in Communications Satellite Payloads.* Workshop J—On GaAs MMIC System Insertion and Multifunction Chip Design, IEEE MTT-S International Microwave Symposium, Boston, MA, June 1991.
[28] A. Meulenburg et al., Total dose and transient radiation effects on GaAs MMICs. *IEEE Trans. Electron. Devices* ED-35, 2125–2132, Dec. 1988.
[29] W. Anderson et al., Heavy ion total fluence effects in GaAs devices. *IEEE Trans. Nuclear Sci.* 37, 2065–2070, Dec. 1990.
[30] R. Gupta, J. Reynolds, M. Fu, and T. Heikkila, Design and modeling of a GaAs monolithic 2- to 6-GHz feedback amplifier. *COMSAT Tech. Rev.* 17, 1–22, Spring 1987.
[31] R. Gupta, *Technology Assessment of C-Band Switching Devices for Transponder Gain Control.* COMSAT Technical Note MWT/85-6028, July 1985.
[32] R. Gupta and B. Geller, Microwave digitally controlled solid-state attenuators having parallel switched paths. U.S. Patent 4,978,932, Dec. 18, 1990.
[33] R. Gupta et al., *A 0.05–14 GHz MMIC 5-Bit Digital Attenuator,* pp. 231–234. 1987 IEEE GaAs IC Symposium, Portland, OR, Oct. 1987.
[34] R. Mott, *A GaAs Monolithic 6-GHz Low-Noise Amplifier for Satellite Receivers.* 1987 IEEE MTT-S International Microwave Symposium, Las Vegas, NV, pp. 561–564, June 1987.
[35] F. Assal, R. Gupta, K. Betaharon, A. Zaghloul, and J. Apple, A wideband satellite microwave switch matrix for SS-TDMA communications. *IEEE J. Selected Areas Commun.* SAC-1(1), 223–231, Jan. 1983.
[36] R. Sorbello et al., *A Ku-Band Multibeam Active Phased Array for Satellite Communications.* 14th International Communications Satellite Systems Conference, Washington, DC, March 1992.
[37] R. Gupta, A. Zaghloul, T. Hampsch, and F. Assal, *Development of a Beam-Forming Matrix Using MMICs for Multibeam Active Phased Arrays.* 1994 IEEE APS Symposium and URSI Radio Science Meeting, Seattle, WA, June 19–24, 1994.

[38] E. Eklman, E. Kohls, A. Zaghloul, and F. Assal, *Measured Performance of a Ku-Band Multibeam High-Power Phased Array.* 1994 IEEE APS Symposium and URSI Radio Science Meeting, Seattle, WA, June 19–24, 1994.
[39] J. Upshur, R. Gupta, G. Estep. and R. Kroll, and D. Rouques, *1-W Ku-Band MMIC SSPAs for Communications Satellite Phased-Array Antenna Applications,* pp. 1013–1024. 15th AIAA International Communications Satellite Systems Conference, San Diego, CA, Feb.–Mar. 1994.
[40] G. Noreen, Mobile satellite communications for consumers, *Microwave J.* 24–34, Nov. 1991.

10
Direct Broadcast Satellite Receivers

Charles Huang
Anadigics, Inc., Warren, New Jersey

10.1 Introduction
10.2 System Descriptions
10.3 Receiver Components
10.4 GaAs MMIC Development for DBS
    10.4.1 Low-Noise Amplifier (LNA)
    10.4.2 Mixer
    10.4.3 Oscillators
    10.4.4 IF Amplifier (IFA)
    10.4.5 LNB Subsystem
10.5 Commercially Available GaAs MMICs
    10.5.1 NEC
    10.5.2 Hitachi
    10.5.3 Mitsubishi
    10.5.4 Fujitsu
    10.5.5 Avantek
    10.5.6 Anadigics
10.6 Designing Receiver Components with MMICs
    10.6.1 Designing an LNB
    10.6.2 Designing a DBS Tuner
10.7 Application Issues
    10.7.1 Mounting and Bonding Considerations
    10.7.2 Biasing the Circuits
    10.7.3 Reliability Studies
    10.7.4 Temperature Stability
10.8 Future Trends
10.9 Summary
    References

10.1 INTRODUCTION

The inherent advantage of the direct broadcast satellite (DBS) system is that it offers instant television programming to a wide area. The small and inexpensive Ku-band DBS receiver system is ideal for direct-to-home broadcast services. It should be able to compete with terrestrial broadcast and cable television for a substantial share of the television audience because of the wider area coverage (when compared to broadcast television) and lower installation cost per household (when compared to cable television).

The DBS system is also suitable for high-definition television (HDTV) and stereo sound transmissions. These additional features should make the DBS system extremely popular.

Direct broadcast satellite television distribution in the United States started in 1975 when Home Box Office (HBO) began using C-band (3.7–4.2 GHz) communication satellites to beam movie programs to cable operators. Soon after, 10′ television receive-only (TVRO) dishes began to sprout up in the backyards of rural homes for satellite direct-to-home television reception. It is estimated that more than 3 million C-band TVRO dishes are in use today.

The cost of C-band TVRO receivers (including dishes and indoor receiver units) is high. It was U.S. $10,000 in 1980 and is about U.S. $2500 today. The high cost and large dish size continues to limit the market growth.

The Ku-band (12 GHz) DBS receiver system uses a small, 1′–2′ dish. The small size dish is possible because of shorter wavelengths and higher satellite power. It is less expensive (U.S. $350) than TVRO receivers and the dish can be installed anywhere in a house or an apartment. Therefore, it has a much better market potential.

Ku-band DBS got started in 1977 when the World Administrative Radio Conference (WARC) allocated orbit slots and frequency channels for high- and medium-power Ku-band DBS systems. The Ku-band DBS system got a head start in Europe and Japan because cable television is not pervasive in those countries and consumers crave more television programs.

The first Ku-band DBS direct-to-home programming began on May 12, 1984, in Japan with the BS2A satellite. On July 4, 1987, a full 24-h service started with the BS2B satellite. Today, there are almost 30 direct-to-home Ku-band DBS channels available in Europe and 3 in Japan. A partial listing of these channels is provided in Table 1.

10.2 SYSTEM DESCRIPTIONS

Communication and broadcast satellites revolve synchronously with Earth, hence the satellites appear stationary from any point on Earth's surface and are ideally suited for television broadcasting. Examples of several Ku-band satellite orbit slots in Europe are provided in Fig. 1.

The television picture quality is determined by the carrier to noise ratio C/N which is a function of satellite's equivalent isotropic radiated power (EIRP) and many other parameters. The carrier to noise ratio is determined by

TABLE 1 Partial Listing of European DBS Channels as of 1991 Illustrates Versatile Programming and Wide Coverage

Satellite	Channel	Main markets	Homes (millions)	Location	Transponder	H	Video	Scramb. sys.
Astra	Screensport	Europe	8.7	19.2 E	1	18	PAL	Clear
Astra	RTL Plus	Germany	15.6	19.2 E	2	20	PAL	Clear
Astra	Children's	UK/Scand.	.9	19.2 E	2W	12	PAL	Clear
Astra	TV 3	Scandinavia	6.2	19.2 E	3	8	D2-MAC	D2-MAC
Astra	Eurosport	Europe	22.6	19.2 E	4	20	PAL	Clear
Astra	Lifestyle	UK	2.2	19.2 E	5	15	PAL	Clear
Astra	Sat 1	Germany	14.2	19.2 E	6	19	PAL	Clear
Astra	TV 1000	Sweden	0.1	19.2 E	7	7	D2-MAC	D2-MAC
Astra	Sky One	UK	1.7	19.2 E	8	17	PAL	Clear
Astra	Teleclub	Switz./Germ.	0.1	19.2 E	9	15	PAL	Clear
Astra	3 SAT	Germany	11.0	19.2 E	10	10	PAL	Clear
Astra	FilmNet	Scandinavia	0.5	19.2 E	11	24	PAL	Satbox
Astra	Sky News	UK	1.7	19.2 E	12	24	PAL	Clear
Astra	RTL 4	Netherlands	4.7	19.2 E	13	18	PAL	Clear
Astra	Pro 7	Germany	7.3	19.2 E	14	24	PAL	Clear
Astra	MTV	Europe	18.3	19.2 E	15	24	PAL	Clear
Astra	Sky Movies	UK	0.8	19.2 E	16	24	PAL	Vidcrypt.
MarcoPolo	Sky 1	UK	2.3	31.0 W	6	19	D-MAC	Eurocypher
MarcoPolo	Movie Channel	UK	0.2	31.0 W	20	13	D-MAC	Eurocypher
MarcoPolo	Sky News	UK	2.3	31.0 W	4	24	D-MAC	Eurocypher
MarcoPolo	Power Stn.	UK	2.3	31.0 W	16	21	D-MAC	Eurocypher

(continued)

TABLE 1 (continued)

Satellite	Channel	Main markets	Homes (millions)	Location	Transponder	H	Video	Scramb. sys.
MarcoPolo	Sports Ch.	UK	2.3	31.0 W	12	13	D-MAC	Eurocypher
TVSat2	Eins Plus	Germany	7.0	19.2 W	18	4	D2-MAC	Clear
TVSat2	RTL Plus	Germany	15.6	19.2 W	2	20	D2-MAC	Clear
TVSat2	3 Sat	Germany	11.0	19.2 W	10	10	D2-MAC	Clear
TVSat2	Sat 1	Germany	14.2	19.2 W	6	19	D2-MAC	Clear
TDF1/2	Canal Plus	France	3.0	19.0 W	17	22	D2-MAC	Eurocrypt
TDF1/2	Euromusique	France	0.2	19.0 W	5	24	D2-MAC	Clear
TDF1/2	La Sept	France	2.7	19.0 W	9	12	D2-MAC	Clear
Tele-X	NRK	Norway	1.5	5.0 E	32	9	D-MAC	Clear
Tele-X	TV 4	Sweden	2.5	5.0 E	26	5	PAL	Clear
Olympus 1	Rai' Sat	Europe	5.2	18.8 W	24	15	PAL	Clear
Olympus 1	Enterprise	Europe	—	18.8 W	20	6	D2-MAC	Clear

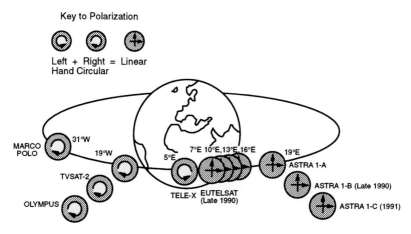

FIG. 1 Illustration showing longitudinal locations of several European K-band DBS and medium-power satellites (as of 1991).

$$\frac{C}{N} = EIRP + \frac{G_r}{T_s} - L - K - B, \qquad (1)$$

where G_r/T_s is a function of antenna gain and the receiver noise figure, L is the sum of the signal losses through free space, rain, atmosphere and incidence to the antenna, K is Boltzmann's constant, and B is the bandwidth of the received signal. In Eq. (1), each term is a quantity expressed in dB. Carrier to noise of at least 30.0 dB is required to provide acceptable picture quality. This level in turn typically requires an EIRP value of 44 dBw.

Figure 2 shows the European footprint where the EIRP of Astra 1A satellite is strong enough for direct-to-home television reception.

Because more and more satellites are occupying orbital slots, they are being spaced closer. Interferences from nearby satellites are additional noise sources and should be taken into consideration in receiver design.

The most common baseband modulation formats for DBS systems are:

A. Frequency modulation (FM) and
B. Multiplexed analog components (MAC).

The frequency modulation format is provided in Fig. 3. The carrier frequencies are modulated by a composite video signal and the subcarrier frequencies are modulated by sound channels. The main problem with FM format is the intermodulations between the luminance and chrominance signal components. In the MAC transmission system, the chrominance and luninance signals are separated, as shown in Fig. 4. There are many different MAC standards, as listed in Table 2.

Recently, transmission by digital compression techniques has gained

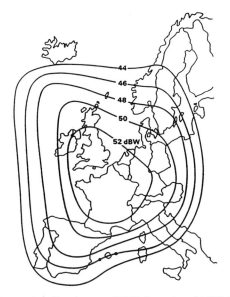

FIG. 2 Equivalent isotropic radiated power (EIRP) footprint of ASTRA 1A satellite showing that a single satellite can effectively cover most of Europe.

popularity. DirecTV, a DBS venture in the United States, has proposed an 8:1 compression technique. Also, many different HDTV formats are being proposed.

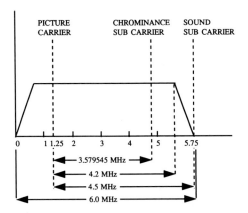

FIG. 3 Composite of frequency modulation (FM) format used in DBS systems.

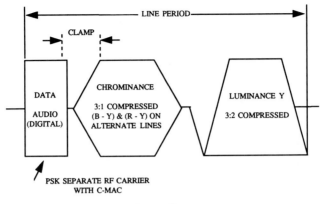

FIG. 4 Multiplexed analog components (MAC) signal showing separation of chrominance and luminance signal components.

10.3 RECEIVER COMPONENTS

Each direct-to-home DBS receiver system consists of an outdoor and an indoor unit. The main body of an outdoor unit is made of a parabolic dish antenna (Fig. 5) that is pointed directly toward the satellite. The dish focuses the signal onto the feedhorn of the low-noise block downconverter (LNB). The antenna gain is a function of the size of the dish. Typically, a 60-cm (2′) dish provides 36.0 dB of gain and a 35-cm (14″) dish provides 31.0 dB of gain.

The LNB receives, amplifies, and converts the Ku-band signals to approximately 1.0 GHz for transmission to the indoor receiver unit. A block diagram of an LNB is provided in Fig. 6. Many of the LNB functions are implemented with gallium arsenide (GaAs) devices.

Key performance parameters for an LNB are noise figure and conversion gain. Generally, noise figures of less than 1.5 dB and conversion gain of more than 50 dB are required for DBS reception.

Three different bands are being used for Ku-band DBS systems:

10.95–11.70 GHz—TV, communications
11.70–12.50 GHz—DBS
12.50–12.75 GHz—telecommunications

Typically, specific LNBs are designed for each band.

To fit more channels into each band, both linear and circular polarizations can be used. A polarizer in front of the LNB is required to select the

TABLE 2 Attributes of Several Different MAC Standards

Attribute	C-MAC	D2-MAC	B-MAC (625)	B-MAC (525)
Line/frame rate	625/25	625/25	625/25	525/29.97
Line frequency, Hz	15625	15625	15625	15734
Time increments/line	1296	1296	1365	1365
Active lines	574	574	574	483
Aspect ratio	4:3	4:3	4:3	4:3
Luminance compress.	3:2	3:2	3:2	3:2
Chrominance compress.	3:1	3:1	3:1	3:1
Sound/data	QPSK/BPSK	Duo binary	Quaternary/binary	Quaternary/binary
Bit rate	3.08 Mb/s	1.54 Mb/s	1.59 Mb/s	1.60 Mb/s
Multiplexing	Separate RF carrier	Packet mux at baseband	Baseband	Baseband
Baseband	5.6 MH$_3$	5.6 MHz	5.0 MHz	4.2 MHz
RC bandwidth, kHz	27.0	27.0	27.0	24.0
Used in	Scandinavia	TV Sat, TDF-1, Europe	Australia	US/Canada

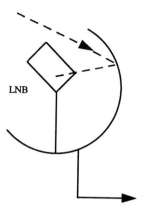

FIG. 5 Outdoor receiver unit consisting of dish antenna and low-noise block (LNB) downconverter.

desired signal. For circularly polarized DBS reception, flat-plate antennas are sometimes used. A 35-cm square flat-plate antenna has an antenna gain of 31.0 dB.

A block diagram of an indoor receiver is provided in Fig. 7. Several functions suitable for GaAs MMIC technology are indicated. The DBS tuner selects a desired channel and an FM demodulator converts it into baseband signals. Baseband signals are used to modulate a radio frequency (RF) signal before it is transmitted to the television set. A descrambler is required if signals received from the satellite are scrambled. A MAC indoor receiver unit is shown in Fig. 8.

A DBS tuner provides power gain, automatic gain control (AGC), and image rejection. Figure 9 shows a block diagram of a DBS tuner.

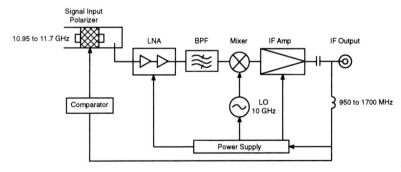

FIG. 6 A low-noise block (LNB) diagram showing functions that can be realized in GaAs MMIC technology.

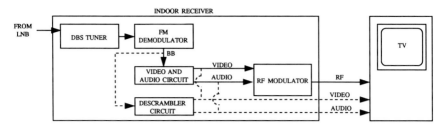

FIG. 7 An indoor receiver block diagram. LNB and DBS tuner functions can be realized in GaAs MMIC technology.

10.4 GaAs MMIC DEVELOPMENT FOR DBS

Among the DBS receiver components, the LNB and DBS tuner are the most appropriate for GaAs monolithic microwave integrated circuit (MMIC) implementation. In this section, an account of the development history is presented.

LNB MMICs have attracted the greatest amount of attention thus far. A comprehensive review on the development history was provided by Konishi in 1990 [1]. Not surprisingly, most of the development work has been done in Japan or Europe, where DBS markets are the most active. The basic elements of a LNB are low-noise amplifier, mixer, oscillator, and intermediate frequency (IF) amplifier. These circuits are briefly described in the following.

10.4.1 Low-Noise Amplifier (LNA)

There have been many attempts [2–7] to develop a MMIC LNA suitable for LNB applications. The most successful LNA was presented by Ayaki *et al.* [6] in 1989, where a four-stage LNA MMIC with 1.76 dB noise

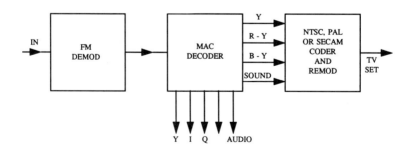

FIG. 8 MAC components receiver.

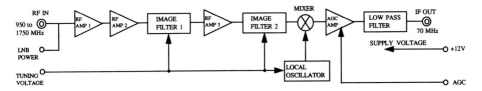

FIG. 9   A DBS tuner block diagram.

figure and 28 dB of gain was reported. Unfortunately, these results are not good enough for practical applications because the Ku-band DBS LNB market requires a noise figure of 1.3 dB or better.

To date, the only way to meet this stringent noise figure requirement is to use a discrete high electron mobility transistor (HEMT) in conjunction with the MMIC. HEMTs with noise figures of 1.0 dB or less are marketed by many Japanese companies. Using these HEMTs, LNBs with noise figures of 1.3 dB or less can be made. It is most likely that LNAs will continue to be made with discrete HEMTs because of the difficulty in monolithically integrating HEMT technology in a cost-effective manner.

There also is an economic reason for using discrete HEMTs. It takes two discrete HEMTs to make an LNA with satisfactory noise figure and with 25 dB of gain. The size of each HEMT chip is 0.0625 mm^2, or a total of 0.125 mm^2 for two chips. By comparison, the LNA MMIC reported by Ayaki *et al.* required an 8.0 mm^2 chip area, more than 60 times larger than the two discrete HEMTs combined. Since the cost of GaAs devices is directly proportional to the chip size, there is no question that using the MMIC is more expensive than using the discrete HEMTs in the production of LNAs.

10.4.2 Mixer

MMIC implementation of both diode and field effect transistor (FET) mixers have been extensively studied [8–10]. For FET mixers, both single-gate metal semiconductor FET (MESFET) and dual-gate MESFET mixers were thoroughly investigated. Unlike LNAs, mixers for LNBs do not have noise figure or other requirements that dictate the use of discrete HEMT devices. As a result, MMIC mixers can be an attractive option compared to discrete diode mixers.

A representative dual gate mixer MMIC was reported by Sugiura *et al.* in 1985 [9]. The chip size is 1.21 mm^2 and a single side band noise figure of 12.3 dB was realized. A much better result was reported by Michels *et al.* in 1990 [10], where a conversion gain of 5.5 dB and a single side band noise figure of 8.5 dB was realized with a tiny 0.45 mm^2 chip. The latter result presented the first glimmer of hope that a high-volume MMIC mixer commercial application is feasible.

10.4.3 Oscillators

Next to the LNA, the oscillator MMIC is the second most actively researched device. Hundreds of papers have been published on the subject [11–15]. Some issues discussed in these papers are:

A. Integration of the tuning varactor diode on the chip,
B. Tuning range of the on-chip varactor diode,
C. Provision of negative resistance oscillator.

A practical approach was presented by Moghe *et al.* in 1987 [16], where a 10.74-GHz oscillator with 16 dBm output and a −80 dBc (10 KHz) single side band phase noise with ±4 ppm/°C temperature stability was reported. The chip size is 0.81 mm^2.

The oscillators can be stabilized by an external dielectric resonator (DRO) or by varactor diode (VCO). The varactor diode can be on-chip or off-chip. A recent article by McNally *et al.* [17], prescribes a high-Q, hyperabrupt, planar ion implanted on-chip varactor with a tuning range of 20% at Ku-band frequencies. However, for block downconversion, no tuning is required and a DRO is commonly used.

For ultrastable and low phase noise operations, phased-locked loop (PLL) stabilized oscillators are needed. An important result on PLL frequency synthesizers was reported by Ohira *et al.* in 1990 [15], incorporating two GaAs chips. The first MMIC contains a Ku-band VCO, a dual output buffer amplifier, a balun, and a prescaler. The second chip contains a dual modulus prescaler, programmable counters, and a phase comparator. The synthesizer exhibited a 1-GHz tuning range and a phase noise of −70 dBc/Hz at 1 KHz offset from the carrier. The chips are 2.4 and 9.9 mm^2, respectively.

A high-speed prescaler is a critical part of a PLL oscillator system. In 1988, a 9.5-GHz monolithic GaAs ¼ prescaler was reported by Takahashi *et al.* [18]. The phase noise is −100 and −120 dBc/Hz at 10 and 100 KHz offsets, respectively.

10.4.4 IF Amplifier (IFA)

The input frequency to an IFA is around 1.0–2.0 GHz, a frequency range at which both GaAs and silicon bipolar devices can operate. Several GaAs IFA papers were reported between 1985 and 1989 [19–21]. Silicon monolithic IFAs have been available commercially from NEC and Avantek since 1983.

10.4.5 LNB Subsystem

In 1983, Hori *et al.* [2], reported the development of a LNB using MMICs. The MMIC chips used are

A. A three-stage LNA
   3.4 dB noise figure
   19.5 dB gain
   4.5 mm² chip size
B. A DRO
   10.0 MW output power
   1.5 MHz temperature stability
   2.3 mm² chip size
C. An IFA
   3.9 dB noise figure
   23 dB gain
   2.3 mm² chip size
D. Mixer
   No MMIC used.

The total cost of such a LNB would be in the range of $30.00 to $40.00, which is too high for practical application.

In 1990, Wallace *et al.* [22] reported the integration of LNA, mixer, DRO, and IFA on a single MMIC chip. Data reported were:

5.0 dB noise figure
36.0 dB conversion gain
2.1 mm² chip size

Thus, for the first time, an economical ($6.00) fully integrated LNB had been developed.

The function of the DBS tuner is similar to that of the LNB; however, two differences are:

A. Tuning of the oscillator is required,
B. The RF input frequency is 1.0–2.0 GHz (12.0 GHz for LNB).

The lower input frequency provides more circuit design options, a key factor in achieving favorable results.

In 1991, Philippe and Pertus reported a fully integrated 2-GHz MMIC downconverter [23]. The on-chip oscillator tunes over a 1.2-GHz range which covers the entire 950-MHz to 2.0-GHz TV band. Local oscillator power leakage at the RF port is −40 dBm and at the IF port is −30 dBm. The circuit operates with a single 5-V power supply with 30 mA current on a 0.70 mm² chip. The conversion gain is 6.2 dB and noise figure is 10.5 dB.

10.5 COMMERCIALLY AVAILABLE GaAs MMICs

Commercial GaAs MMIC products are available from NEC, Hitachi, Mitsubishi, Fujitsu, Avantek, and Anadigics for DBS receiver applications.

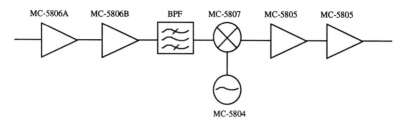

FIG. 10 NEC's LNB GaAs MMICs virtually cover all LNB functions; cost is approximately $12.00.

10.5.1 NEC

Figure 10 shows NEC's GaAs MMIC product offerings for outdoor receiver units. Key characteristics of each part are listed in Table 3; these characteristics are suitable for practical application and have been used in hundreds of units. NEC also offers silicon MMICs for indoor receiver units. The part number and key performance characteristics are listed in Table 4. In late 1991, NEC introduced a GaAs MMIC (μPC 2721) for indoor receiver units. The block diagram and key performance characteristics are detailed in Fig. 11 and Table 5. NEC also offers GaAs prescalers as indicated in Table 6.

10.5.2 Hitachi

Hitachi has two GaAs MMICs for indoor receiver units. HA21005 is a preamplifier and HA21002MS is a mixer circuit. The key performance parameters for these products are shown in Tables 7 and 8. Both products are offered with surface mount packages.

10.5.3 Mitsubishi

Mitsubishi offers GaAs MMIC amplifiers that can be used for indoor receiver units. Key data are provided in Table 9.

TABLE 3 NEC's Ku-Band LNB GaAs MMICs Have Performance That Meets DBS Requirements

Function	Part No.	Specification
LNA-1	MC-5806A	G_a = 16 dB, NF = 2.2 dB
LNA-2	MC-5806B	G_a = 16 dB, NF = 3.0 dB
MIX	MC-5807	L_c = 6 dB
OSC	MC-5808	P_o = 15 mW, F_o = 10.678 GHz
IF Amp	MC-5805	G_a = 16 dB, NF = 4.0 dB

TABLE 4 NEC's Si MMICs for Indoor Receiver Unit

Function	Part No.	Specification	Package
LNA	μPC1659	$G_a = 16$ dB, NF = 5 dB, BW = 1.8 GHz	8-Pin mini flat
Prescaler	μPB581C	+2, $F_{max} = 2.2$ GHz	8-Pin DIP
	μPB582C	+4, $F_{max} = 2.2$ GHz	8-Pin DIP
AGC amp	μPC1476	$CG_a = 25$ dB, $F_{max} = 650$ MHz, AGC: 40 dB	16-Pin DIP
FM demodulation (PLD)	μPC1477C	DG:DP = 2° 2%, $F_{max} = 650$ MHz	16-Pin DIP
QPSK demodulation	μPC1478C		16-Pin DIP
IF amp	μPC1652G	$G_a = 16$ dB,	8-Pin mini flat
	μpC1656C	NF = 4 dB, BW = 1 GHz	8-Pin DIP
PCM demodulation	μPD9301G	Input block	44-Pin flat
	μPD9302G	Output block	64-Pin flat

10.5.4 Fujitsu

Fujitsu offers GaAs monolithic prescalers for DBS applications. The data are provided in Table 10.

10.5.5 Avantek

Avantek offers a series of silicon MMICs for IFA applications. Data are provided in Table 11. Additionally, silicon MMICs are also available for mixer/amp and prescaler applications, as listed in Tables 12 and 13.

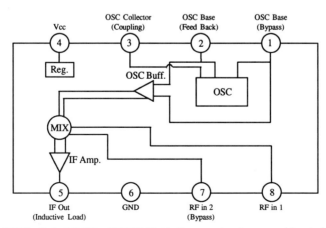

FIG. 11 NEC's GaAs MMIC μPC2721 block diagram showing several key indoor receiver functions integrated on a single chip and packaged in a standard 8-pin SOIC configuration.

TABLE 5 NEC's μPC2721 Specification Showing Good Compliance with DBS Requirements at an Affordable Price (~$2.00)

Parameter	Symbol	Test condition	Min	Typ	Max	Unit
Current	1cc	No signal	25	35	45	mA
Conversion Gain 1	CG1	f_{RF} = 900 MHz, IF = 402.8 MHz	17	20	23	dB
Conversion Gain 2	CG2	f_{RF} = 2.0 GHz, IF = 402.8 MHz	17	20	23	dB
Noise Figure 1	NF1	f_{RF} = 900 MHz, IF = 402.8 MHz		11	15	dB
Noise Figure 2	NF2	f_{RF} = 2.0 GHz, IF = 402.8 MHz		11	15	dB
Power Out 1	$P_o(sat)1$	f_{RF} = 900 MHz, IF = 402.8 MHz	2	5		dBm
Power Out 2	$P_o(sat)2$	f_{RF} = 2.0 GHz, IF = 402.8 MHz	2	5		dBm

Note. Specification (T_2 = 25°C, V_{cc} = 5 V).

TABLE 6 NEC's GaAs Prescalers

Part No.	Function/divide ratio	Useable frequency range (GHz)
UPG501B, P	÷4	1.5 to 5
UPG502B, P	÷2	1 to 5
UPG503B, P	÷4	3.5 to 9
UPG504B, P	÷2	2 to 9
UPG506B, P	÷8	6 to 14
UPG507B, P	÷2	6 to 14

GaAs prescalers

Part No.	Function/divide ratio	Useable frequency range (GHz)
UPB588G, P	÷64	1–5
UPG501B, P	÷4	1.5–5
UPG502B, P	÷2	1–5
UPG503B, P	÷4	3.5–9
UPG504B, P	÷2	2–9
UPG506B, P	÷8	6–14
UPG507B, P	÷2	6–14

TABLE 7 Hitachi's Indoor Receiver HA21005 GaAs MMIC Preamplifier Specifications Showing Good Compliance with DBS Requirements at an Affordable Price (~$1.00)

Parameter	Test condition	Specification
Gain	f = 950 MHz	12 dB typ
VSWR	f = 950 MHz	2 typ
Noise figure	f = 950 MHz	7 dB typ
Third-order intermod	f = 990 MHz, −25 dBm, 2 tone	55 dB typ

TABLE 8 Hitachi's Indoor Receiver HA21002MS MMIC Mixer Specifications Showing Good Compliance with DBS Requirements at an Affordable Price (~$1.00)

Parameter	Test condition	Specification
Gain	f = 990 MHz, f_{local} = 480 MHz, P_{local} = 4 dBm, V_{AGC} = 5 V	30 dB min
AGC range	f = 990 MHz, f_{local} = 480 MHz, P_{local} = 4 dBm, V_{AGC} = 5 V → 2 V	45 dB min
Noise figure	f = 990 MHz, f_{local} = 480 MHz, P_{local} = 4 dBm, V_{AGC} = 5 V	10 dB typ
Third-order intermod	f = 990 MHz, f_{local} = 480 MHz, P_{local} = 4 dBm, GR = 50 dB, −25 dBm 2 tone	45 dB min

TABLE 9 Mitsubishi's Indoor Receiver GaAs MMIC Amplifiers Showing Good Compliance with DBS Requirements at an Affordable Price (~$1.00)

Type	f (GHz)	NF (dB) Typ	GP (dB) Typ	Out line
MGF7003	0.2–1.8	2.5	8	GD-4
MGF7006	0.2–1.8	3.0	17	GE-1
MGF7201A	14.0–14.5	n/a	23	GF-15

TABLE 10 Fujitsu's Monolithic Frequency Dividers Showing Good Compliance with DBS Requirements at an Affordable Price (~$2.00)

Part No.	P_{out} typ. (dBm)	Divide by	f (GHz)	V_{ss} or V_{oo}[a] (V)	I_{ss} ro I_{oo} typ. (mA)	Package type
FMM104FG	4	4	1.0–4.2	−5 or +5	120	FG
FMM106HG	4	8	2.0–6.5	−5 or +5	140	HG
FMM110HG	4	8	0.6–12	−5 or +5	120	HG
FMM201AG[b]	1.6 V(V_{out})	64/65 or 128/129	0.5–1.3	+5	7	AG
FMM202AG[b]	1.6 V(V_{out})	128/129 or 256/257	1.0–2.3	+5	20	AG

[a]Only one supply voltage required.
[b]Two modulus.

10.5.6 Anadigics

Anadigics offers fully integrated downconverter MMICs for LNB applications. The MMIC incorporates LNA, band passed filter (BPF), mixer, low pass filter (LPF), IFA and local oscillator (LO) functions on a single GaAs chip (2.0 mm²). The circuit block diagram of the downconverter is shown in Fig. 12, and the performance characteristics are listed in Table 14.

The design approach of the downconverter MMIC is as follows [22].

A. The key concern is to keep the current consumption as low as possible (<150 mA).
B. LNA key characteristics are:
   Two-stage amplifiers
   Common source resistor bias
   Inductive loads
   Reactive matching
   Parasitic absorption.
C. Mixer key characteristics are:
   Dual-gate FET mixer
   Inductive load
   No LO buffer used.
D. Filter key characteristics are:
   Spiral inductor and MIM (metal–insulator–metal) capacitors
   Low Q (<10)
   Impedance matched
   Small size (<0.1 mm²).
E. IFA key characteristics are:
   Two-stage amplifier

TABLE 11 Avantek's Intermediate Frequency Amplifiers Showing Good Compliance with DBS Requirements at an Affordable Price (~$1.00)

Part No.	$\|S_{21}\|^2$ gain at 0.1 GHz (dB)	$\|S_{21}\|^2$ at 1.0 GHz (dB)	Noise figure at 1.0 GHz (dB)	P_{1dB} at 1.0 GHz (dBm)	Maximum power supply voltage (V)	Device voltage (V)	Device current (mA)	Case type
MSA-0100	19.0	17.0	6.0	1.5	7	5.0	17	Chip
MSA-0200	12.5	12.0	6.5	4.5	7	5.0	25	Chip
MSA-0300	12.5	12.0	6.0	10.0	7	5.0	35	Chip
MSA-0170	19.0	16.5	6.0	1.5	7	5.0	17	70-Mil stripline
MSA-0270	12.5	12.0	6.5	4.5	7	5.0	25	70-Mil stripline
MSA-0370	12.5	12.0	6.0	10.0	7	5.0	35	70-Mil stripline
MSA-0135	19.0	16.5	6.0	1.5	7	5.0	17	Micro-X
MSA-0235	12.5	12.0	6.5	4.5	7	5.0	25	Micro-X
MSA-0335	12.5	12.0	6.0	10.0	7	5.0	35	Micro-X
MSA-0104	18.5	15.0	6.0	1.5	7	5.0	17	145-Mil plastic
MSA-0185	18.5	15.0	6.0	1.5	7	5.0	17	85-Mil plastic
MSA-0204	12.5	11.0	6.5	4.5	7	5.0	25	145-Mil plastic
MSA-0285	12.5	12.0	6.5	4.5	7	5.0	25	85-Mil plastic
MSA-0304	12.5	11.0	6.0	10.0	7	5.0	35	145-Mil plastic
MSA-0385	12.5	12.0	6.0	10.0	7	5.0	35	85-Mil plastic
MSA-0186	18.5	15.0	6.0	1.5	7	5.0	17	Surface mount plastic
MSA-0286	12.5	12.0	6.5	4.5	7	5.0	25	Surface mount plastic
MSA-0311	11.5	11.0	6.0	9.0	7	4.7	35	SOT-143 plastic
MSA-0386	12.5	12.0	6.0	10.0	7	5.0	35	Surface mount plastic

TABLE 12 Avantek's Double-Balanced Active Mixers Showing Good Compliance with DBS Requirements at an Affordable Price (~$1.50)

Part No.	RF–IF RF and LO frequency (GHz)	Third-order IF frequency (GHz)	Minimum conversion gain (dB)	Intercept point (dBm)	LO–RF isolation (dB)	Power supply voltage (V)	Device current (mA)	Case type
MSF-8635	0.1–2.0	DC to 0.5	9.0	7	0	5	16	Micro-X
MSF-8670	0.1–2.0	DC to 0.5	9.0	7	0	5	16	70-Mil stripline
MSF-8685	0.1–2.0	DC to 0.5	8.0	7	0	5	16	85-Mil plastic
MSF-8835	0.5–8.0	DC to 2.0	9.0	16	0	10	36	Micro-X
MSF-8870	0.5–8.0	DC to 2.0	9.0	16	0	10	36	70-Mil stripline
MSF-8885	0.5–8.0	DC to 2.0	8.5	16	0	10	36	85-Mil plastic

Note. Two-port self-oscillating mixers (typical specifications at 25°C case temperature).

TABLE 13 Avantek's Static Prescalers Showing Good Compliance with DBS Requirements at an Affordable Price (~$2.00)

Part No.	f_{in} minimum (GHz)	f_{in} maximum (GHz)	Dividing factor	Input sensitivity at $f = 1.0$ GHz (dBm)	Output level (dBm)	Minimum power supply voltage (V)	Device current (mA)	Case type
IFD-50010	0.05	5.0	+4	−30	−13	5	25	100-Mil stripline
IFD-50210	0.05	3.5	+4	−30	−13	5	25	100-Mil stripline

Note. Typical specifications at 25°C case temperature.

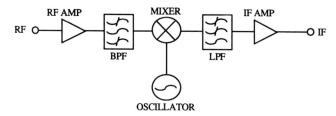

FIG. 12 Anadigics' AKD12000 downconverter block; all functions are realized on a single GaAs MMIC (2.0 mm²).

Dual supply with self bias
Resistive feedback
Active load
Source follower output.

Anadigics markets products for four different bands. Part numbers for each band are listed in Table 15.

Anadigics also has a fully integrated MMIC, ADC20010, for DBS tuner applications. The circuit block diagram is provided in Fig. 13. Key performance specifications are listed in Table 16.

10.6 DESIGNING RECEIVER COMPONENTS WITH MMICs

10.6.1 Designing an LNB

A straightforward way to design an LNB converter is provided in Fig. 14, where two low-noise HEMTs, a bandpass filter, a MMIC downconverter, and an external dielectric resonator are used.

TABLE 14 Anadigics' AKD12000 Specification Showing Good Compliance with DBS Requirements at an Affordable Price (~$6.00)

Device performance	
Conversion gain	36 dB
Noise figure	5 dB
Output power at 1 dB compression point	5 dBm
LO leakage at IF port	−9 dBm
LO leakage at RF port	−20 dBm
LO phase noise	−70 dBc at 10 KHz
Supply voltage	$V_{dd} = 6$ V
	$V_{ss} = -5$ V
Supply currents	$I_{dd} = 100$ mA
	$I_{ss} = -3$ mA
Chip size	2.0 mm²

TABLE 15   Anadigics' Downconverter MMIC Part Nos.

Model	Bandwidth (GHz)	Application
AKD 12000	10.95–11.70	Communications/DBS
AKD 12010	11.70–12.20	DBS
AKD 12011	11.70–12.50	DBS (full spectrum)

An LNA circuit diagram with two HEMTs is provided in Fig. 15 [2]. The matching circuits are designed to optimize the noise figure of the first-stage amplifier and gain of the second-stage amplifier. The source resistor, Rs, and RF-bypassed capacitor, Cs, are incorporated for single power supply operation. DC-blocking capacitors are used between the stages.

Two-stage LNAs with a noise figure of 1.0 dB and associated gain of more than 20.0 dB are routinely manufactured with 0.25-μm gate length HEMT devices.

The bandbass filter is used for image and out-of-band signal rejections and also for reduction of the LO RF leakages. Typically, 40 dB of image rejection and 20 dB of leakage reduction is desired [25]. A photograph of a MMIC downconverter (AKD12000) is shown in Fig. 16. The circuit block diagram is shown in Fig. 12. The MMIC incorporates RF amplification, filtering, mixing, oscillator and RF amplification functions on a single GaAs chip.

The high level of functionality greatly simplifies LNB manufacturing. However, the proximity of many signal processing functions on an IC does introduce some unique problems, notably cross talk from LO signals through substrates or package pins.

Leakages to the RF port usually returns to the antenna and are particularly troublesome. Luckily, the bandpass filter, in front of the MMIC downconverter, can reduce the leakages. However, the presence of the filter affects the conversion gain, which is explained below.

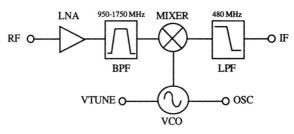

FIG. 13   Anadigics' ADC20010 GaAs MMIC DBS tuner chip block diagram.

TABLE 16 Anadigics' ADC20010 Downconverter Specification

Parameter	Min	Typ	Max	Units
Frequency				
RF	950		1750	MHz
LO[a]	1430		2230	MHz
IF		480		MHz
Conversion gain	7	9		dB
SSB noise figure		8		dB
Output power at 1 dB gain compression		−2		dBm
Third-order IMD				
−25 dBm tones 5 MHz apart		−60		dBc
−15 dBm tones 5 MHz apart		−40		dBc
Second-order IMD (−25 dBm input level)		−35		dBc
LO leakage[a,b]				
RF port			−35	dBm
IF port			0	dBm
LO output level for PLL[c]		−5		dBm
Tuning voltage[d] (V_c)	0.5		24	Volts
VCO phase noise				
10 KHz offset		−65		dBc/Hz
100 KHz offset		−100		dBc/Hz
Input VSWR		2:1		
Output VSW		2:1		
Power supply current (+5 V)		60		mA

Note. Electrical specifcations (packaged unit, $T_A = 25°C$, $V_{dd} = +5$ V).
[a]Measured in Anadigics' Test Fixture with suitable off-chip varactor (e.g., Siemens BB811).
[b]Includes external coupling through test fixture.
[c]Oscillator output for external PLL.
[d]Measured with a BB811 off-chip varactor.

The bandpass filter is reflective for out-of-band signals. Therefore, the LO signal is reflected and is amplified before being reintroduced at the RF port of the mixer. The reflected LO recombines with the original LO at the

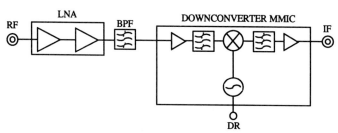

FIG. 14 Designing an LNB with an AKD12000 GaAs MMIC downconverter.

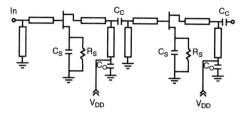

FIG. 15 A circuit diagram of a two-stage LNA designed to optimize the noise figure of the first-stage amplifier and the second.

mixer. It either increases or decreases the conversion gain, depending on whether the two signals are in phase or out of phase.

To realize maximum conversion gain, the electrical length between the filter and the RF pin of the MMIC downconverter needs to be carefully optimized.

To operate the LO, the LO pin is connected by a 50-Ω microstrip line on a personal computer (PC) board to a 50-Ω resistor termination. A dielectric resonator (DR) is placed on a spacer between the LO pin and the termination and placed inside a cavity (a mechanical housing) to resonate the oscillator, as shown in Fig. 17.

The position of the DR, relative to the LO pin and the 50-Ω microstrip line, greatly affects the downconverter performance. The X and Y axes are defined in Fig. 18. Figures 19 and 20 show typical gain and noise figures as a function of the DR position.

Note that the performance contour provides a good starting point but, for best results, each individual DR, spacer, cavity and printed circuit board needs to be carefully considered.

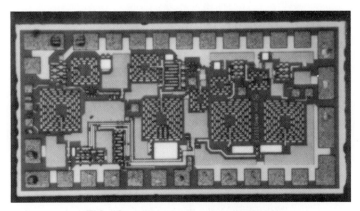

FIG. 16 A photograph of an AKD12000.

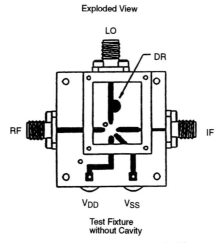

FIG. 17 Placing of a dielectric resonator inside a test fixture.

10.6.2 Designing a DBS Tuner

A functional block diagram of a DBS tuner MMIC (ADC20010) was shown earlier in Fig. 13 [25]. The chip is composed of a single-stage LNA, a mixer, and a voltage-controlled oscillator. Additionally, there are filters for image and LO rejections. Conversion gain is distributed as follows:

| LNA | 11 dB |
| Mixer | −2 dB |

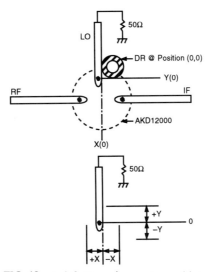

FIG. 18 A definition of resonator positions.

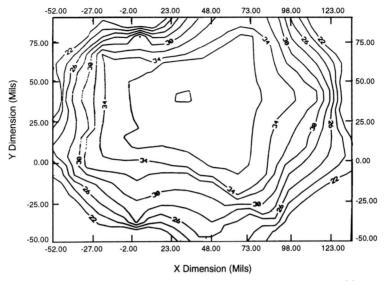

FIG. 19 AKD12000 MMIC downconverter gain versus resonator position.

Filter   −1 dB
Total   8 dB

The MMIC requires an external resonator, two coupling capacitors, two bypass capacitors, and an RF choke to operate, as shown in Fig. 21. Additionally, good grounding is required.

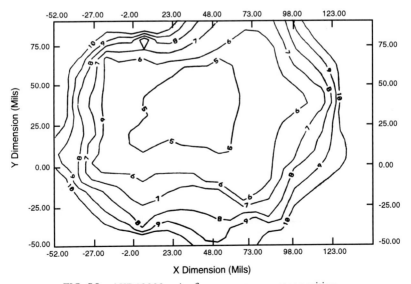

FIG. 20 AKD12000 noise figure versus resonator position.

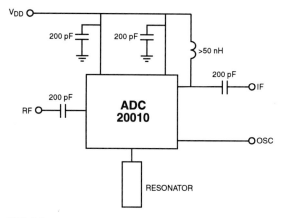

FIG. 21 External circuit for the ADC20010 tuner MMIC.

A typical DBS tuner using the MMIC is shown in Fig. 22. In addition to the MMIC, a fixed high-pass filter and a tracking low-pass filter are needed at the RF input. These filters are chosen, rather than a tracking bandpass filter alone, because the combination offers better insertion loss.

The three-pole, high-pass filter ($f_c \sim 800$ MHz) serves two functions. First, it prevents IF noises from reaching the MMIC. Second, it provides a capacitive impedance loading for the IF frequency at the RF input pin to maintain good noise figure performance. The high-pass filter must be very close electrically to the RF pin to satisfy this requirement. The tracking low-pass filter is used for image and LO rejections.

10.7 APPLICATION ISSUES

Key application issues are:

A. Mounting and bonding considerations
B. Biasing the circuits

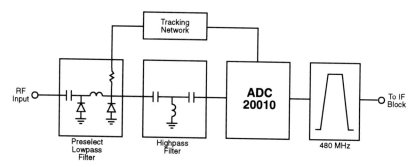

FIG. 22 Designing a DBS tuner with ADC20010 MMIC chip.

C. Reliability studies
D. Temperature stability.

10.7.1 Mounting and Bonding Considerations

DBS receiver MMICs are available in TO cans, ceramic metal packages, plastic dual-in-line (DIP) and small outline integrated circuit (SOIC) packages and in bare chips. Precautions are needed to guard against electrostatic discharge (ESD) and thermal damage when mounting or bonding devices. ESD is caused by people and apparatus that are not grounded properly. Careful grounding should be applied at work benches, operator's clothing and anything that generates static electrical charges [26]. A temperature rating curve for mounting a plastic package device is provided in Fig. 23 [27].

Because the operating frequency is high, it is important not to allow extra lead lengths when mounting or bonding devices, especially on the ground leads.

10.7.2 Biasing the Circuits

Biasing for DBS receiver MMICs is straightforward. Unlike silicon bipolar devices, most commercially available GaAs MMICs are made with FETs. The bias current for FET devices decreases with temperature increases. Therefore, the thermal runaway problem, more commonly associated with bipolar devices, is not a problem here.

10.7.3 Reliability Studies

Reliability for DBS receiver MMICs is no different than the well documented reliability of GaAs MMICs that are used for military/defense applications. Mean-time-to-failure of 10 million h, at an operating channel temperature of 140°C, is routinely achieved.

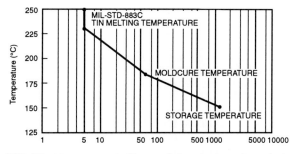

FIG. 23 Temperature/solder graph for plastic package devices.

10.7.4 Temperature Stability

Temperature stability of the local oscillator is of critical importance to DBS receiver applications. The temperature stability of ± 1.0 MHz over the operating temperature range has been achieved.

10.8 FUTURE TRENDS

Currently (1995), there are approximately 26.0 million Ku-band DBS receivers (dishes and indoor receiver units) in service. Approximately 6.0 million in Japan and 20.0 million in the U.K., Germany, Scandinavian, and East European countries.

The Ku-band DBS service has been slow to develop in the United States due to pervasive availability of cable television services. Recently, however, Primestar (1994) and Hughes Communications Inc. DirecTV/United States Satellite Broadcasting, Inc. (USSB) (1994), started to provide Ku-band DBS services.

To spur market growth, it is important that high-quality receivers be obtainable at low prices. The availability of more MMICs for DBS receivers helped to make this happen.

Future MMICs presumably will incorporate more and more functionality on a chip and will be available in inexpensive, easy to use plastic, surface mountable packages. The DBS receiver will continue to be an important market for GaAs MMICs.

10.9 SUMMARY

In this chapter, the Ku-band DBS system was described. The development and the availability of GaAs MMICs for DBS receivers were discussed. Design and application information on selected MMICs were provided. Finally, the future development of DBS receiver MMICs was reviewed.

REFERENCES

[1] Y. Konishi, GaAs devices and the MIC applications in satellite broadcasting. *IEEE 1990 Microwave and Millimeter-Wave Monolithic Circuits Symp. Dig.* 1–6.
[2] S. Hori et al., GaAs monolithic MICs for direct broadcast satellite receivers, *IEEE Trans. Microwave Theory Tech.* **MTT-31**, 1089, Dec. 1983.
[3] E. M. Bastida and G. Donzelli, Airbridge gate for GaAs monolithic circuits. *IEEE Trans. Microwave Theory Tech.* **MTT-33**, 1585–1590, Dec. 1985.
[4] H. Honjo et al., X-band low-noise GaAs monolithic frequency converter. *IEEE Trans. Microwave Theory Tech.* **MTT-33**, 1231–1235, Nov. 1985.

[5] T. Kato et al., A super low-noise self-aligned multi-layer gate MESFET using a P-LDD structure for MMICs. *1989 National Convention Record, IEICE (Japan)*, C-358.
[6] N. Ayaki et al., A 12 GHz-band super low-noise MMIC amplifier using self aligned multi-layer gate MESFET. *IEICE Tech. Rep.* MW 89-29, 1989.
[7] R. Dean Eppich et al., A monolithic variable gain Ku-band LNA. *IEEE MTT-S Dig.* 529–532, 1989.
[8] T. Takenaka et al., A miniaturized, broadband MMIC mixer. *1989 GaAs IC Symp. Tech. Dig.* 193–196, 1989.
[9] T. Sugiura et al., 12 GHz-band GaAs dual-gate MESFET monolithic mixers. *IEEE Trans. Microwave Theory Tech.* **MTT-33**, 105–110, Feb. 1985.
[10] R. Michels et al., A high performance, miniaturized X-band active mixer for DBS receiver application with on-chip IF noise filter. *IEEE Trans. Microwave Theory Tech.* **38**, 1249–1251, Sept. 1990.
[11] T. Mekata et al., Plastic packaged 10 GHz band MMIC oscillator. *1988 Spring National Convention Record, IEICE (Japan)* SC-8-4.
[12] T. Mekata et al., Very small BS converter module. *IEICE Tech. Rep.* MW 89-26, 1989.
[13] M. Madihian and K. Honjo, GaAs-monolithic ICs for an X-band PLL stabilized local source. *IEEE Trans. Microwave Theory Tech.* **MTT-34**, 707–713, June 1986.
[14] T. Ohira et al., A Ku-band MMIC PLL frequency synthesizer. *1989 MTT-S Dig.* 1047–1050, 1989.
[15] T. Ohira et al., Dual-chip GaAs monolithic Integration Ku-band phase-lock-loop microwave synthesizer. *IEEE Trans. Microwave Theory Tech.* **38**, 1204–1209, Sept. 1990.
[16] S. B. Moghe and T. J. Holden, High performance GaAs MMIC oscillators. *IEEE Trans. Microwave Theory Tech.* **MTT-35**, 1283–1287, Dec. 1987.
[17] P. J. McNally et al., Ku- and K-band GaAs MMIC varactor-tuned FET oscillators using MeV ion-implanted buried-layer back contacts. *1990 IEEE MTT-S Dig.* 189–192, 1990.
[18] M. Takahashi et al., 9.5 GHz commercially available ¼ GaAs dynamic prescaler. *IEEE Trans. Microwave Theory Tech.* **36**, 1913–1919, Dec. 1988.
[19] T. Onoda et al., GaAs MMIC for 50 MHz–3 GHz broadband amplifier. *NEC Tech. Rep.* 38(2), 1985.
[20] T. Yoshimasu et al., GaAs broadband monolithic amplifiers. *1987 National Convention Record, IECE (Japan)* 219.
[21] E. Jarvinen, Low noise GaAs monolithic L-band E/D-amplifiers with low power consumption, pp. 1276–1281. 19th European Microwave Conference, 1989.
[22] P. Wallace et al., A low cost high performance MMIC low noise downconverter for direct broadcast satellite reception. *IEEE 1990 Microwave Millimeter-Wave Monolithic Circuits Symp. Dig.* 7–10, 1990.
[23] P. Philippe and M. Pertus, A 2 GHz enhancement mode GaAs downconverter IC. *IEEE 1991 Microwave Millimeter-Wave Monolithic Circuits Symp. Dig.* 61–64, 1991.
[24] Anadigics Technical Brief, *AKD Series Ku-Band GaAs MMIC Downconverters.*
[25] Anadigics Technical Brief, *ADC20010 GaAs MMIC 950-1750 MHz Broadcast Satellite Tuner.*
[26] *Avantek Microwave Semiconductors Data Book*, p. 6–3.
[27] *Avantek Microwave Semiconductors Data Book*, p. 6–4.

Appendix A

Military Electronic Equipment Nomenclature [1,2]

The U.S. Military uses a standard electronic equipment designation which consists of the letters AN, a slant bar, and three additional letters appropriately selected to indicate where the equipment is installed (aircraft, missile, water, surface craft, etc.), the type of equipment (radar, countermeasures, radio, etc.), and the purpose of the equipment (weapon control, navigation, detection, etc.). Table A.1 lists the nomenclature. Following the three letters are a dash and a numeral. The numeral is assigned in sequence for that particular combination of letters. Thus the designation Radar Set AN/SPS-48 is for a search radar designed for installation on ship. The number 48 identifies this particular equipment and indicates that it is the 48th in the SPS. A suffix letter (A, B, C, etc.) follows the original designation for each modification that does not affect interchangeability of the sets or systems as a whole. These modification letters are assigned only if the frequency coverage of the unmodified equipment is maintained.

The civil radars used in the Air Traffic Control System of the U.S. Federal Aviation Agency (FAA) utilize the following nomenclature:

ARSR: air route surveillance radar
ASDE: airport surface detection equipment
ASR: airport surveillance radar
PAR: precision approach radar (not to be confused with phased-array radar denoted by PAR in this book).

As with the AN nomenclature, the numeral following the letter designation indicates the particular radar model of that type.

REFERENCES

[1] M. I. Skolnik (Ed.), *Radar Handbook*, Chap. 1, McGraw–Hill, New York, 1970.
[2] E. A. Wolff and R. Kaul, *Microwave Engineering and Systems Applications*, Chap. 3, Wiley, New York, 1988.

TABLE 1 Summary of Joint Electronic-Type Designation System Nomenclature

First letter: Platform or installation	Second letter: Type of equipment	Third letter: Purpose or function
A Airborne (installed and operated in aircraft)	A Infrared	A Auxiliary assemblies (not complete operating sets used with or part of two or more sets or sets series)
B Underwater mobile, submarine	C Carrier (wire)	
D Pilotless carrier	D Radiac	
F Fixed ground	E Nupac	
G Ground, general ground use (includes two or more ground-type installations)	F Photographic	B Bombing
	G Telegraph or teletype	C Communications (receiving and transmitting)
	I Interphone and public address	
	J Electromechanical (not otherwise covered)	D Direction finder and/or reconnaissance
K Amphibious	K Telemetering	
M Ground, mobile (installed as operating unit in a vehicle which has no function other than transporting the equipment)	L Countermeasures	E Ejection and/or release
	M Meteorological	G Fire control or searchlight directing
	N Sound in air	
	P Radar	H Recording and/or reproducing (graphic meteorological and sound)
	Q Sonar and underwater sound	
	R Radio	L Searchlight control (inactivated, use G)
P Pack or portable (animal or man)	S Special types, magnetic, etc., or combinations of types	M Maintenance and test assemblies (including tools)
S Water surface craft		
T Ground, transportable	T Telephone (wire)	
U General utility (includes two or more general installation classes, airborne, shipboard, and ground)	V Visual and visible light	N Navigational aids (including altimeters, beacons, compasses, racons, depth sounding, approach, and landing)
	W Armament (peculiar to armament, not otherwise covered)	
	X Facsimile or television	
	Y Data processing	
V Ground, vehicular (installed in vehicle designed for functions other than carrying electronic equipment, etc., such as tanks)		P Reproducing (inactivated, do not use)
		Q Special, or combination of purposes
		R Receiving, passive detecting
W Water surface and underwater		S Detecting and/or range° and bearing
Z Piloted–pilotless airborne vehicle combination		T Transmitting
		W Control
		X Identification and recognition

Appendix B

Radar Terms Definition [1]

Detection. The act or action of discovering, and sometimes locating, something or some event. For example, special-purpose detection radars might be used to detect an intrusion (unidentified object entering guarded area or airspace), a nuclear blast, a rocket launching, etc.

Ranging. The act of determining the distance to an object of interest, usually by means of the Doppler effect. For example, police radars usually operate solely in this mode.

Search. The action of looking for an object or for objects of interest. A search radar ordinarily determines the range and azimuth of objects within its area of detection.

Height-finding altitude sensing. The action of determining the height of an airborne object. A height-finder radar usually scans a horizontal fan beam in elevation to determine the elevation angle of the target.

Surveillance. The continual observation of an area or airspace. Surveillance usually implies the observation of familiar territory.

Reconnaissance. The examination or observation of an area or airspace to secure information regarding the terrain, the location or objects of interest, or any other desired information regarding the situation. Reconnaissance usually implies the observation of unfamiliar territory or of territory not accessible to continual observation (surveillance).

Acquisition. The action by which a radar locates a target for the purpose of tracking.

Tracking. The action of continually observing and following the movements of a target. A tracking radar usually "locks onto" the return signal and automatically tracks in angular coordinates and in range.

Track-while-scan. The action of observing and following the movements of a target while continuing to scan in a search or acquisition mode. New position coordinates are obtained with each scan.

Navigation. The action of determining the location of distinctive terrain features, navigational aids (buoys, beacons, etc.), and other objects of interest (nearby ships, aircraft, etc.). The acquired information is useful when steering, maneuvering, or controlling a vehicle.

Mapping. The action of systematically collecting data which allows for the display of a representation of a portion of Earth's surface. Mapping of synthetic aperture radars is concerned primarily with terrain features and with major cultural targets.

Homing. The action of self-direction toward a given spot or target. A homing radar provides the relative location coordinates of the target.

Terrain-following. The action of controlling an aircraft's altitude so as to fly a path which closely follows the terrain profile along a predetermined course. A terrain-following radar normally scans ahead in elevation to determine the upcoming terrain profile.

Terrain-avoidance. The action of controlling an aircraft's altitude and course so as to fly a path which closely follows the terrain along a generally predetermined course. A terrain-avoidance radar normally scans a solid forward sector in order to sample the upcoming three-dimensional profile. For example, with a terrain-avoidance capability, an aircraft might change course to fly between two mountains rather than fly directly over one of them.

REFERENCE

[1] T. M. Miller and H. A. Corriher, Jr., *Basic Radar Principles, Workshop Notes*, Georgia Institute of Technology, Atlanta, GA, 1987.

Appendix C

Units and Symbols

C.1 SI UNITS AND THEIR SYMBOLS

In 1960 the International System of Units was established as a result of a long series of international discussions. This modernized metric system, called SI, from the French name, Le Systéme International d'Unités, is now, as a general world trend, to replace all former systems of measurement, including former versions of the metric system.

In the SI system, four physical quantities are classified as fundamental: length, mass, time, and charge. For practical purposes, temperature is included here as a basic unit. In Table C.1 the first five are basic quantities and the rest are derived quantities, i.e., their dimensions can be expressed as a combination of the first five.

C.2 METRIC PREFIXES

The nomenclature in this decimal structure is derived from a system of prefixes, which are attached to units of all sorts. For example, the prefix "kilo" means 1000, hence kilometer, kilogram, and kilowatt mean 1000 meters, 1000 g, and 1000 W, respectively. Most of our everyday experiences with metric units will involve some of the prefixes listed in Table C.2.

C.3 DECIBEL UNITS

The ratio of signals between the output and input ports of a network is expressed in decibels and its absolute powers are measured in dBm or dBW.

The Decibel (dB)

The decibel is a logarithmic unit of power ratio, although it is commonly also used for current ratio and voltage ratio. If the input power P_i and the output power P_o of a network are expressed in the same units, then the network insertion gain or loss is

$$G = 10 \log \frac{P_o}{P_i} \quad \text{dB.} \qquad (C.1)$$

TABLE 1 SI Units and Their Symbols

Quantity	SI Unit	Symbol	Dimensions
Length	Meter	m	Basic
Mass	Kilogram	kg	Basic
Charge	Coulomb	C	Basic
Time	Second	s	Basic
Temperature	Kelvin	K	Basic
Frequency	Hertz	Hz	1/s
Energy	Joule	J	kg $\times$ m²/s²
Force	Newton	N	kg $\times$ m/s²
Power	Watt	W	J/s
Pressure	Pascal	Pa	N/m²
Electric current	Ampere	A	C/s
Electric potential (voltage)	Volt	V	J/C
Electric field	Volts/meter	V/m	J/m/C
Resistance	Ohms	Ω	V/A
Resistivity	Ohms-meter	Ω-m	V-m/A
Conductance	Siemens	S	A/V
Capacitance	Farad	F	C/V
Permittivity	Farads/meter	F/m	F/m
Magnetic field	Amperes/meter	A/m	A/m
Inductance	Henry	H	V $\times$ s/A
Permeability	Henrys/meter	H/m	H/m

For example, if $P_i = 5$ W and $P_o = 20$ W, then $G = 10 \log 4 = 6$ dB, i.e., power gain of 6 dB. If $P_i = 5$ W and $P_o = 2.5$ W, then $G = 10 \log 0.5 = -3$ dB and the network is said to have a power loss of 3 dB.

The dBm and dBW

The absolute power levels of a network are expressed in dBm, which is defined as the power level P in reference to 1 mW; i.e.,

$$P(\text{dBm}) = 10 \log \frac{P(\text{mW})}{1 \text{ mW}}. \tag{C.2}$$

Thus $P = 1$ mW $= 0$ dBm, $P = 100$ mW $= 20$ dBm, and $P = 0.5$ mW $= -3$ dBm. If power unit reference is 1 W, the decibels are expressed in dBW.

C.4 PHYSICAL CONSTANTS AND OTHER DATA

Permittivity of vacuum, $\epsilon_0 = 8.854 \times 10^{-12} \simeq (1/36\pi) \times 10^{-9}$ farad/m
Permeability of vacuum, $\mu_0 = 4\pi \times 10^{-7}$ henry/m

TABLE 2 SI Prefixes

Prefix	Symbol	Factor by which the unit is multiplied
Exa	E	10^{18}
Peta	P	10^{15}
Tera	T	10^{12}
Giga	G	10^{9}
Mega	M	10^{6}
Kilo	k	10^{3}
Hecto	h	10^{2}
Deca	da	10^{1}
		10^{0}
Deci	d	10^{-1}
Centi	c	10^{-2}
Milli	m	10^{-3}
Micro	μ	10^{-6}
Nano	n	10^{-9}
Pico	p	10^{-12}
Femto	f	10^{-15}
Atto	a	10^{-18}

Impedance of free space, $\eta_0 = 376.7 \simeq 120\pi$ ohms
Velocity of light, $c = 2.998 \times 10^8$ m/s
Charge of electron, $e = 1.602 \times 10^{-19}$ coul
Mass of electron, $m = 9.107 \times 10^{-31}$ kg
$\eta = e/m = 1.76 \times 10^{11}$ coul/kg
Mass of proton, $M = 1.67 \times 10^{-27}$ kg
Boltzmann's constant, $k = 1.380 \times 10^{-23}$ joule/K
Planck's constant, $h = 6.547 \times 10^{-34}$ joule-s
10^7 Ergs = 1 joule
1 Joule = 0.6285×10^{19} electron volts
1 Electron volt = energy gained by an electron in accelerating through a potential of 1 V
Energy of 1 electron volt = equivalent electron temperature of 1.15×10^4 K
Electron plasma frequency, $f_p = \dfrac{e}{2\pi}\left(\dfrac{N}{m\epsilon_0}\right)^{1/2} = 8.97\sqrt{N}$ Hz, where N is the number of electrons per cubic meter
Electron cyclotron frequency, $f_c = eB/2\pi m = 28{,}000 B$ MHz for B in webers per square meter; $f_c = 2.8B$ MHz for B in gauss
10^4 Gauss = 1 weber/m²
Conductivity of copper, $\sigma = 5.8 \times 10^7$ S/m
Conductivity of gold, $\sigma = 4.2 \times 10^7$ S/m

Index

Acousto-optic receiver, 220
Active aperture arrays, 29
Active filters, 10
Active-phase array antenna (APAA), 44
Active Phased-array radar (APAR), 32, 92, 94–96, 98, 99, 103–106, 133
Adaptive control circuitry, 161
Adaptive polarization jamming, 159
Advanced air-to-air missile (AAAM), 34, 40
Advanced air traffic control (AATC), 32
Advanced medium range air-to-air missile (AMRAAM), 34, 40
Advanced mobile phone service (AMPS), 260, 268
Advanced tactical fighter (ATF), 39
Advanced tactical surveillance (ATS), 32
Aegis radar, 90
Agile receive/transmit module (ARTM), 207, 209–213

Airborne adverse weather weapon system (AAWWS), 32
Airborne early warning and control system (AWACS), 32
Airborne multifunctional guided missile radar, 81
Airborne shared aperture (ASA), 32, 33
Airborne systems, 5
Airport surveillance radar (ASR), 81
AM to PM conversion, 102
Analog/digital (A/D) converter, 44, 61, 221, 244
Angle of arrival (AOA), 179
Antenna radar cross section, 106
Anti-ship cruise missile (ASCM), 96
Application specific integrated circuit (ASIC), 268
Attenuator, 6, 10
Automatic gain control (AGC), 234, 248, 331

Index

Automobile speed detection police radar, 81
Automotive navigation, 43
Automotive object detection, 30

Balanced mixer, 10
Ball grid array (BGA), 287
Ballistic missile early warning system (BMEWS), 88
Balun, 185, 216, 334
Bandwidth time (BT), 264
Beam forming matrix (BFM), 314, 316
Bidirectional-phase shifter, 116
Binary-phase shift keying (BPSK), 268, 272, 283
Bipolar CMOS (BiCMOS), 16, 63–67, 69–71, 74, 77
Bipolar junction transistor (BJT), 100, 226, 227
Bit error rate (BER), 262, 264, 265
Bragg cell receiver, 180, 182, 220
Broadband balun, 240
Buffered FET logic (BFL), 61

Capacitor diode FET logic (CDFL), 61, 63
Car collision avoidance system, 51
Carrier-to-noise ratio, 324
Carry look ahead (CLA), 67
Cellular digital packet data (CDPD), 256, 259–266
Cellular telephone, 46, 49
Cellular Telephone Industries Association (CTIA), 258
Central processing unit (CPU), 57, 65, 71, 75
Channelized receivers, 175, 193, 197
Circulators, 10
Clutter, 80
Cobra Dane, 90
Code division multiple access (CDMA), 260, 284
Cohernt signal processing, 80
Collision avoidance radar, 81
Communication intelligence (COMINT), 183, 221
Communication satellite, 324
Complimentary metal-oxide-semiconductor (CMOS), 60, 62–66, 69–71, 74, 77, 156
Compressive receivers, 143, 220
Computer and networks, 40, 48

Computer-aided design (CAD), 3, 4, 11, 24, 31, 36, 37, 39, 74, 95
Computer-aided manufacturing (CAM), 11, 36, 39
Computer-aided testing (CAT), 11, 36, 39
Condon lobes, 159
Connectionless network protocol (CLNP), 260
Control circuits, 6
Cooperative engagement capability (CEC), 39
Coplanar waveguide, 2
Counter battery radar (COBRA), 39
Crystal video, 180, 182, 183
Crystal video receiver (CVR), 175

Defense and missile guidance system, 81
Defense satellite communication system (CSCS), 33, 39
Demun/mun, 156
Depletion mode FET (DFET), 62
Detector log video amplifier (DLVA), 199
Device under test (DUT), 225
Dielectric resonator oscillator (DRO), 100, 101, 187, 190, 215, 218, 334, 335
Diffusion, 2
Digital/analog (D/A) converter, 61, 145, 151, 165, 238, 239
Digital attenuators, 164
Digital circuits, 6
Digital Communication System (DCS), 268
Digital European Cordless Telephone (DECT), 268, 273, 275
Digital RF memory (DRFM), 143, 149, 155, 156, 171, 172
Digital signal processing (DSP), 200, 284
Diode phase shifter, 97
Direct analog synthesis, 235, 237
Direct broadcast satellite (DBS), 30, 41, 49, 50, 231, 296, 323, 324, 328, 329, 331–333, 335, 337–340, 344, 345, 348, 350, 352
Direct-coupled FET logic (DCFL), 58, 62, 63, 152
Direct digital synthesizer (DDS), 149, 165, 207
Direct digital synthesis, 235, 238, 239, 280
Directional detector, 234
Double balanced mixer, 10, 195, 197, 201, 203, 216, 273
Downconverter specifications, 346

Downconverters, 10
Dual-gate MESFETs, 4, 12, 122, 306, 309, 333
Dual-polarized antenna, 159
Dynamic dividers, 237
Dynamic range, 104, 107, 108, 126, 183, 185, 194, 195, 306
Dynamic random access memory (DRAM), 69

E-beam lithography, 4
Effective isotropic radiated power (EIRP), 295, 298, 302, 310, 324, 327, 328, 332
Effective radiated power (ERP), 140, 141, 157, 161, 171
Electrical mobility, 7
Electronic counter measures (ECM), 33, 94, 99, 137, 139, 149, 179, 181, 182, 206, 220
Electronic counter-counter measures (ECCM), 33, 92, 220
Electronic intelligence (ELINT), 179–182
Electronic support measures (ESM), 33, 179, 181–183, 220
Electronic warfare (EW), 29, 31, 34, 39, 90, 137–139, 140–143, 146, 149, 150, 157, 159–161, 163, 164, 171–175, 177, 179–181, 193, 197, 199, 200, 206, 213, 220, 221
Electronically scanned radar, 84, 85
Electronically steered arrays, 80
Electronics intelligence, 221
Electrostatic discharge (ESD), 351
Ellipso, 298
Emitter-coupled logic (ECL), 62–66, 69–71, 74, 152, 156
Enhancement mode FET (EFET), 62
Epitaxial growth, 2, 4, 61
Evaporation, 2
Exciter, 98, 100
Expendable decoys, 33

FET phase shifter, 97
FM demodulator, 331
Fast hopping frequency synthesizer, 149
Fast-settling frequency synthesizer, 151
Federal Communications Commission (FCC), 255, 257, 258
Ferrite phase shifter, 97

Fiber distributed data interface (FDDI), 74
Fiber optic communication, 74
Fire control radar (FCR), 34
Fixed satellite services (FSS), 295
Fixed target indicator, 90
Fixed tuned oscillator, 100
Frequency converters, 10
Frequency dividers, 10
Frequency-hopping spread spectrum, 280
Frequency modulation (FM), 259, 327
Frequency modulation/continuous wave (FM/CW) radar, 10, 35, 51
Frequency scanned radar (FSR), 85
Frequency shift keying (FSK), 183, 268, 272

Gaussian distribution, 183
Gaussian filter, 264
Gaussian minimum shift keying (GMSK), 259, 261–264, 267
Geosynchronous communications satellite, 43
Gilbert cell active mixer, 272, 275
Global positioning satellite (GPS), 30, 41–44
Global system for mobile (GSM), 268
Globalstar, 45, 283, 298, 319
Groundbased radar (GBR), 39
Gunn diodes, 3
Gunn diode oscillator, 100

Helicopter all-weather fire control (HAWFC), 32
Hellfire anti-armor missile (HAAM), 34, 40
Heterostructure bipolar transistor (HBT), 2, 8, 9, 23, 58, 61, 100, 104, 133, 226, 278, 285, 287
High definition television (HDTV), 324, 328, 352
High electron mobility transistor (HEMT), 2, 8, 9, 23, 35, 53, 60, 61, 100, 112, 113, 133, 226, 227, 232, 269, 287, 333, 344, 345
High probability of intercept, 183
High power switches, 119, 120
High speed anti-radiation missile (HARM), 34, 40
High speed digital processing, 80
Home Box Office (HBO), 324
Hybrid approach, 2
Hyper abrupt varactor, 190, 223

IMPATTs, 3
InP, 7, 8
I–Q modulator, 10
Identification friend or foe (IFF), 90
Image rejection filter, 266
Impact avalanche transmit-time (IMPATT) diode oscillator, 100
In-orbit test transponder (IOTT), 294, 303
Indirect synthesis, 235, 236
Indirect synthesis PLL synthesizer, 150
Industrial, scientific, and medical (ISM) bands, 258, 268
Integrated electronic warfare systems (INEWS), 33
Integrated injection logic, 61
Integrated componication navigation identification avionics (ICNIA), 33
Intelligent-vehicle highway systems (IVHS), 40, 47, 50, 51, 53
Intercontinental ballistic missiles (ICBM), 91
Intermediate frequency (IF) amplifier, 83
Intermodulation distortion (IMD), 247, 248
Internal-voltage-controlled attenuator, 266
International Maritime Satellite Organization (Inmarsat), 46, 297
International Telecommunications Satellite Organization (INTELSAT), 294, 295, 298, 299, 302, 306, 309, 310
Internet protocol, 260
Instantaneous frequency measurement (IFM), 143, 175, 180, 181, 199–201, 203, 204, 220
Ion implantation, 2, 4, 17
Iridium, 283, 298, 319
Iridium communication system, 44
Isolators, 10

Japanese Digital Cellular (JDC), 268
Joint surveillance target attach radar system (JSTARS), 90
Joint tactical information distribution system (JTIDS), 33
Junction field effect transistor (JFET), 60–61

Klystrons, 96

$1/f$ corner frequency, 226, 227, 232
Lange coupler, 204, 246

Large scale integration (LSI), 58, 63
Lattice matching, 2
Light emitting diodes (LEDs), 60, 133
Limiters, 10
Limiting amplifier, 203
Liquid-phase epitaxy, 17
Local area network (LAN), 268
Local multipoint distribution service (LMDS), 50
Local oscillator (LO), 83
Local oscillator leakage, 207, 210
Logarithmic amplifier, 10, 156
Low earth orbit (LEO), 41, 298
Low noise amplifiers (LNA), 7, 10, 44, 83, 104, 108, 109, 111–113, 126, 130, 208, 266, 268, 269, 272, 274, 279, 287, 332–335, 337, 340, 345, 348
Low-noise block (LNB), 296, 320, 328, 329, 332–335, 340, 344
Low power switches, 119
Low radar cross section, 159
Low temperature cofired ceramic (LTCC), 287

MMIC packages, 21
MMIC technology, 1
Magnetrons, 96
Manchester code data, 261
Marine navigation, 43
Mean time between failure (MTBF), 197
Mean time to failure (MTTF), 307, 308, 351
Medium scale integration (MSI), 58, 63
Metal matrix, 132
Metal oxide semiconductor FET (MOSFET), 60, 64
Microstrip, 2
Microwave and millimeter wave monolithic integrated circuits (MIMIC), 31, 35, 95, 142, 170
Microwave packages, 20
Microwave switch matrix (MSM), 301, 309, 310
Military electronic equipment nomenclature, 31, 355
Missile control function, 90
Missile seekers, 35
Missile seeker radars, 81
Mixer, 7, 12
Mixer diodes, 3
Mobile communication, 30, 46, 298

Index

Mobile data base station (MDBS), 260
Mobile intercept resistance radio (MISR), 33
Mobile radio, 49
Mobile satellite services (MSS), 297
Mobile satellite systems (MSS), 41
Modulation doped FETs, 2
Molecular-beam epitaxy, 17
Molecular electronics for radar application (MERA), 88
Monopulse radar, 84, 159
Monopulse tracking, 80
Multibeam phased-array antennas, 294
Multibit phase shifter, 116
Multichannel multipoint distribution service (MMDS), 50
Multifunction self-aligned gate (MSAG), 15–20, 125
Multiple discriminators, 200
Multiple-launched rocket systems and terminally guided warhead (MLRS/TGW), 34, 40
Multiple option fuse for artillery (MOFA), 34, 40
Multiplexed analog components (MAC), 327, 331
Multiplexers, 7
Multipliers, 10

n-channel metal-oxide semiconductor (NMOS), 62
Nonrecurring engineering, 3, 22, 30
Nyquist sampling theory, 238

Odyssey, 46, 298
Optical lithography, 18
Optoelectronic feed, 133

Parasitics effects, 2
Passive balun, 216
Patriot radar, 39, 90
Pattern generator, 248–249
Pave Paw Radar, 88
Pentium, 73
Personal communications, 30, 41
Personal communication network (PCN), 255, 268
Personal communications services (PCS), 47, 51, 53, 255–259, 268, 269, 274, 282–285, 287, 298
Personal computers (PCs), 71, 265, 346

Personal Computer Memory Card International Association (PCMCIA), 258–262, 265–267
Personal handy phone (PHP), 47
Phase comparator, 334
Phase correlators, 200, 201
Phase detector, 10, 151, 154
Phase discriminators, 200
Phase lock loop (PLL), 149–151, 153, 236–238, 264, 266, 267, 334
Phase-locked oscillator, 98
Phase noise, 227
Phase shifters, 6, 10
Phased-array antenna (PAA), 141, 157, 310, 313
Phased-array radar (PAR), 39, 79, 87, 88, 90, 91, 96, 97, 99, 102, 104, 133
Pin diodes, 3, 6, 120, 164, 233, 309
Polarimeter, 160, 161
Polarization, 106
Power amplifiers, 113
Power-aperture product, 99
Power splitters, 10
Precision guided weapons (PGW), 34
Prescalar, 334, 337
Process control monitor (PCM), 37
Programmable array logic (PAL), 145
Programmable attenuators, 122
Programmable counter, 334
Programmable read-only memory (PROM), 145, 151
Pulse compression, 80
Pulse generator, 252
Pulse-repetition frequency (PRF), 139

Quadrature IF mixer (QIFM), 143–146, 148
Quadrature phase shift keying (QPSK), 267, 268, 283
Quenchable oscillator, 10, 190, 191

RF Radio, 258
Radar homing and warning receiver (RHWR), 179–181
Radar systems, 6
Radar terms definition, 257
Radar warning receiver (RWR), 179, 180, 206
Radiation hardness, 7
Radio frequency identification (RFID), 41, 47

Random access memory (RAM), 105, 155, 172
Read-only memory (ROM), 165, 238, 265
Received signal strength indicator (RSSI), 267
Receiver, 6, 10
Receiver sensitivity, 183
Reduced instruction set computer (RISC), 65, 75
Reliable advanced solid-state radar (RASSR), 88

SOIC packages, 21, 22, 351
SP2T switch, 148, 164
Sample and hold circuit, 152
Satellitebased personal communication services (SPCS), 44
Satellite cellular telephone, 44
Satellite services, 295
Satellite transponders, 294, 302, 319
Saturation electrical velocity, 7
Schottky-barrier diodes, 3, 4, 8
Schottky diode mixer quads, 203
Search lock oscillator (SLO), 143, 146, 148
Semi-insulating substrate, 2
Sense and destroy armor (SADARM), 34, 40
Signal intelligence (SIGINT), 221
Signal-to-noise ratio (SNR), 183
Silicon carbide imbedded alumina, 132
Silicon-on-sapphire, 7
Single-channel advanced man portable (SCAMP), 39
Single pole double throw (SPDT), 120, 123, 128, 251, 268, 279, 306
Single sideband (SSB), 283
Single-tone intermodulation suppression, 185
Small computer system interface (SCSI), 74
Small kinetic energy weapon, 35
Smart weapons, 31, 32, 34, 39
Smart target activated fire and forget (STAFF), 40
Solid-state power amplifier (SSPA), 294, 314, 316, 317
Source-coupled FET logic (SCFL), 61–63
Space-qualification of MMICs, 294, 307
Spectral efficiency, 264
Spectrum analyzer, 232, 244, 245
Sputtering, 2
Statistical process control (SPC), 294
Stealth aircraft, 141, 161

Superhet receiver, 181, 184
Surface acoustic wave (SAW), 100
Surfacebased weather radar, 81
Switches, 6, 10
Switch matrix, 10
Synchronous optical network (SONET), 48, 74
Synthetic aperture radar (SAR), 90

TDMA/TDD, 267, 275
T/R block, 267
T/R module, 3, 30, 33, 39, 44, 53, 92, 94–96, 98, 99, 102, 105–110, 112, 113, 116, 125, 128, 133, 141, 157, 159, 161, 206, 268
T/R routing switch, 287
T/R switch, 109, 130, 133, 268, 270, 279
Tangential sensitivity, 183
Telecommunications Industries Association, 257
Teledesic, 46
Telemetry intelligence, 221
Television receive only (TVRO), 41, 49, 324
Thermal coefficient of expansion (TCE), 131, 132
Thermal conductivity, 7
Third-order distortion, 273
Third-order intercept (TOI), 104, 106, 108
Time division multiple access (TDMA), 53, 279, 283, 284
Transceivers, 6
Transimpedance amplifier, 10
Transmit/receive (T/R) switch, 120
Traveling wave amplifier, 241, 248
Traveling wave tube (TWT), 80, 96, 141, 143, 148, 159, 160, 163, 171
Transistor-transistor logic (TTL), 148, 152, 164

Units and symbols, 359
Unmanned aerial vehicle, 31
Upconverters, 10, 275, 276

Vapor-phase epitaxy, 17
Variable attenuators, 227, 233
Very large scale integrated (VLSI), 44, 62–63, 75, 105
Very small aperture terminal (VSAT), 41, 298, 320